The Ice Age in Britain

The Ice Age

1

in Britain

B. W. SPARKS

R. G. WEST

METHUEN
LONDON AND NEW YORK

First published in 1972

This edition reprinted in 1981 by
Methuen & Co Ltd
11 New Fetter Lane, London EC4P 4EE
Published in the USA by
Methuen & Co
in association with Methuen, Inc
733 Third Avenue, New York, NY 10017

Printed in the United States of America

British Library Cataloguing in Publication Data

Sparks, B. W.
The Ice Age in Britain.
(Methuen library reprints)
1. Glacial epoch — Great Britain
I. Title   II. West, R. G.
551.7'92'0141   QE 617

ISBN 0-416-32160-7

Contents

List of Figures

List of Plates

[*Plates fall opposite page 302*]

Acknowledgements

We are grateful to the following for permission to base diagrams on existing material: to C. Embleton and C. A. M. King and Edward Arnold (Publishers) Ltd for fig. 3.21; to J. B. Sissons and Oliver and Boyd for figs. 3.4, 3.19, 3.20, 3.31 and 3.36; to J. B. Sissons, R. A. Cullingford and D. E. Smith and the Institute of British Geographers for fig. 5.20; to D. E. Sugden and the Institute of British Geographers for figs. 3.14, 3.15 and 3.16; to the Institute of British Geographers for figs. 3.28 and 3.29; to W. B. Wright and Macmillan London and Basingstoke for figs. 3.11 and 5.2; to the Linnean Society for figs. 7.1, 7.4 and 7.5; to K. P. Oakley and The Geologists' Association for figs. 8.1F, G and fig. 8.2; to F. W. Shotton and The Geologists' Association for fig. 8.3; to The Geologists' Association for figs. 5.16, 5.17 and 7.2; to K. P. Oakley and the Trustees of the British Museum (Natural History) for figs. 8.5, 8.6C, E, F, G and 8.8A, E, F, G, H; to W. B. Wright and G. Bell & Sons Ltd for fig. 8.7; to Paul Woldstedt and Ferdinand Enke Verlag for fig. 1.5; to F. H. Hatch and R. H. Rastell and George Allen & Unwin Ltd for fig. 2.1; to H. Godwin and Cambridge University Press for fig. 6.16; to Reid Moir and Cambridge University Press for fig. 8.6B; to R. F. Flint and John Wiley & Sons Inc. for fig. 5.1; to Grahame Clark and Thames and Hudson Ltd for fig. 8.4; to The British Association for fig. 3.27; to Cambridge University Press for figs. 6.11 and 6.19; to H. Godwin and P. A. Tallantire and Blackwell Scientific Publications Ltd for fig. 6.4, and to Blackwell Scientific Publications Ltd for fig. 7.3; to The Royal Society for figs. 5.13, 5.14, 6.3, 6.8 and 10.1; to F. W. Shotton and The Royal Society for fig. 5.22; to C. Turner and The Royal Society for fig. 6.7; to H. Godwin and The Royal Society for fig. 9.8; to J. Turner and The Royal Society for fig. 8.11; to Ingrid U. Olsson and Almqvist & Wiksell for fig. 9.3; to N. Stephens and R. E. Glasscock and *Irish Geographical Studies*, Department of

Geography, The Queen's University of Belfast, for fig. 8.10; to H. Godwin and Blackwell Scientific Publications Ltd for fig. 8.12; to G. C. Simpson and the Royal Meteorological Society for fig. 2.3; to E. H. Willis and the New York Academy of Sciences for fig. 5.15; to J. K. Charlesworth for fig. 3.24; to H. M. van Montfrans, J. Hospers and Koninklijk Nederlands Geologisch Mijnbouwkundig Genootschap for fig. 9.5; to Panstwowe Zaklady Wydawnictw Szkolnych for figs. 6.10 and 6.17; to W. E. Le Gros Clark and The University of Chicago Press for fig. 8.1A; to M. A. Stokes and T. A. Smiley and The University of Chicago Press for fig. 9.1B; to K. S. Sandford and The Clarendon Press for fig. 5.16; to G. H. Dury and the Royal Scottish Geographical Society for fig. 3.32; to F. M. Synge and the Royal Scottish Geographical Society for fig. 5.19; to the Geological Society of London for figs. 5.7 and 5.12; to P. F. Kendall and the Geological Society of London for fig. 3.34; to the Institute of Geological Sciences for figs. 5.9 and 5.21; to J. R. Mackay and the Government of Canada for fig. 4.3; to Ordnance Survey for figs. 3.13 and 3.22; to J. K. Brierley and Heinemann Educational Books Ltd for figs. 8.1B, C, D, E; to P. A. Mellars and *Antiquity*, St John's College, Cambridge for figs. 8.8B, C, D; to the Prehistoric Society, Faculty of Archaeology and Anthropology, Cambridge for figs. 8.8I, J; to the National Museum of Wales for figs. 8.8K, L; to the Wiltshire Archaeological & Natural History Society for figs. 8.9C, D; to Thames & Hudson Ltd for figs. 8.9E, F; to Piggott and Cambridge University Press for figs. 8.9A, B; to C. D. Ovey and the Royal Anthropological Institute of Great Britain and Ireland for fig. 8.6D; and to R. B. G. Williams, F. G. Davies and The Royal Society for figs. 4.4 and 4.5.

We would also like to thank Dr J. K. S. St Joseph, Director in Aerial Photography at the University of Cambridge, Sir Harry Godwin, Mr R. C. Bailey and Mr A. Barlow for the photographs they have let us use; and also Mr Michael Young and the staff of the drawing office of the Department of Geography, University of Cambridge, for producing the line diagrams.

Preface

The Ice Age, the period specialists refer to as the Pleistocene, is the last, and a relatively insignificant, appendage to geological time. Yet a continuous stream of writing on the Ice Age is poured out by geologists, botanists, geomorphologists, archaeologists and zoologists, and there is no sign of the stream drying up. Indeed, the amount of publication on the Ice Age probably nearly equals that on the rest of geological time.

This is very unfortunate. Not unfortunate in an absolute sense, but unfortunate because the non-specialist finds it ever more difficult to make any sense of the increasing mass of literature on various aspects of the Ice Age. Yet the Ice Age is man's pre-history; it has affected and still affects his environment in a multitude of ways, so that he must be excused for being interested in it.

So, perhaps in extenuation of the fact that we, too, have contributed our quota to the total of specialist writing on the Ice Age, we have tried to write an account of the Ice Age in Britain which will be comprehensible to many. We have kept the number of detailed references to a minimum, and, perhaps in a very unscholarly fashion, have directed the attention of readers to texts which themselves subsume the detailed references. We hope that it may interest the person at the end of school or at the beginning of university study, and the person of education and intelligence who at any stage of life happens to be interested in the Ice Age. Perhaps to the botanist the botany may be too simple, to the geomorphologist the geomorphology too generalized, and to the zoologist the zoology too naïve. But the problem of the Ice Age as a whole must be presented and we hope that we may be excused any shortcomings which may result from such a broad study.

Cambridge, November 1971

I

The General Geography of the Ice Age

Introduction

The term Ice Age may in fact give a wrong impression. The period which geologists know as the Quaternary or the Pleistocene and which occupied the last 1·5 to 2·0 million years was not one long continuous glaciation, but a period of oscillating climate with ice advances punctuated by times of interglacial climate not very different from the climate experienced now. Ice sheets derived from an ice cap centred on northern Scandinavia reached south in Britain at their maximum almost to the Thames–Bristol line, and in Europe almost to the line of the Hercynian block mountains that stretch eastwards from the Rhine block through the Harz to the Bohemian massif. At the same time the glaciers of the Alps and the Carpathians also expanded. Beyond the margins of the ice sheets climatic oscillations affected most of the rest of the world; for example, in the deserts periods of wetter conditions (pluvials) contrasted with drier, interpluvial periods.

Although the time involved is so short, about 0·04 per cent of the total age of the earth or 0·3 per cent of the time during which fossils have been deposited in rocks, the attention devoted to the Pleistocene has been incredibly large. This is probably because of its immediacy, because of the way in which the interest of so many people has been aroused, and because the period largely coincides with the occurrence on earth of man and his immediate ancestors. Naturally, a chronology of events was highly desirable for the various glacial advances and retreats and to date the prehistory of man.

Even now there is no reliable way of dating much of the Ice Age. Geological dates are usually obtained by measuring the rates of decay of various radioactive minerals into others. Some of these geological clocks are suitable for very old dates but involve increasing errors when they are used for young rocks; others are suitable for very young rocks and the errors increase rapidly in older rocks. Most of the Ice Age lies between two such systems (see Chapter 9). The eight-day clock, as it were, is provided by the decay of Potassium 40 into Argon 40 and with this dates of 1·5–2·0 million years can be obtained for the very earliest part of the Pleistocene, but their accuracy may not be very high. The stop-watch is provided by the decay of naturally occurring radiocarbon (^{14}C) (see Chapter 9). The limit of this method is normally about 40,000 years before the present (B.P.), though some laboratories have produced dates to 70,000 B.P. with lesser certainty. Between these two methods the bulk of the Ice Age at present remains uncertainly dated.

Apart from radiometric dating geologists use zone fossils for correlating rocks. In rocks of Jurassic age, which have the best type of zone fossils – namely ones which evolve rapidly and hence have short vertical time ranges but which are nevertheless widely dispersed – the average length of a zone is of the order of 1 million years, although under ideal conditions subzones of the order of $\frac{1}{4}$–$\frac{1}{2}$ million years may be possible. Even this, which in terms of pre-Pleistocene geology is an ultra-refined time-scale, is far too coarse a framework to provide the dating required for the Pleistocene. Furthermore, Pleistocene deposits are usually land and freshwater (continental) deposits which are notoriously poor providers of good zone fossils. The fossil changes from bed to bed are more likely to reflect changes in climatic and environmental conditions than the changes in evolution which limit the time ranges of good zone fossils.

Yet, in spite of these known difficulties of correlation, Pleistocene researchers have at times pinned their hopes on to all sorts of more or less fanciful model schemes of how they would have arranged the Ice Age had they been in charge of events. For example, an early classification of Alpine glaciation developed by Penck and Brückner suggested the existence there of four glaciations, the famous (or infamous) Günz, Mindel, Riss and Würm. This succession was based primarily on a series of deposits and events not directly related to glacial and interglacial periods rather than on the more usual modern method of the study of biological remains included in interglacial beds themselves interstratified within glacial deposits. Yet the succession was forced willy-nilly by later workers on to the glaciated parts of northern Europe (and of other continents), where there are partial successions of true glacial ground moraines (see Chapter 3) and interglacial deposits, which one hopes can ultimately be pieced together from area to area to provide a complete Pleistocene succession. The same scheme was also applied to river valley successions of alternations of gravel deposition and dissection, which are notoriously liable to local variation. The eradication of the Alpine nomenclature, which should never have been applied widely in the first place, has proved and indeed is still proving a Herculean task.

Again, it is possible to suggest that if certain astronomical features are subject to rhythmic variations the combined curve of their effects will produce a rhythmic pattern of variations of radiation received at the earth's surface. This could give rise to cold periods (glacials) separated by warmer periods (interglacials), with perhaps shorter mild periods (interstadials) within the glacial periods. Such exercises in the construction of models should really be modified with the discovery of real events, whereas the interpretation of real events has too often been modified to fit the model.

A chronology of man's stone-working industries has been evolved, much of the relative dating being based upon some other model scheme fitted to the Pleistocene, and then this chronology has been used to date events relative to each other. One argues in a circle.

In spite of the way in which animal assemblages vary with local conditions as well as with time, attempts have been made to characterize certain faunas as time indicators. There is, for example, a cold, arid land snail fauna characterized by *Columella columella*, *Vertigo parcedentata* and *Pupilla muscorum*, which for a long time was thought to indicate the last cold phase in southern Britain. It proved to be yet another environmentally determined fauna – in geological terms, a facies fauna. Of course, as the deposits of the last glaciation are nearest to the surface and unlikely to have been eroded away by later glaciations, the odds are that the vast majority of such deposits will prove to be of that age, but all of them are not. The same is probably true of other attempts to introduce the fossil zone concept into a geological period in which it is unlikely to be applicable, for example the idea that the last interglacial in Britain is characterized by *Hippopotamus* and *Corbicula fluminalis*, a freshwater bivalve shell confined now to sub-tropical regions of the world.

The application of pollen analysis in the last half century to the study of the deposits of the Ice Age (Chapter 6) has opened up great possibilities. By this means it is possible to evolve a picture of continuous change through a deposit. It has been discovered that the various interglacials are characterized by consistent differences in their general vegetation successions, though some of these are subtle and their interpretation requires the skill of a Pleistocene botanist. Even so, the differences are not so clear that every thin deposit can be attributed to its correct place in its correct interglacial. Nor are all the characteristics of an interglacial necessarily revealed in the first half dozen deposits studied. It was thought at one time that the last interglacial in Britain was characterized by the absence of alder (*Alnus*), which seemed to be thoroughly characteristic of the middle parts of the interglacial before. Recently, a last interglacial site has been discovered near Stoke Ferry in south-west Norfolk which has not only an abundance of alder pollen but also of alder cones and fruits, thus indicating the proliferation of the tree in the vicinity of the deposit. The flora is thus like the fauna in being strongly affected by local conditions, though probably to a lesser degree, so that, even with pollen analysis, care in interpretation and, it might be added, geographical imagination, are both required.

Lately, deep borings in East Anglia have become available for analysis, and careful study of the succession of plant and animal remains through them has shown that there have been at least six cold periods in Britain. In only three of these did an ice sheet actually reach Norfolk and in only two was the county submerged beneath the ice. But we must not commit earlier faults and try to fit the six cold-phases scheme to Pleistocene deposits everywhere in case future work demonstrates the existence of more than six cold phases.

The present stage of Pleistocene studies is one of building up the succession from a careful study of the field evidence in various areas, and not from the application of a model scheme based on limited evidence, yet nevertheless, very rigidly held. Progress will almost certainly be slow and laborious. The only likely short cut would seem to lie in the discovery of a method of radiometric dating, wide in its applicability, to bridge the gap between the ^{14}C and ^{40}K methods. These comments do not apply to the period since the last glaciation, the Devensian or Weichselian glaciation: the ^{14}C clock covers this period so that correlation and a common nomenclature can be introduced for this phase. This general state of affairs in Pleistocene studies may not appeal to those who like to see order at the expense of truth in Nature, but the study of the Ice Age cannot again tolerate the chaos that results from the over-enthusiastic and premature application of generalized schemes.

What then do we know for certain about the Ice Age and in particular about the Ice Age in Britain?

Firstly, there was more than one glaciation. There have, in fact, been those who believed in only one, monoglacialists they are called. This method of thinking has been of two types. Some held this view by maintaining that all cold faunas were of one date and, hence, there had been only one cold period. Others, more sophisticated, saw that apparently complex series of deposits were explicable in terms of one glaciation as a result of the variety of environmental conditions likely to be encountered in the decay of an ice sheet. Such observations were a salutary correction to the interpretation of every minor change in deposits as signifying a glacial, an interglacial or an interstadial. Nevertheless, the truth was almost certainly distorted. Complex successions at given localities in Britain are often to be interpreted as the result of one glaciation, but to extend this to the suggestion that there has been only one glaciation sets on one side all the incontrovertible botanical evidence of interglacials. So we may believe in the six cold periods shown by the East Anglian glacial sequence and the deep borings below it. We suspect that this downsinking area has preserved an essential key to the unravelling of Pleistocene stratigraphy.

Secondly, there is no conclusive evidence about the relative length, relative complexity – i.e. number of interstadials – and the relative warmth and cold of the various glacials and interglacials. There may have been a simple rhythm; there may not. Indeed, it is conceivable on a continental scale that one glaciation may have been more profound in one place, another in another It depends on a whole range of topographic and climatic factors. We know, for example, that the last glaciation but one, the Saale of north European or the Wolstonian of British terminology, generally reached farthest south, but locally in Essex the glaciation before that, the Elster or Anglian, approached the Thames most closely and also probably exceeded the Wolstonian limit in places in the southern Midlands.

Thirdly, the first three of the six cold phases did not result in glaciation in Britain as far as is known, though evidence may have been removed by later glaciations. The fourth and fifth, the Anglian and Wolstonian, were extensive glaciations comparable in size with each other. The sixth, the Devensian, involved a somewhat lesser glaciation of Britain. It reached only Cheshire, York and the north Norfolk coast compared with the approximately Gloucester–London limit of the previous two advances, although it did extend practically as far south in Wales as the earlier glaciations.

One might be tempted to infer from this that the peak of the Ice Age has been passed. Such a conclusion would be foolhardy in the face of our expressed scepticism about a neat periodicity in Pleistocene affairs. We do not know whether we live in a postglacial period, although we call it that, or an interglacial period. The former would be a comforting belief; the latter more likely to be correct. Indeed, the chill truth seems to be that we are already past the optimum climate of postglacial (Flandrian) time. Studies of the pollen of certain temperate plants in Denmark, notably *Hedera* (ivy), *Ilex* (holly) and *Viscum* (mistletoe), suggest decreases of a degree or two in both summer and winter temperatures. The same is suggested by the fossil distributions of *Corylus* (hazel) and *Trapa* (the water chestnut) in Denmark and Sweden. So we may be in the declining climatic phase leading to glaciation and extinction. But again, if one does not believe in simple periodicities, we may not. Perhaps there will be a reversal of trend. The present could be merely an oscillation. The final truth is hardly likely to appear in our lifetime, and, in any case, one cannot insure against it.

Ice caps

An ice cap will accumulate when the summer melting of ice and snow fails to keep pace with the winter accumulation. The conditions under which this might happen are considered in Chapter 2. For the present it is sufficient to note that accumulation may be related not only to increasing cold but also to increasing precipitation. The latter might well be effective in a region of predominantly winter precipitation where small glaciers are just about in balance. If the precipitation in winter in the form of snow increases, summer melting, unless conditions change in this season as well, will be unable to melt the excess, accumulation will result, the glaciers will expand and will ultimately merge into an ice sheet.

The process is more complex than this in practice, because changes usually trigger off other changes which either accentuate or counteract the first change. Such repercussions are often spoken of as feedback, the term being borrowed from electronics. If a change induces another change which tends to offset the first, it is known as negative feedback: for example, if the movement of ice becomes very rapid it may reduce the surface gradient of the ice sheet, the effect of which will be to reduce the velocity. If a change induces another change which accelerates the first, it is termed positive feedback. A good example is provided by our growing ice cap. Because the surface area of ice increases and because ice reflects a much higher proportion of incoming solar radiation than rocks do, the melting effects of summer heat may well be reduced. Yet, on the other hand, the greater reflection of radiation may promote a local cooling of the atmosphere and the tendency for a local anticyclone to develop. This in turn would reduce precipitation and decrease the rate of snow accumulation; an example of negative feedback. But at the same time the local anticyclone would also promote cooling and so lead to less thawing; a simultaneous example of more positive feedback. Practically all natural events of the type we are discussing promote a series of interlocked events and one has to be careful in thinking about the likely results.

To return to our two different causes for the accumulation of ice: a very good example of the first type, i.e. caused by cooling, is provided by the Antarctic ice cap. In the southern hemisphere there are very few places where increased precipitation is likely to push up the rate of accumulation of snow and ice: the southern Andes of Chile and Argentina and the Southern Alps of South Island, New Zealand, are two exceptions. Snow and ice accumulate in Antarctica because of low temperatures and not because of excessive precipitation. The area seems to be usually occupied by a high pressure system, but there are occasional frontal interactions when Antarctic air flows out and milder air flows in giving precipitation. However, precipitation is generally slight and accumulation depends on there being little or no summer melting.

Conditions were very different in the northern hemisphere during the Ice Age. It is true that the Greenland ice cap increased in size, but no polar ice cap developed. In middle high latitudes there are great landmasses, unlike the southern hemisphere, and these include large areas of high ground. There were, as a result, developments of ice caps over the western cordilleras of North America, over the eastern part of the Laurentian Shield, over Scandinavia (with a subsidiary centre in Scotland and minor centres in other parts of the British Isles), over the Alps and so on. In all of these increased precipitation was probably combined with a relatively slight cooling: fluctuations of the mean annual temperature of the order of 5–8°C seem to be generally agreed.

As ice caps develop they thicken, their surface gradients increase and ice flows outwards. In an area such as the Antarctic it is mostly split off as icebergs when it enters the deep water surrounding the continent and so does not expand far. But in Europe in the Ice Age the concomitant fall in sea level, to be discussed later in this chapter, allowed the Scandinavian ice sheets to cross a dry North Sea floor, to invade the North European lowlands, and to impinge on the east coast of Britain where they came into contact with local ice sheets deriving from the British centres. The limits they reached were governed by the relative rates of supply, resulting from flow, and melting, or ablation, caused by the warmer climate. When equilibrium is reached the ice front becomes stationary. In detail, of course, it is never really static, because variations in ablation and variations in accumulation, and hence in flow, may not coincide, so that their interaction produces an ice front oscillating back and forth within fairly narrow limits.

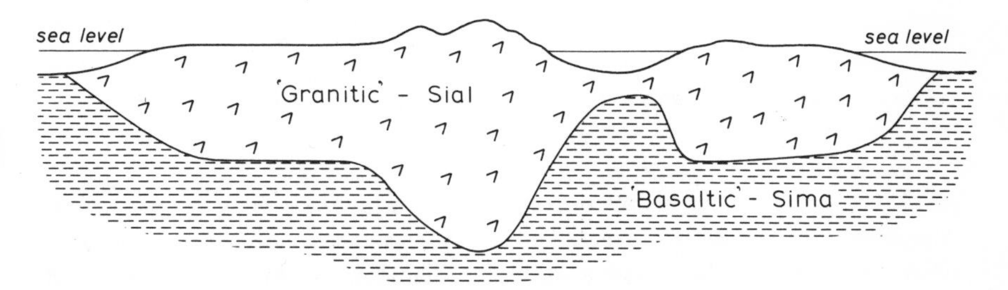

FIG. 1.1 Isostatic equilibrium in the earth's crust.

One other effect needs to be mentioned in connection with ice caps. It concerns the balance or isostatic effects of loading the earth's crust. Briefly, the theory of isostasy envisages the earth's crust more or less floating in a denser substratum, projections of mountains upwards being compensated by projections of roots downwards, rather as depicted in fig. 1.1. The relative density of the sial is about 2.7 and that of the sima about 3.3. In practice the gradation between the two is not as clearcut as this. Much of the earth's crust is in balance, though there are areas where compression probably holds the crust down and where uplift will result when the forces are released. If thick ice is loaded on to a block of the earth's crust (fig. 1.2A) it becomes no longer in balance and tends to sink as a compensating effect (fig. 1.2B). The amount of the sinking depends on the ratio between the density of ice, about 0·9, and that of the sima, about 3·0 to 3·3. Thus, depression should be between about one quarter and one third of the total thickness of the ice. Indeed, this ratio in reverse has been used to calculate the likely thickness of ice at the centre of the Scandinavian ice sheet: a figure of 2650 m has been suggested, assuming a total crustal downwarp of some 730 m.

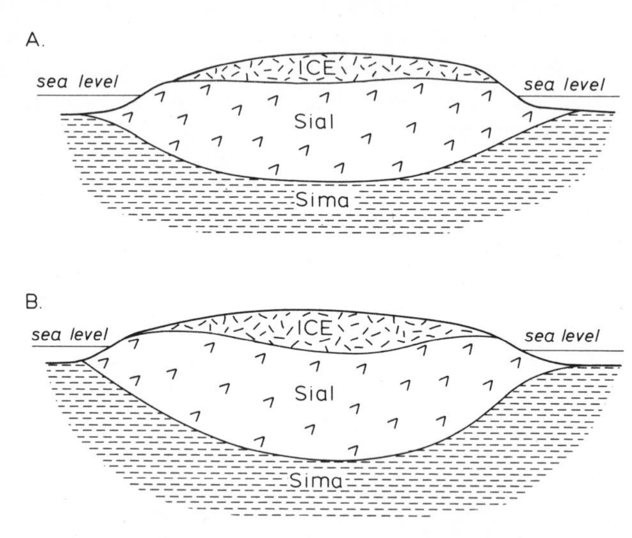

FIG. 1.2 The effect of an ice cap on the equilibrium of the earth's crust.

There seems to be a limiting size below which ice caps do not cause depressions of the earth's crust. Daly (1934) has shown that in those cases where warping can be demonstrated the diameter of the ice cap exceeded 500 km and that its thickness was probably greater than 1 km. In European terms this means that the Scottish and Scandinavian ice caps caused warping but not the Welsh one.

When the ice sheet melts the earth's crust recovers but usually with a considerable lag effect. It seems that recovery is not yet complete either in Scandinavia or in Scotland: the head of the Gulf of Bothnia, which was the site of the centre of the Scandinavian ice cap, is rising at the rate of about 9 mm per year and the centre of Scotland at about 3 mm per year. The time taken for complete recovery is unknown and may well vary considerably: it will probably range up to 20,000 years in the case of Scandinavia.

It may seem incredible that areas of old crystalline rocks, heavily fractured and folded, could behave in the elastic fashion outlined above. It must be realized that a total depression of 730 m for the centre of the Scandinavian ice cap is quite small compared with the distance involved. Copenhagen is at about the limit of the warping effect and the gradient from here to the centre of downwarping is about 60 cm per km, i.e. 1 in 1650 approximately.

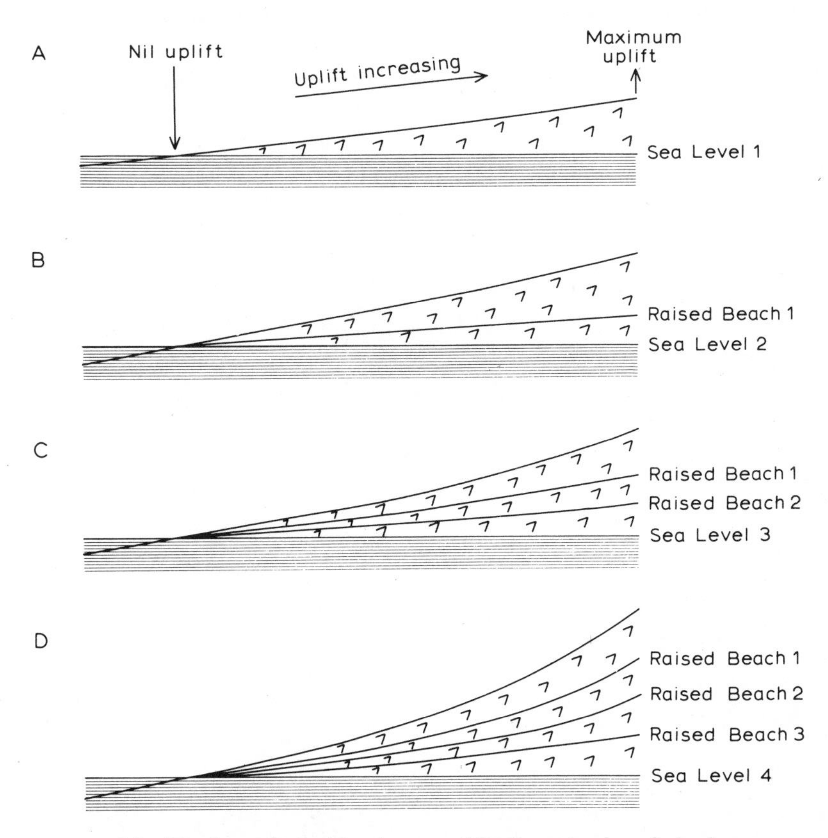

FIG. 1.3 The form of raised beaches caused by the updoming of a landmass.

The recovery results in a complex pattern of raised beaches, and this is rendered all the more confused by the fact that world-wide sea level was rising at the same time due to the melting of the ice. But let us eliminate the second complication for the moment and consider only the first. In highly simplified form fig. 1.3 shows a sea lapping against a landmass and cutting beaches, which are then uplifted as recovery takes place. If one can date the raised beaches and contour them, then a method is available for estimating the downwarp from the amount of recovery which has taken place. The amount of recovery still to take place can be ascertained by geophysical methods. If there is an excess of light crustal material still downwarped it will affect the value of gravity and from the isostatic gravity anomaly the likely upwarp can be estimated. The current amount of upwarping can be estimated by measuring the relative elevation of sea and land in a number of places by means of tide gauges.

But, at the same time as upwarping has been taking place mean sea level has been rising irregularly through the general melting of the world's ice. Thus, we may imagine the sea cutting a bench at level 1 (fig. 1.4A) and then both the land and the sea rising to the levels shown in fig. 1.4B. Near the edges of the upwarped region the sea has risen faster than the land so that beach 1 goes below beach 2 as can be seen in fig. 1.4C, where more upwarp of

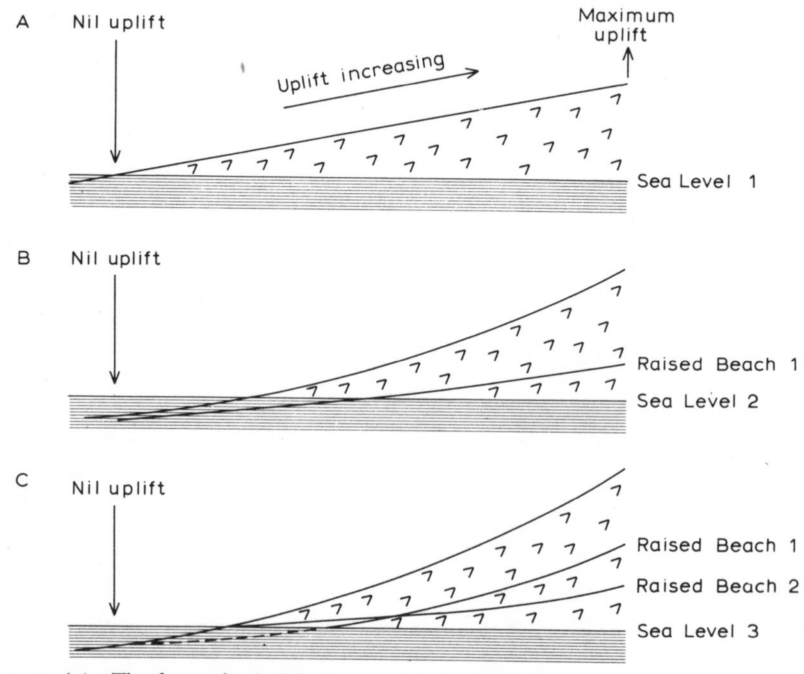

FIG. 1.4 The form of raised beaches caused by the updoming of a landmass and a rising sea level.

the land has occurred but there has been no further rise in sea level. Whether beach 1 will be preserved below the level of beach 2 or obliterated is a difficult question to answer. An example of such a type of movement in Scotland in the last 12,000 years is shown in fig. 5.20.

This all adds up to a very complex problem of interpretation, especially as the beaches are only preserved in sections and not continuously. As they are warped it is very difficult to be sure that bits of one beach are not being correlated with bits of another. Hence, there are still discussions in Scandinavia as to whether there are discontinuities in the uplift and especially whether there are hinge-lines separating areas of more rapid and less rapid

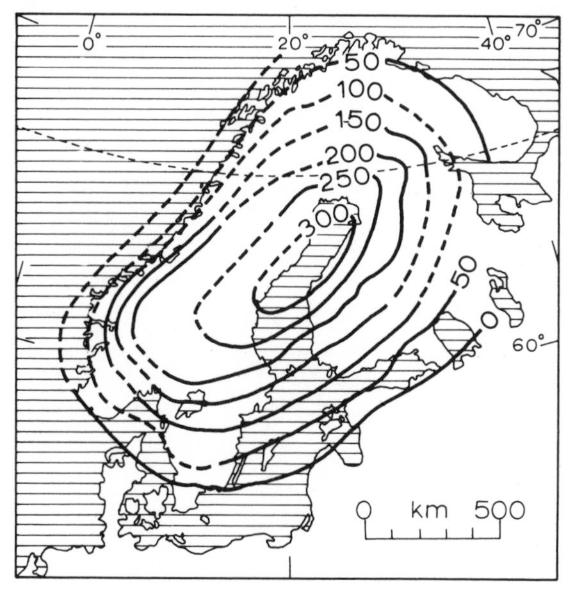

FIG. 1.5 Isostatic recovery (m) of Scandinavia in the last 10,000 years (after Woldstedt, 1954, fig. 119).

uplift. An idea of the general amount and distribution of uplift of Scandinavia in the last 10,000 years is given in fig. 1.5, and of the present annual vertical movements affecting north-western Europe in fig. 1.6.

World changes of climate

When the areas of cold glacial climates enlarged there must have been repercussions on the distribution of the rest of the world's climatic belts. It is generally accepted today that the cold periods in the northern and the southern hemispheres were synchronous, so that all the climatic belts must have been squeezed towards the equator. The exact form of the squeezing is a matter for discussion. It is unlikely that the equatorial belt could have been completely eliminated, for a general zone of convergence there would seem to be an essential part of the world's atmospheric circulation. Yet evidence of larger glaciers on tropical and equatorial mountains in the Pleistocene must imply some cooling and/or increased precipitation even there. It seems

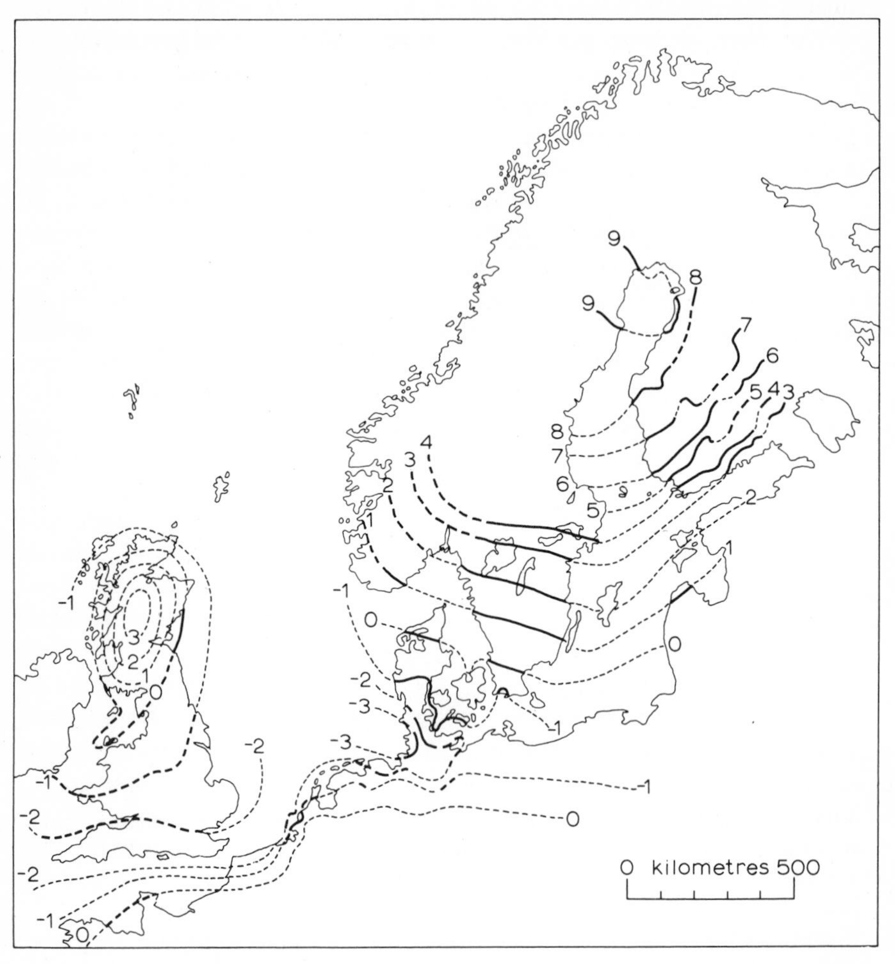

FIG. 1.6 Annual vertical movements in north-west Europe (from West, 1968, fig. 8.2; after E. Fromm, 'Map of Glaciation and Changes of Level', in *Atlas over Sverige*, Stockholm, Generalstabens Litografiska Anstalts Förlag, 1953, and H. Valentin, 'Gegenwärtige Niveauänderungen im Nordseeraum', *Petermann's Geographische Mitteilungen*, 1954, Vol. 98, 103–8). The change is plotted in mm/year (positive or negative):

—— fairly certain
– – – less certain
- - - - by interpolation

most likely that the steepening of the 'climatic gradient' between the glaciated areas and the equator was accompanied by a consequent narrowing of some of the climatic zones.

The effects of the spread of the ice sheets in Britain is the general subject of this book and various aspects are dealt with in later chapters, so that only some general points will be made in this introduction.

One can too readily imagine that the present climate of the Arctic frozen wastes of the northern continents, the tundra region, was transferred bodily to Britain and to Europe just beyond the limits of the ice sheets. It is probably true that general temperatures were similar and that landforming processes, as well as plant and animal communities, were not all that different. But the climate was not identical. As far as is known the position of Britain on the earth's surface and the general attitude of the earth to the sun have not changed since the beginning of the Pleistocene, apart possibly from slight changes in the angle which the earth's axis of rotation makes with the plane in which it rotates elliptically round the sun. This relationship, the obliquity of the ecliptic, is considered in Chapter 2 in connection with the cause of Ice Ages, but the variation is only a degree or two on either side of the present approximately $23\frac{1}{2}°$ from the vertical. The maximum variation could not alter the fact that all those features governed by general astronomical factors, such as length of seasons, length of days and angle of incidence of sun's rays were the same in periglacial Britain in a glacial phase as in temperate Britain today. Thus, although Britain may have had in general terms a tundra climate, it was not the land of the midnight sun at any time of the year and the much higher angle of the sun's rays, especially when coupled with slopes facing south – whether these were major escarpments or merely minute local features – meant probably surprising degrees of local warmth. One should therefore be prepared for anomalies in processes, and in plant and animal life as well. There are, in fact, no exact homologues on the surface of the earth today of the climates Britain experienced in the cold glacials of the Ice Age.

This statement has been made with specific reference to Britain, the main subject of this book, but it is also true of any other shift of climatic belts; one cannot change the very important latitudinal factors, so that, although climates may be approximately comparable with certain present climates, they are never exactly the same.

The other major Pleistocene variant was the intervention of pluvial periods in the world's hot desert belt. A pluvial period could have been caused either by the spread of the wet winter, Mediterranean climate from the poleward side or of the wet summer, Savanna climate from the equatorial side, or by a general weakening of the subtropical high pressure systems allowing more frequent incursions of depressions bringing rainfall from either side. A model scheme of shifts of climatic belts towards the equator would seem to imply the system outlined in fig. 1.7, in which a pluvial on the poleward side of the desert would appear to correspond with a glacial period and a pluvial on the equatorial side of the desert with an interglacial. Perhaps these are the equivalents of what Butzer (1965) has called the cold pluvials and the warm pluvials, the one represented by increased stream activity, the other by soil reddening. It is conceivable that they could have overlapped in some parts of the Sahara. If, however, the deserts were narrowed and the Mediterranean and equatorial zones simultaneously extended, pluvials could

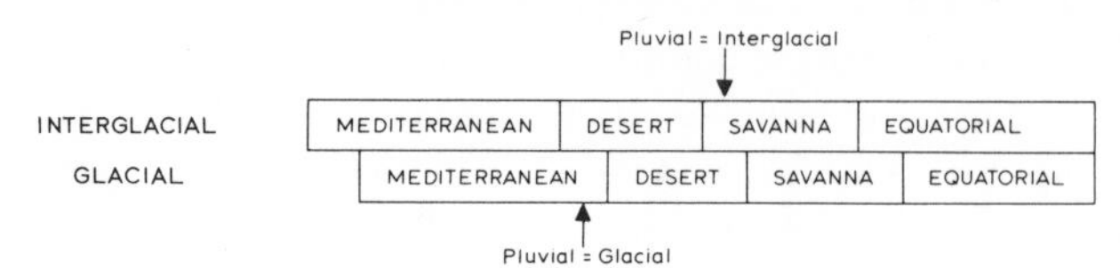

FIG. 1.7 Scheme of glacial and interglacial climatic shifts in tropical areas.

be synchronous on both sides of the desert zone. It would be interesting to have an expert meteorological opinion on the physical probability of this. The unravelling of the sequence of pluvials in deserts will require more rigorous pollen correlations and radiometric dates than are at present possible.

Whatever the chronology of the pluvials and interpluvials, it is patently obvious that the Sahara was appreciably wetter at some stage or stages in the Pleistocene, and the same is true of many other deserts elsewhere. Grove and Warren (1968) have advanced good physical evidence for a mega-Chad lake anything up to forty times its present area. There is evidence of former widespread lakes and stream action in wadis in the form of commonly-occurring faunas of freshwater Mollusca, which are of such a type as to imply freshwater bodies of some permanence. It must not be assumed that the Sahara in the Pleistocene was equatorial in its rainfall, but large parts of it were appreciably wetter than they are now, although the Libyan Sand Sea on the margins of Egypt and Libya probably remained a true desert.

World-wide changes of sea level

When the amount of ice in the world is increased the only final sources of the water involved are the oceans. Hence, an Ice Age must be accompanied by a general fall in sea level. The amount will obviously depend on the size of the ice sheets and is a matter for estimation rather than accurate calculation. The extent of former ice sheets can be estimated from the geological evidence left at their margins, provided this has not been overrun and destroyed by a later ice sheet. An estimate of the gradient of the surface of an ice sheet and hence of its volume can be obtained by the application of physical theories concerning ice flow. Finally, an allowance can be made for the isostatic depression likely to have occurred beneath the ice sheets. There will be room for differences of estimation in all these calculations, so that it is not surprising that estimates of sea-level depression in the last glaciation have varied, usually from about 100 m to about 140 m. There is thought to be enough water still locked up in ice at the earth's surface to cause a further oceanic rise of about 50 m, over three-quarters of this being held in the Antarctic ice sheet.

Changes of sea level worked out by calculating the amounts of ice to be formed and melted are not simply made by converting the ice to water and adding this volume to the present surface area of the oceans, because the general surface area must change as sea level rises. If one takes this factor into account, the calculated falls in glacial periods will increase and the rises in complete deglaciation will diminish: for example, there is enough ice at present to add approximately 90 m to the present area of the oceans, but sea level would probably rise little more than 50 m because of the change in area.

It seems unlikely that all the ice would have melted in the interglacials, so that variations of the order of 120 m in sea level might be expected between the glacials and the interglacials. World-wide movements of sea level are known as eustatic; those caused by the fluctuations of ice sheets are known as glacio-eustatic, for it is possible for them to be caused by major earth movements, which could produce changes in the volume of the ocean basins. Indeed, something has complicated the position of sea level in the Pleistocene apart from the glacio-eustatic effect.

If glacio-eustasy alone operated, the curve for glacial and interglacial periods would be as on fig. 1.8A, on which the peaks represent interglacials, for example B and C, and the troughs, *a*, *b* and *c*, represent glacials. The levels are not all shown as the same height or depth because of possible variations in the size of the world's ice sheets. The distance, X, would then represent the difference between preglacial and present sea levels, which would also be the same as the rise in sea level likely to be caused by the melting of all the world's ice, i.e. 50 m or so. The real position is more like that shown in fig. 1.8B in which the distance *y* to the sea level of interglacial A is about double that which could be achieved by melting all the world's ice. The true preglacial sea level was higher still.

No one knows the causes of these facts: some would even dispute the evidence, but, although a lot of this is dubious, there is too much for it to be set completely aside. The evidence consists of the identification and correlation of terraces formerly cut by the sea and rivers during interglacial periods in order to derive approximate figures for sea levels at those times. The difficulties really reside in the fact that there are no foolproof methods of fossil or radiometric correlation and dating such terraces, so that they are very often correlated on height alone, but this implies no earth movement, which is more or less the hypothesis one is trying to test. Over-enthusiastic workers will probably correlate everything; under-enthusiastic ones will probably attribute all terraces at different heights to differential earth movements.

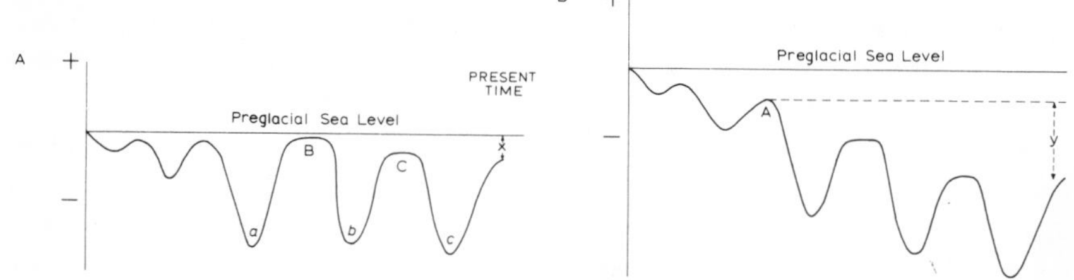

FIG. 1.8 Ice Age sea-level changes. A, theoretical changes with glacio-eustasy; B, changes as recorded by raised beaches.

It is possible to get into further difficulties through such technical errors as confusing the origin of terraces and hence not being in a position to know which heights on the terraces are the significant ones; lack of knowledge of the amount – and it often varies – of dissection since the terrace was formed; inaccuracies in the estimation or measurement of heights; failing to consider the effects of variable tidal ranges on the possible heights of marine terraces. A full treatment of all the technicalities is beyond the scope of the present discussion. The upshot of all this is that an honest survey of the terraces of any region usually produces a scatter of fragments at a variety of elevations, with a tendency for them to be clustered at certain levels. These tendencies to concentration probably represent genuine former sea levels. While people argue over details and which terrace is to be fitted to what, there is little doubt that fig. 1.8B represents a generalization of the truth, the full content of which is specialist work.

From the point of view of the general geography of the Ice Age the importance of the fluctuating sea level is that at different times the rivers were graded to different levels. In glacial times they were graded to lower sea levels, so that their lower valleys were flooded in the higher sea levels of interglacial periods, which are now represented by terraces on the valley sides. In the classic examples the drowning of the lower parts of valleys produced in unglaciated areas, such as south-west Ireland or north-west Spain, long embayments known as rias: in glaciated areas such as Norway, western Scotland, southern Chile or South Island, New Zealand, the rise is responsible in part for the formation of fjords (see Chapter 3).

But this does not mean that drowned valleys will always be formed. Imagine an area such as the lands around the North Sea, including eastern

England. A fall in sea level in an interglacial does not mean that the sea suddenly or even slowly drops 100 m at Cromer. It obviously cannot without carving 99 m of its bed away. The strand-line retires across the floor of the North Sea until it reaches a position on the sea floor gradient 100 m lower, probably well north of the Dogger Bank. Over this emerged North Sea floor the rivers of eastern England, the Low Countries and northern Germany will flow. The gradient may well be equal to that of the rivers in an equilibrium state, i.e. that in which the gradient of the stream is adjusted to the power required for transporting the amount of debris supplied to it by the weathering of the land. If this is so, there is no reason why they should cut down, so that when the sea level rises in the next interglacial there will be virtually no valleys to flood.

Again, if the stream flowing into the head of a drowned valley is carrying a lot of sediment it may fill the drowned section of the valley with this material. Sediment can be brought in at such a rate that the valley is filled as fast as the sea rises, or the valley may be filled after the sea has risen and become stabilized at its interglacial level. One or the other happened in the case of the rivers of Sussex, the Arun, the Adur, the Ouse and the Cuckmere, which in their gaps through the South Downs meander over a sedimentary fill which hides the fact that the solid rock floors are at a considerable depth below, in the case of the Arun, the deepest of them, some 30 m below.

The oscillating sea level occasioned by a series of glacial and interglacial periods has been one of the main causes, but not the only one, of island formation in the Pleistocene. The mechanism is in essence simple. Imagine a peninsula (fig. 1.9A) with two small streams, A and B, in the preglacial period. The same state of affairs is shown in section in fig. 1.9B. The original width of the peninsula is here xy. In a glacial period the sea level will fall, and the rivers, rejuvenated, will cut down to the glacial sea level. When the sea rises in the following interglacial the width of the peninsula between the heads of the drowned estuaries will have reduced to $x'y'$. The repetition of this process, coupled with marine erosion, may finally isolate the end of the peninsula. The Chalk ridge between Purbeck and the Isle of Wight was probably breached in this way. Maybe the eastern part of the English Channel, including the Straits of Dover, was also formed in the same way, but it has also been suggested that this was ripped out by the overflow of a lake impounded by one of the ice sheets in the southern North Sea.

Even in tropical seas the rising and falling sea levels had important effects, apart from the formation of estuaries and islands, according to Daly's (1934) theory of the formation of coral atolls. Although a full examination of the theory and its merits would be too lengthy, it does seem reasonable that low glacial sea levels and lower sea temperatures near the poleward margins of the coral belt would mechanically destroy reefs after the corals had been

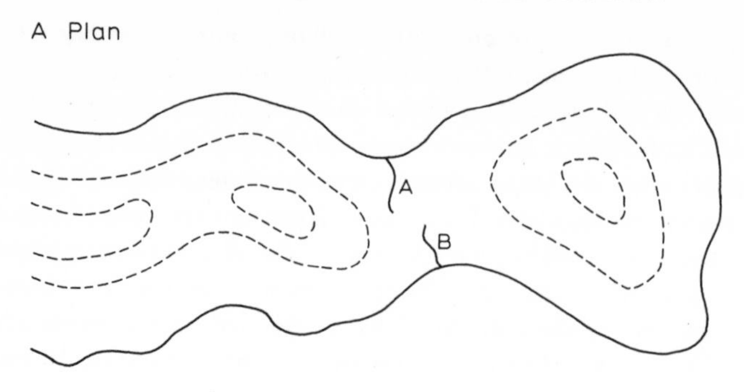

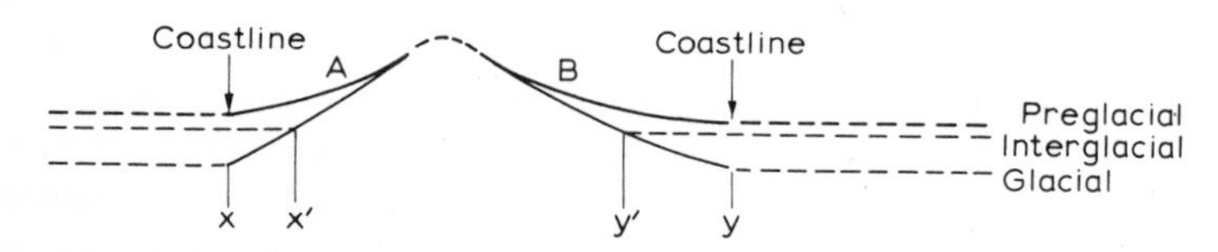

FIG. 1.9 Formation of an island by an oscillating sea level.

killed off or weakened. These reefs would have to re-form in every interglacial period and this may explain the tendency for reefs to be comparatively narrow except in some positions close to the equator where they survived through glacials and interglacials alike. But it is less likely that alternating glacial and interglacial sea levels provide a complete explanation for the formation of coral atolls: just as local structural movements complicate the succession of interglacial raised beaches, so do vulcanicity and earth movements affect the history of coral islands. Indeed, they may well dominate it.

The glaciated areas and their margins

Broadly speaking, material was eroded from the centres covered by ice sheets and ultimately dumped near the margins of the ice either by the ice itself or by meltwater streams flowing away from it. This transport may not have been accomplished in one haul, but debris may have been carried and deposited, picked up again and redeposited in one form or another until it got to its final destination.

Again one must bear in mind the succession of events which in sum add up to a glaciation. At the present day snow lingers in a couple of north-facing gullies in the highest Scottish mountains all the year round. If there were a general refrigeration of climate the Scottish Highlands might easily develop a tundra climate and the physical processes would then become periglacial (see Chapter 4). Ultimately snow accumulation should bring a small ice cap into existence, from which glaciers would flow away down some of the valleys. At such a stage the main agent of denudation would be the splitting of rocks caused by water getting into cracks, freezing there, expanding, and so exerting great destructive force. All this would take place mainly above the level of the ice. When the ice cap completely covered the area the main actions would be the grinding and abrasion exerted by the ice on the rocks beneath. The likely extent of this is a controversial question to be discussed in Chapter 3.

Out of the English lowlands a similar sort of climatic succession would unfold, although in soft rocks freeze–thaw would probably effect a more total destruction of the rocks than it could in the hard rocks of the highlands. The glaciation would ensure the later submergence of most of the land beneath the ice except for some of the higher hills, such as the North York Moors in the last glaciation.

The deglaciation in either area would not have been the mirror image of the development of ice sheets. For glaciation to occur cold has to develop to such an intensity that ice sheets build up. Thus the land may have been subjected to very low temperatures indeed in the pre-glaciation periglacial period – some authors have attributed specific effects to this cause. On thawing, however, it is obvious that the land will not be exposed through the ice sheet until temperatures are high enough for the ice sheet to have been substantially melted. At the same time the very fact of melting will mean that water action will play a massive and increasing role, whereas in the corresponding pre-glaciation period it played a minor and decreasing role.

This sequence of events provides a spectrum of conditions of denudation, which, together with interglacial conditions, must have been repeated several times over. Thus, it may become extremely difficult to attribute particular effects to particular phases of glacial history.

The essentially unvegetated landscape of debris washed out from and deposited in front of the ice sheet is exposed to attack and redistribution. Outwash streams swollen by summer thawing will transport and spread vast quantities of coarse gravels along stream courses, in lakes and so on. Such bare areas are wide open to wind attack and sand dunes may be formed from the finer material, while the very finest material is often winnowed by the wind and redeposited at the margins of the periglacial zone as loess.

In any part of the areas covered by ice sheets considerable modification of the drainage patterns may occur. These are considered in more detail in Chapter 3. Apart from the disintegrating effects of irregular scour and deposition, they involve the transfer of drainage over watersheds into other basins by a number of processes, so that the post-Ice Age pattern may be substantially different from the pre-Ice Age pattern. Without Ice Ages and with crustal stability drainage tends to develop in conformity with the basic rock pattern of any region given a long enough period of time, so that our present patterns, at least in glaciated areas, may be quite atypical.

In addition, there are some less widely appreciated effects. The glacial periods of the Ice Age allowed the production of vast quantities of waste, much of it by freeze–thaw weathering. This material was widely distributed by ice sheets which carted it over watersheds from one drainage basin to another. Much of it was dumped on areas which are now submerged by the sea and form parts of the Continental Shelf. Thus, we probably have a much more generalized distribution of weathered waste and the availability of larger quantities than had previously been the case in and around the British Isles. The accumulation on the Continental Shelf was probably accentuated in glacial periods, when the land beyond the ice sheets was more bare and exposed to denudation and the rivers at least seasonally more powerful, so that large quantities of material tended to be transported on to the Continental Shelf by the rivers as well as by the ice sheets. Thus in our present postglacial phase the sea has more debris than has been available before to build up into constructional features such as bars, spits and cuspate forelands. One has only to think of the coast of East Anglia: Scolt Head Island, Blakeney Point, Winterton Ness, Yarmouth spit, Benacre Ness, Orford Ness to name some of the most prominent features. This proliferation of coastal forms may well be an unusual feature directly dependent on the Ice Age.

General effects on plants and animals

Very little probably lives when an ice sheet covers most of the land. On the other hand, the climates of the interglacial periods were as genial as our present climate if not more so. The flora and fauna that existed at the end of the Tertiary period in Britain immediately prior to the Ice Age were very different from those we have today. The general effect of a succession of glacial periods has probably been to impoverish the variety of our plant and animal life, although we have unfortunately no continuous record because rocks representing the 30 million years before the Pleistocene are virtually absent from Britain.

One would expect a series of climatic oscillations to lead to an impoverishment. Refrigeration means extinction for those plants and animals which cannot survive it. When the climate reverts to an interglacial one the formerly glaciated area will be repopulated, usually from the nearest glacial refuge of the plant or animal concerned (fig. 6.12). Glacial refuge must be understood to mean that locality where a given plant or animal survived during a glacial period. One must avoid the picture of plants and animals deliberately seeking shelter in inclement weather: unfortunately there is some implication of such an image in our general use of the word refuge.

In a theoretical example one can imagine 100 species poised to repopulate Britain. The refuges of some may be in the unglaciated parts of Britain; others may have survived only on the Continent. Their general success in the reoccupation of the territory climatically suited to them will to some extent depend on the nearness of the refuge but also on their speed of movement, voluntary or involuntary. Freely-moving vertebrate animals, especially birds, are ideally adapted for reoccupation; so are plants with light wind-borne seeds, especially those with wings, but trees with fruits such as acorns or chestnuts must spread more slowly. Yet anomalies are provided by involuntary transport. Imagine a bird that paddles in mud and eats elderberries. Provided that its digestive processes are slow it might travel some distance before the ingested seeds are deposited, often with their chances of germination improved by having their hard outer coat weakened and set, as it were, in their own manured seed bed. That same bird may have in the mud on its feet a couple of minute water snails, whose speed of reoccupying territory under their own steam would be negligible. Both elderberries and snails may fail to survive where they are deposited, but, if there is an intermittent bombardment with involuntary immigrants, colonization success will be achieved at some stage or another.

However, of our 100 species, perhaps half a dozen may fail to reoccupy their former territory, either through sheer bad luck or because they had not crossed the site of the English Channel before sea level rose to flood it and severed Britain from the Continent. Of the 94 left and forced out later by the next glaciation, again half a dozen may fail to reoccupy their former areas, and so on through the various glaciations that have occurred. It does not seem very likely that our theoretical decrease of six species per glaciation will be compensated by six new species not known in Britain before, because their glacial refuges will presumably be farther away still, although there have undoubtedly been examples of new introductions. The latter eventuality is especially likely if one interglacial happens to be a lot longer and warmer than another and should also occur at the end of the Ice Age when the richness of the flora and fauna could in the absence of man be expected slowly to increase.

The astonishing thing is how involuntarily mobile, immobile creatures can be. Perhaps this is because we tend to visualize major earth events, such as the isostatic recovery of Scandinavia or the repopulation of Britain with plants and animals, in terms of diagrams with exaggerated scales. This point has been stressed above in calculating the real gradient involved in what at first thought appears to be a very great upwarping. We should think of the time-scale available for repopulation in the same way; in terms of thousands and not of tens of years.

It was mentioned in discussing climate that many of the Ice Age climates had no exact present-day homologues and that because of the high altitude of the sun, there may have been quite warm local climates in regionally cold climates. We may expect to see southern species occurring occasionally in generally cold environments. There may be considerable differences between end-of-interglacial and beginning-of-interglacial conditions. The occupation at the beginning of an interglacial will reflect climatic suitability, distance of glacial refuge, and speed of migration. The distribution towards the end of an interglacial will be governed by climatic refrigeration and the ability of the species to survive in the deteriorating climatic conditions.

Over and above all it must be remembered that the size and habit of organisms controls the detail of the climate to which they are subjected. A snail living in the humid atmosphere of decaying leaf litter takes a molluscan view of climate, not a human one: large free-standing plants such as trees are more likely to be affected by the same aspects of climate that affect humans than are small plants growing low down in the general vegetation cover where the micro-climate would be appreciably different: winged, flying insects are more likely to be affected by the general regional climate than are those which spend their time crawling on the ground. We shall revert to these considerations in discussing the meaning of fossil plant and animal assemblages in Chapters 6 and 7.

Man has probably had as much effect on plants and animals as the oscillating range of Pleistocene climates, and this effect has reached its peak in postglacial times (Chapter 8).

The largest effect on vegetation has been caused by the clearance of land for cultivation and the extreme of this can be seen in the abuse of clearance which has led to soil deterioration and erosion. The resulting change of vegetation patterns is enormous. There have, too, been considerable effects on individual species; either through their desirability as elements in gardens or conservatories, or through human magpie instincts, or because of lack of interest shown in trying to preserve threatened species.

On animals the effects have probably been even greater. Whether for profit or sport or both, animals have been hunted and killed with the result that the present distribution of larger vertebrates bears little relation in many cases to their potential climate distribution. Some disappeared long ago, such as wolves in Scotland: others are in process of extinction now, such as the tiger in India and the rhinoceros in Africa, though for different reasons. It might be thought that this applied only to larger animals, but this is not so. Direct hunting of smaller animals is limited to such irritants as wasps and mosquitoes and such offending, if not offensive, creatures as fleas and bugs. Man's alteration of vegetation involves great changes in micro-climate. The general clearance for cultivation, for example, creates micro-aridity, because of bare soils, where micro-humidity, because of the humus layer, had previously prevailed. The resulting changes in the population of land snails was one of the greatest changes affecting these animals in the Pleistocene. One dreads to think what effects the present chemical warfare associated with farming will have on many of the groups of small creatures which do not directly arouse general human interest and hence interference.

Thus the distribution of plants and animals, whether directly through climate or indirectly through man, was greatly influenced by the Ice Age.

Some economic repercussions

Although this book is primarily concerned with natural history, the impact of the Ice Age upon many natural features has clearly affected their economic value to man, sometimes enhancing them, sometimes impoverishing them.

In the glaciated highlands it might seem that the impact was adverse. The ice sheets tended to scrape away weathered waste mantles and soils and to leave bare rock outcrops. Areas of disorganized and disintegrated drainage patterns became areas of raw, acid peat bog. Any slight agricultural assets the regions may once have had – and they can never have been very fertile given the type of rocks and climate prevailing there – were destroyed in the Ice Age. But there are some advantages, even if they do not fully compensate.

The usual glaciated U-shaped valley is more efficient as a dam site. The same area of masonry is needed to impound the same volume of water, other things being equal, but the same volume of water can be impounded with a smaller surface area, and hence with a lower evaporation loss, in a U-shaped than in a so-called, typical, V-shaped river valley. Furthermore, many glaciated valleys have gentler and even occasionally reversed gradients so that greater volumes of water can be ponded up for a given height of dam. These and such glaciated features as corries and hanging valleys offer more potential for hydro-electricity development than do unglaciated mountains, though our highlands are poor relations of the Alps and Scandinavia in their hydro-electric potential. Finally, in an age which is aesthetically thrilled and not repelled by grim crags and bare rocks, glaciated mountains have more tourist, mountaineering, rock climbing and winter sport potential than unglaciated mountains with their smoother outlines and more complete waste and soil mantles. Again, the income-earning potential of the Highlands of Scotland does not equal that of the Alps, but it is being exploited, most recently in the development of winter sports in the Cairngorms.

In lowland Britain the widespread deposition of drift has in places enhanced, and in others, reduced the natural fertility of the local soils. Where chalky boulder clay overlies such heavy intractable clays as the London Clay and the Gault Clay it may provide a soil which is very little easier to work but which is a great deal more fertile because of its calcareous content. Where it overlies Chalk, as in much of the western part of East Anglia, it provides a soil superior to the natural, thin, dry Chalk soils that one might otherwise expect. Where great expanses of sandy and gravelly outwash or other lithologically similar preglacial deposits occur, infertile, dry, stony, leached soils usually develop. Fortunately such areas, which are common in the North European lowland and give rise to extensive heathland, are rare in Britain: the Breckland of East Anglia, a region of conifer plantations, aerodromes, heaths and the greatest density of rye cultivation in Britain, is probably our best example. Beyond the glaciated parts the deposition of wind-borne silt has created one of the richest farming zones known, the famous loess belt. In Britain, its attenuated representative, the brickearth of the Sussex coastal plain south of the South Downs, was once the second most valuable stretch of farmland in Britain and much of it has even greater value now as non-urban building land. The most valuable farming land is in the Fens, themselves an accumulation of mild humus resulting from the last phase of Pleistocene variations of sea level, when with sea level rising a great swampy aggradation took place in the area and ultimately provided the enormously fertile peat soils, which are a wasting agricultural asset of modern Britain.

Glaciation in various ways has breached watersheds (see Chapter 3) and this effect was important in communications, especially when railways were dominant since gradients are much more of a problem in railway than in road construction. The Ferryhill gap south of Durham is a well-known example of the use of a glacial overflow channel by a main trunk railway, the old main Great Northern route to Scotland. Another example is the use of the breach between Loch Eil and Loch Shiel by the line from Fort William to Mallaig; if not a trunk line, it is a route which would have been much more difficult without the breach. The A5 trunk road from Bettws-y-Coed to Bangor uses a breached watershed near Llyn Ogwen. Even the Cambridge to Liverpool Street line uses a glacially modified gap through the Chalk.

It might be thought that the oscillating sea level resulting in, and partly responsible for rias and fjords was another example of an Ice Age benefit. But unfortunately most of these – Norway, British Columbia, southern Chile, South Island, New Zealand, south-west Eire, north-west Spain and the Dalmatian coast – are backed by mountains, a necessary condition for the formation of good rias and fjords, so that the resulting harbours have very poor hinterlands. It is true that cruises in the Norwegian fjords and among the Dalmatian islands are tourist attractions and hence economically profitable, but this cannot apply yet to southern Chile. It is also true that the Norwegian fjords were ideal for hiding German capital ships in the 1939–45 war, but such usages are fortunately not regular. Recently, however, it seems that these remote harbours may have one real advantage in the provision of oil terminals, where giant tankers can discharge their cargoes in accessible deep-water harbours far from the main ports for which they are rapidly becoming too large. From the oil terminals it is now practicable to construct pipelines to centres of consumption and to refineries, if these are not on the spot.

Enough has been said to indicate the very great and very varied effects of the Ice Age on the world around us. We live in an altogether special period of time, in the shadow of past and possibly also future glaciation.

2

The Causes of Ice Ages

Most Pleistocene research workers are satisfied to try to decipher the meaning of the Pleistocene record in terms of its changing geographical environment. Occasionally some of them may pause to wonder what mechanism lay behind the changing geography. Sometimes workers in other fields, climatology for example, have become interested in the question of basic causes. It is not a field in which many people can dwell comfortably for a long time because it is almost entirely speculative. From time to time increase in knowledge about the sequence of events in the Ice Age may involve the rejection of older theories which explain too small a proportion of the facts and the birth of new theories which better explain those facts. There is probably no 'right' theory to explain the cause of Ice Ages, only a number of more or less probable ones.

One must bear in mind the following facts which require explanation:

(1) Ice Ages have occurred before the Pleistocene Ice Age. Evidence for these lies in the rocks but it is by no means unequivocal. The main form of ground moraine in the Pleistocene is till or boulder clay (see Chapter 3, p. 76). Essentially this consists of an unbedded mixture of clay and stones. Mixtures indistinguishable from such Pleistocene till have been reported from various geological horizons (fig. 2.1). But to say that these are fossil tills leaves open the more fundamental question of whether tills are always indistinguishable from certain other deposits of different origin. In the belief of some authorities they are not, for unsorted clay–stone mixtures may be formed in other ways. Turbidity currents generated on the continental slope, for example by earth tremors, may lead to a flow of unsorted waste down into greater ocean depths, provided that the ratio of solid to liquid is high.

Such slumped sediments, known as turbidites, are recognized in rocks formed in subsiding oceanic troughs (geosynclines) in the past, for example in the Lower Palaeozoic rock succession of Wales. Such turbidities have been confused with tillites (as fossil till deposits are called). It is also possible that some terrestrial mudflows, again provided that they remain very viscous, could simulate till. If the water content of any form of flow becomes too high it is most likely that the deposit will show bedding to some recognizable extent.

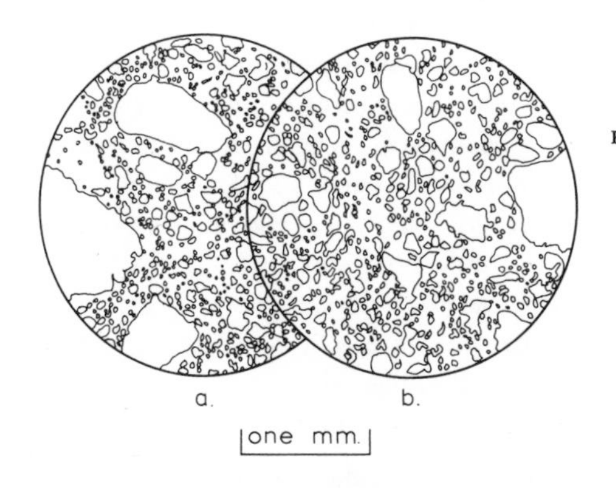

FIG. 2.1 Glacial deposits in thin section. a, Pleistocene till from Leeds; b, Carboniferous tillite from South Africa (after F. H. Hatch and R. H. Rastell, *Petrology of the Sedimentary Rocks*, London, Allen & Unwin, 1965, fig. 9).

Because of the possibilities of confusion in a too cursory consideration of the lithology, Flint (1957) has suggested that other features should be present as well. In addition to the unsorted lithology, he recommends that the deposits should include striated and faceted stones, and that there should also be evidence of a striated and smoothed rock surface below. Striation is a characteristic glacial feature (see Chapter 3, p. 59) which affects not only the rock below a glacier but also the stones within the till, which are usually striated parallel to their longest axes. The chalk pebbles of the chalky tills of eastern England are often striated in this manner, but they have to be cleaned for the striation to be seen.

If these criteria are applied, there seems to be strong evidence from South Africa, India and Australia of glaciation in the Permo-Carboniferous rocks. This, incidentally, was one of the pieces of evidence on which the theory of Continental Drift rested. All three areas are old shield areas, which are thought to have been formerly joined together as a continent known as Gondwanaland, which later broke up and drifted apart to the present positions. Such a mechanism would explain the presence of glacial deposits of the same age in such widely separated areas. Strictly speaking, they are not necessarily of the same age, as the Carboniferous and the Permian together cover a span of some 125 million years and precise correlation is very difficult. Similar evidence occurs in Scandinavia, especially in north-eastern

Norway, in the rocks known as sparagmites: these are either late Pre-Cambrian or early Cambrian in age and are thought to be the equivalent of the Torridonian Sandstone of western Scotland. There is less good evidence of glaciation at about the same level in the United States. A certain amount of evidence exists at other geological levels but it is not conclusive.

(2) The Ice Age was apparently preceded by a gradual cooling of the climate through the Tertiary period, which lasted some 60 million years. Of course, some of this cooling may have been due to shifting of the continents relative to the poles, but a similar pattern of very slow cooling has been deduced for both Europe and North America and steady cooling over the last 30 million years has been demonstrated from a series of temperature determinations on shells in ocean cores. These facts suggest a secular pre-Pleistocene cooling irrespective of position on the earth's surface. Similarly, the evidence of the study of palaeomagnetism, which fixes the position of the poles in the past, is against variations in pole positions in the Pleistocene apart from the well-known systematic variation between the magnetic and geographic poles. It should be noted that the rate of temperature change invoked for the Tertiary period is very much slower than that required for the onset of glacial conditions after an interglacial within the Ice Age.

(3) The Ice Age consisted, or consists, of a number of alternations between glacial and interglacial conditions with minor fluctuations imposed on this general pattern. Most of the minor variations seem to have involved the glacial periods. These are best known in the last glacial period in which a number of interstadials have been suggested. It is difficult to define precisely the difference between an interglacial and an interstadial. An interglacial period normally involves the development in the climatic optimum of a vegetation at least as temperate as that we have in Britain at present, i.e. mixed oak forest. An interstadial period implies something less than this: usually it means a climax dominated by a conifer such as pine or, in the case of a shorter amelioration, a phase of birch woodland between phases of Arctic treeless vegetation. Obviously it is theoretically possible to have a genial climatic interlude very difficult to designate either as an interglacial or as an interstadial. Fortunately there seem to be few such cases, at least in Britain. They are more likely to occur in warmer latitudes where the total interglacial change is less, so that the extent of the change is less easy to distinguish.

Apart from knowing that there have been fluctuations, we know very little about them in detail, especially about their rhythm. Before the presence or absence of a rhythm can be established, we must have an independent form of correlation, namely some method of radiometric dating. Without this it is difficult to see how our knowledge can advance. If we think of one of the older glacial periods, we can imagine a faunal fluctuation towards warmth enclosed between two cold deposits. At the start we do not know whether it is a reflection of a relatively long-term, general phenomenon or something very short-lived and local. It is impossible to say whether it is synchronous with a suspected similar variation in the same glacial period in another locality. The last thing that we must do is to have a model rhythm for major and minor climatic oscillations in the Pleistocene and force our evidence into that pattern. After all, what we are trying to do is to discover whether, in fact, there is a pattern of major and minor rhythms.

Hence, a test for any hypothesis of the cause of the Ice Age is that it should provide for variations between warmer and colder climatic conditions. Beyond that we cannot at present go and this is a serious limitation to our ability to provide a stringent test to any hypothesis.

(4) Any hypothesis must account for the deduced range of climatic change. One cannot say the observed range, for the extent of climatic change is a matter for expert and inspired guesswork. The estimates are probably very near the truth, but it must not be forgotten that they are estimates. Hence they cannot be held as a complete defence against a theory which seems to demand somewhat smaller or somewhat larger changes.

The conditions generally agreed by expert estimates seem to be as follows:

(a) Climatic changes, which affected the whole world, affected both hemispheres simultaneously. The second assumption has not always been accepted in the past, especially by those predisposed towards astronomical theories involving alternate glaciations in northern and southern hemispheres. However, the study of ocean cores and the advance of radiocarbon dating make it more likely than not that glaciations in both hemispheres were synchronous.

(b) The variation of mean annual temperature (not, let it be noted, a very significant figure because of the way in which difference between summer and winter temperatures lies hidden within it) was of the order of 8°C in temperate regions and 6°C in tropical regions between glacials and interglacials. This is a field where the inference from biological and landform evidence is most involved: neither is easy nor foolproof.

(c) The range of climate led to a variability of the snow line of as much as 1300 m in places. It should be noted that this could be partly a direct climatic effect and partly a feedback of the type suggested in Chapter 1 resulting from increased precipitation rather than greatly increased cold.

To explain the facts and inferences described above specialists have had recourse to a variety of ideas and suggestions involving changes in various features of the earth and its atmosphere, changes in the earth's attitude as a planet, or cosmic changes which have repercussions on the earth's climate. Usually the possible changes are split into five or six different categories, but these can probably be reduced to three main classes:

(A) *Changes in the geography of the earth's surface.* Such changes may affect:

(i) The positions of the climatic belts in relation to the landmasses on the surface of the earth. Obviously if these have changed different parts of the lands could have experienced glacial conditions. Generally all such hypotheses involve only a variation in climatic distribution, whereas point (2) above demanded a world-wide slow cooling before the Ice Age and point (4) seemed to imply whole world cooling and warming within the Pleistocene to explain the facts as they are at present interpreted. So, hypotheses of this sort are not likely to satisfy one of the essential requirements.

(ii) The general relief of the earth's surface. The usual idea involved here is that a very large increase in elevation can bring about the formation of ice caps by increasing the amount of relief precipitation and especially by increasing that proportion of it which falls as snow, because of the lower temperatures of high elevations.

(iii) The gas content of the atmosphere. The composition of the atmosphere affects the absorption and reflection of solar energy both as a whole and selectively at different wave-lengths.

(B) *Changes in the earth's attitude as a planet.* Involved here are the characteristics of the earth's path around the sun and the attitude of the earth to the plane in which it orbits.

(C) *Changes in general cosmic conditions.* The majority of ideas in this class are concerned with the total emission or radiation from the sun and the way in which it may vary.

However, more fundamental than the above classification, which is really a classification of the location of causes and could hence be termed geographical were it confined to the earth, is a simple twofold classification into:

I. Those hypotheses which attempt to explain why Ice Ages have occurred only occasionally at the earth's surface. They could be called fundamental hypotheses.

II. Those theories which attempt to explain the alternation from cold to warm in the Pleistocene – it can hardly be called a periodicity for this is one of the questions which remain to be answered.

A number of ideas which have been put forward are discussed below. At the outset we must be clear that there is no basic reason why the Permo-Carboniferous glaciation of parts of the present southern hemisphere should necessarily be due to the same causes as the Pleistocene Ice Age, though it would be better to try to find a cause applicable to all major Ice Ages before adopting unique hypotheses. Similarly, there is no reason why one single idea should explain all aspects of the Pleistocene and, indeed, some of the theories discussed below are combinations of ideas.

It may well be that we require a basic hypothesis to explain why any Ice Age came into existence and another to explain the variations within it. The basic idea could well be of the sort which earth scientists usually call catastrophic, i.e. one representing a unique and non-repetitive departure from normal conditions. Ideas of this sort were common enough in the early days of earth science when, for example, the features we now attribute to glaciation were largely held to be the work of the Biblical flood. Some catastrophic event of the sort envisaged could have allowed a secondary rhythm to come into effect: for example, periodic variations in the earth's attitude as a planet, which were not enough to cause more than minor climatic fluctuations themselves, may have been superimposed on some general cooling caused by a basic catastrophic cause and so have given rise to glacials and interglacials. Similarly, a pattern of climatic variations produced by solar energy variations may have been coupled with a general increase in the world's surface elevation to give an Ice Age and its climatic oscillations. In addition to these one needs to consider possible feedback effects. These may augment changes brought about by other causes for a while, i.e. positive feedback, but later be self-defeating by the generation of negative feedback. Thus they may augment or supplement the variations due to other causes.

Examples of Ice Age theories

GEOGRAPHICAL CHANGES AT THE EARTH'S SURFACE

The distribution of the lands in relation to the climatic belts can be brought about by two main mechanisms: either the lighter continents have drifted over the denser sima below, or the poles of the earth have changed their positions.

The hypothesis is hardly tenable for the Pleistocene Ice Age. It may help a lot in explaining the present position of the evidence of Permo-Carboniferous glaciation in the southern hemisphere, assuming the disjointed areas to have been glaciated at the same time. One needs a mechanism to explain how the areas separated and, further, why they are so far from present possible glacial conditions. Continental drift and pole movements could have provided these causes.

Continental drift originated as a hypothesis to explain inferred climatic distributions in the geological past and certain morphological and structural fits on opposite sides of oceans, the Atlantic providing the best case. Erected primarily for the purposes of climatology, it came under fire as being physically impossible and was relegated for many years to the position of a museum piece, which was occasionally brought out, discussed in a passionate symposium, and then put back into its case. However, in the last twenty years it has been shown to be likely that rocks preserve the magnetism imposed on them by the earth's magnetic field at the time of their formation. If this palaeomagnetism is measured, the position of the earth's magnetic field and hence of the magnetic poles at the time of their formation can be assessed. These positions for older rocks are very different from the present pole positions. Certain reasonable assumptions have to be made about the relationship between the magnetic and geographic poles remaining constant. If one accepts these, it is difficult to escape the conclusions of pole movement and continental drift to explain the deduced pole positions.

Yet, at the same time as the study of palaeomagnetism makes continental drift and pole wandering a much more likely possibility, it also makes it clear that there has been very little alteration in the Pleistocene and hence as causes of the Ice Age these explanations are excluded. Furthermore, although this is the type of explanation that could explain the catastrophe as a whole it cannot explain the variations between glacial and interglacial conditions without recourse to extensive continental or polar movement within the Pleistocene, for which there is no evidence.

A theory using pole movement as its starter mechanism and then one of the feedback mechanisms described above to explain the alternations of climate in the Ice Age is that advanced in the later 1950s by Ewing and Donn (1958). According to this theory the North Pole is alleged to have reached its present position in the Arctic Ocean and the South Pole its position in Antarctica about the beginning of the Pleistocene. At that time the Arctic Ocean would have been ice-free. Being ice-free it contributed much moisture to the atmosphere and hence to precipitation on the surrounding lands. Once the North Pole had moved to its Arctic position cooling would have begun with the result that an increased proportion of the precipitation fell as snow, accumulated and ultimately led to the formation of ice caps. At this stage a positive feedback system started because the increased proportion of solar radiation reflected from the growing ice surface (the percentage of the radiation which is reflected is known as the albedo) resulted in an acceleration of the cooling. This was reinforced by another factor. With an ice-free Arctic Ocean the circulation of the atmosphere there would tend to be cyclonic, with inblowing winds transferring moisture and water from the Atlantic to the Arctic, but once colder conditions had become established there was a tendency for the formation of an anticyclonic system with outblowing winds repelling Atlantic influences and so accelerating the onset of glacial conditions. As the ice caps grew, sea level fell and the exchange of water over the Atlantic–Arctic sill decreased. The Arctic became colder and froze over. Here is the self-destroying phase, because, as the Arctic froze, the supply of precipitation to the surrounding lands declined and solar radiation was able to start a slow melting of the ice caps. As they thawed sea level rose, more Atlantic–Arctic exchange took place and this further accelerated the thaw. In fact a positive feedback system working in the opposite sense was initiated, until, according to the theory, there was so much precipitation that an increased part of it fell as snow, ice caps started to increase and things were back where they were originally with the initiation of another cold phase.

In effect, the system proposed is almost one of perpetual motion: phases when the speed of change in one direction is accelerated by positive feedback lead ultimately to a negative feedback which reverses the trend and leads to an acceleration of change in the opposite direction. The idea is ingenious, it involves no recourse to extra-terrestrial sources, and by a combination hypothesis it attempts an explanation of all features of the Pleistocene changes. However, it has met with considerable criticism. First, it does not seem very likely that the Poles reached their present positions at the beginning of the Pleistocene. Second, although many examples are known of positive feedback systems in earth science which ultimately lead to a counteracting factor which brings them slowly to a halt, the idea of a reversal and an acceleration in the opposite direction seems to demand too much unless some external factor intervenes.

Pole wandering and continental drift might be described as belonging to the geophysical level of geographical hypotheses. It is possible to imagine that Ice Ages are due to variations in the amount of heat transmitted from the interior of the earth. Heat is generated in the earth by the decay of radioactive minerals and its transfer to the surface is indicated by the fact that there is a thermal gradient within the earth. If we assume that there have been periodic or aperiodic variations in the generation of such heat, we have another type of geophysical hypothesis which might be used to explain the Ice Age. But in this case it is very much arguing in a circle, for we deduce our variations in radioactivity from an assumed knowledge of the glacial–interglacial fluctuations and then use the deduced variations in radioactivity to explain the climatic fluctuations. The idea is incapable of verification – but it could be true.

The general class of geographical hypotheses also includes those which refer to increased altitude or increased continentality as causes of Ice Ages. Now, in general terms it is true that the Pleistocene Ice Age followed the mid-Tertiary Alpine orogeny and hence we live in an unusually mountainous phase of earth history. This would provide a good fundamental hypothesis for the Ice Age and incidentally a basic cause for the Permo-Carboniferous glaciation which followed the earlier Hercynian orogeny. It need not follow that all orogenies (major phases of mountain building) were followed by Ice Ages, because some mountain building episodes may have occurred in the Tropics where glaciation would be very unlikely. After all, mountain building is located where great thicknesses of sediment accumulated in geosynclines have been crumpled up and the location of geosynclines does not seem to follow any clear pattern. So the hypothesis needs to be tested not only by reference to the time of mountain building but also by reference to its location in relation to the pole. The two factors may have to coincide for the formation of an Ice Age.

Furthermore, it is well known that mountain building cycles do not suddenly start and as suddenly cease. Although the climax of Alpine folding was probably late Oligocene, folding in different areas had occurred a long time before and, if this had resulted in increased elevation, it might help to explain the very slow cooling through the Tertiary period.

So far an impressive case seems to be building up for the elevation hypothesis. But there are snags. The Pleistocene Ice Age was delayed 25 million years or so after the climax of the Alpine orogeny, though not necessarily after the maximum elevation reached by the mountains, and then its onset and fluctuations were seemingly rapid. It is true that there were substantial amounts of uplift in the Pliocene and early Pleistocene. Embleton and King (1968) cite uplifts of 2000 or 3000 m in this period for various mountain ranges. Here we are very near the same circular argument dangers we have met before. It is easy to fall into the trap that the explanation of the Ice Age demands that very late Tertiary uplift shall have taken place and, as a result, moulding the evidence to fit the theory. Further, although general elevation consequent upon mountain building can provide the fundamental cause of an Ice Age, it cannot explain the alternation from cold to warm in the Ice Age except by a colossal and rapid waxing and waning of elevation, an improbability which none of the adherents of the hypothesis have suggested.

Finally, among the geographical hypotheses is included variation in the gas content of the atmosphere. To understand how this might work involves some knowledge of the insolation–radiation balance at the earth's surface. Incoming radiation from the sun is essentially short-wave radiation, which is far less absorbed by the constituents of the earth's atmosphere than is long-wave radiation. Outgoing radiation from the earth, on the other hand, is long-wave radiation beyond the red end of the spectrum and not visible as light: such radiation is much more susceptible to absorption by the components of the atmosphere, especially water vapour and carbon dioxide. Here is the basis for a possible cooling mechanism: decrease the water vapour or the carbon dioxide and more outgoing radiation can get into space and so upset the insolation–radiation balance. Hence, a decrease in water vapour or carbon dioxide might lead to a general cooling of the earth and so produce glaciation.

The problem here lies in suggesting a reason why the proposed decrease should ever have happened. It is much more likely to be part of a general feedback effect resulting from a glaciation but not causing it. Given a glaciation, there might have been two possible effects. The solubility of carbon dioxide is increased at low temperatures, so that much more could have been dissolved in the earth's oceans and seas, which, however, would have been decreased in volume. At the same time the decrease in vegetation in cold regions would have led to the reduction in the amount of carbon dioxide supplied to the atmosphere by vegetation decay and humus development. Similarly, if the atmosphere cooled appreciably, the water vapour content may well have decreased as well. But, let it be repeated, these are results, not causes of an Ice Age, though it is conceivable that they may have accentuated the cooling.

Another atmospheric cause is the possible absorptive effect of clouds of volcanic dust. The 1883 eruption of Krakatao furnished much dust and consequent magnificent sunsets, and such eruptions might decrease the insolation received at the earth's surface. However, this is unlikely to have a global effect, nor could it account for the alternating warm and cold climates of the Pleistocene. Further, there have been geological periods in the past with paroxysmal volcanic activity and no Ice Ages. It has also been observed that recent eruptions have had little measurable effect. Finally, it is arguable whether the dust would interfere more with incoming than with outgoing radiation: if the latter it could lead to a warming and not to a cooling.

CHANGES IN THE EARTH'S ATTITUDE AS A PLANET

There are a number of effects here which might cause variations in the heat received at the earth's surface and hence promote glaciations. The effects, which require explanation, are:

(1) The precession of the equinoxes.
(2) The obliquity of the ecliptic.
(3) The eccentricity of the orbit.

Let us consider each of them in turn.

The earth revolves in an elliptical orbit round the sun (fig. 2.2A), the nearest position to the sun being known as perihelion and the farthest position as aphelion. At the same time the earth rotates about its axis which is inclined to the plane of the ecliptic (fig. 2.2B). The inclination is about $23\frac{1}{2}°$ from the vertical at present and in its present position the northern hemisphere summer occurs in aphelion and the southern summer in perihelion (fig. 2.2B) with consequent results on intensity of insolation. But the earth's axis also revolves, rather like a wobbling top, around the direction normal to the plane of the ecliptic, so that in time the attitude of the earth will change from that shown on the left of fig. 2.2C to that shown on the right by a rotation shown in polar view in fig. 2.2D. This means that in time the northern summer will occur in perihelion and the winter in aphelion with consequent variations in heating and cooling. The effect is known as the

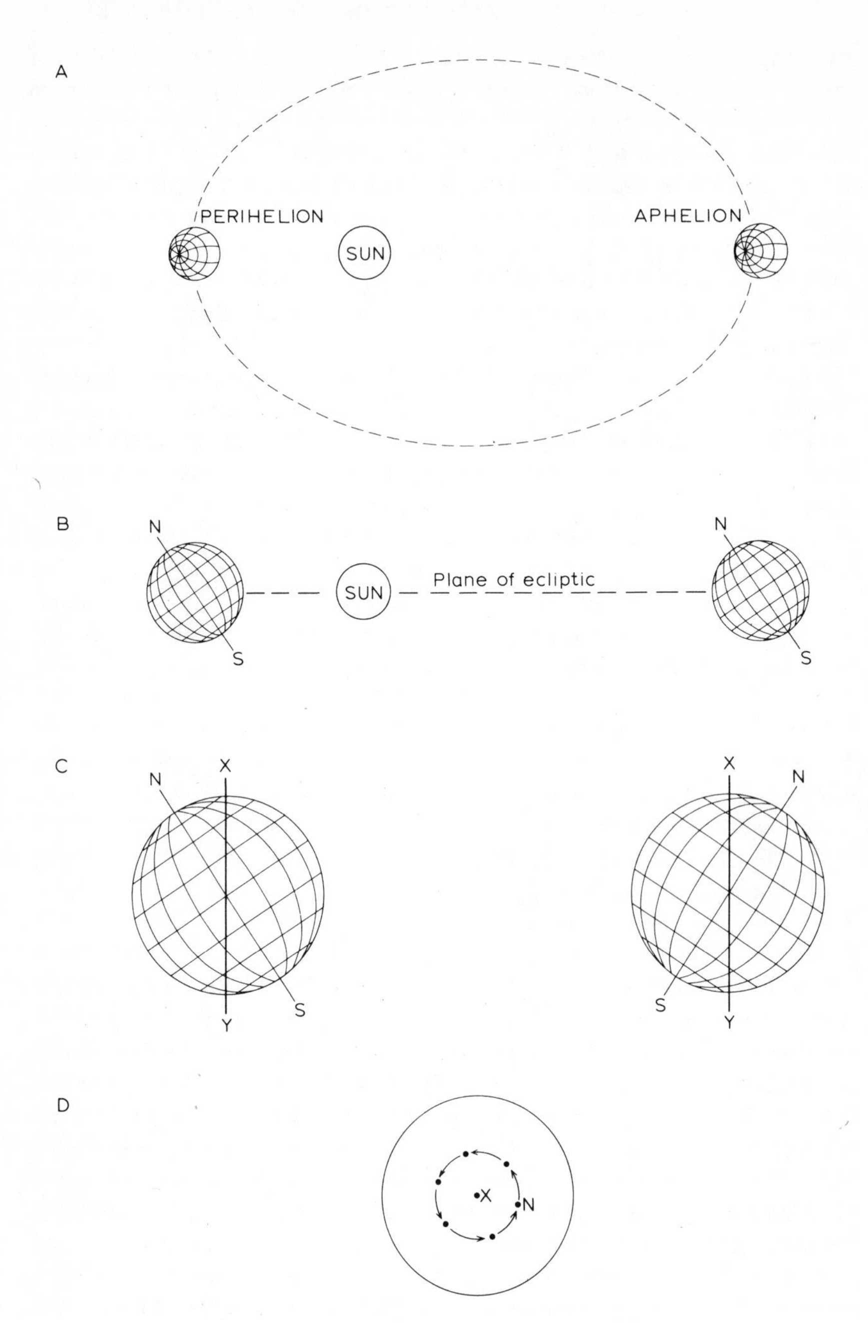

FIG. 2.2 Variations in the earth's orbit. For explanation see text.

precession of the equinoxes and has a periodicity of 21,000 years. If we believe that short hot summers and long cold winters (the position when the hemisphere concerned has its summer in perihelion) lead to ice accumulation and vice versa (the position when summer is in aphelion), we have a mechanism giving alternating cooling in different hemispheres. It has been alleged that the effect would be a sufficient lowering of temperature to usher in a glacial period.

Added to this is the obliquity of the ecliptic, that is, the angle between the NS and XY axes on fig. 2.2C. This can vary from about $21\frac{1}{2}°$ to $24\frac{1}{2}°$ and obviously if a maximum occurs when the hemisphere in question has summer in perihelion it can heighten the seasonal contrast. The periodicity is 40,000 years.

Finally, owing to the perturbation caused by other planets, notably Venus, the eccentricity of the orbit can change from almost circular to pronouncedly elliptical (fig. 2.2A). This again will obviously affect the relative intensity and duration of the seasons in different hemispheres. The periodicity is 92,000 years.

The importance of the precession of the equinoxes formed the basis of a famous nineteenth-century theory put forward by Croll, and, with the other two effects, of a radiation curve calculated by Milankovitch and used in this century to account for the variations of climate in the Ice Age. Milankovitch's curve, which combines the three effects, was used as the basis of a general scheme, notably by Zeuner (1959), for classifying the events of the Ice Age.

The combined curve does not require glaciation alternately in different hemispheres, but it does not allow for synchronous glaciation. It is not completely periodic because of the interaction of the three factors. However, it runs into a number of difficulties. It can only explain the alternation of colder and warmer conditions and hence is not a fundamental explanation of the cause of the Ice Age. It can only explain a variable distribution of heat on the earth's surface and not a total increase or decrease. Therefore, it does not explain the pre-Pleistocene general cooling nor the overall cooling in the glacial phases of the Pleistocene. Indeed, if it implies cooling in one area it must imply corresponding warming elsewhere, a suggestion which does not seem to be borne out by the evidence. Furthermore, it has been alleged that so many simplifications have been made in radiation calculations that these are probably not very accurate and are in any case insufficient to account for the changes evidenced by the deposits.

By providing a predetermined pattern of warmer and colder periods it yields a pigeon-hole classification, into which one is tempted to put events which cannot yet be independently dated. Schemes such as this appeal strongly to scientists who crave a framework for their data at all costs.

CHANGES IN COSMIC EFFECTS

Suggestions have been put forward about cosmic causes that require unprovable catastrophes, such as the passage of the earth through cold regions of space or through clouds of cosmic dust. But most cosmic hypotheses have turned upon variations in solar radiation, which Flint (1971) regards as the most reasonable cause since the sun is the ultimate source of energy and since most present climatic changes seem to be related to solar phenomena.

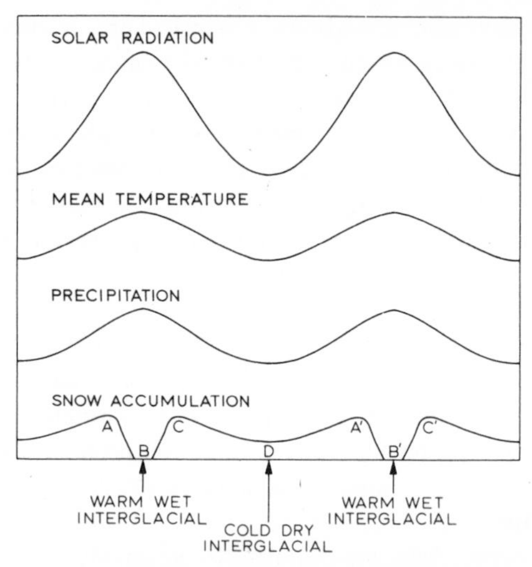

FIG. 2.3 Simpson's theory of glacials and interglacials (after Simpson, 1934, fig. 3).

An ingenious interaction between solar radiation and glaciation was suggested by Simpson (1934). It is illustrated graphically in fig. 2.3. According to this theory increased solar radiation results in greater storminess on the earth and hence more rapid and vigorous interchange between equator and poles of large masses of moisture-laden air. Thus precipitation increases in the polar regions and a large part of it still falls as snow because the earth has not yet greatly warmed up. Therefore snow accumulates until at point A on fig. 2.3 increased temperature offsets the snow accumulation and there is rapid melting leading to a warm and wet interglacial at B. With decreased radiation but still with fairly high precipitation snow will accumulate again to a maximum at point C. After that a fall in precipitation results in a slow decrease in snow accumulation to give a long, dry, cool interglacial at D. Then the cycle begins to repeat itself through A′B′C′. Thus two cycles of solar radiation are invoked to explain four glacials, A, C, A′ and C′, two short, warm, wet interglacials, B and B′, and one long, cold, dry interglacial,

D. This hypothesis would ensure synchronous glaciation in both hemispheres, and it produces a pattern which fitted very well into earlier model schemes for the Pleistocene with four glacials and one long middle interglacial. It implies, however, a high mean temperature in glacials, which does not seem to accord with the evidence, and also two types of interglacials for which there is no biological evidence.

A combined hypothesis making use of solar radiation variations is that advocated by Flint. His hypothesis suggests a comparatively slow and not very great fluctuation of solar radiation that normally would not cause a great enough climatic change to start a series of glacials and interglacials. But in periods when the earth possessed large areas of abnormally high land, the radiation and relief factors combined to cause the Ice Age. Flint produced a lot of evidence to support a Pliocene and early Pleistocene increase in elevation in many mountainous areas. Much of this is evidence of uplift, and not necessarily of mountain building. It could well be that in the phase of intense mountain building compression holds the ranges down and they only rise into isostatic equilibrium when the pressure is released.

Not all hypotheses have been covered in this brief survey, but a fair sample of the range of more or less probable ones has been included. There is no perfect hypothesis and until we know a lot more about the detailed chronology of Pleistocene variations – and this means awaiting new and better means of radiometric dating – we lack the primary means of testing any hypothesis. At present one can infer a cause from limited evidence and then fit the evidence into the inferred cause only too readily.

3

Glacial Landforms in Britain

In this chapter we shall consider the type of landforms impressed on Britain by the Ice Age irrespective of their part in the general succession of the Ice Age. The latter will be discussed in Chapter 5.

Snow and ice

When more precipitation falls as snow in winter than can be thawed out and drained away as water in summer, accumulation exceeds ablation, which is the general term for the transformation of ice into water and vapour. There are a number of smaller factors which may intervene but they do not greatly affect the gross balance outlined above. For example: snow may be blown away by wind; direct sublimation from snow to water vapour is possible; snow may thaw and refreeze within the snowfield; rime may be precipitated directly from the air. The economy or mass balance of a snowfield is a very important study, which bears very closely on future glacier movement with all that this means in terms of landforms.

When snow falls it is in the form of crystals with a hexagonal symmetry. It is a matter of common observation that the size of snowflakes varies considerably. When temperatures are not far removed from freezing point the flakes are large because they consist of aggregates of crystals. Such snowfall occurs usually when warm fronts or warm occlusions are pushing against an anticyclone centred over Europe: the result is usually heavy snowfall well to the west in Britain (fig. 3.1A). Such fronts often have only a very small component of the pressure gradient at right angles to them and remain almost stationary or even retreating. Precipitation falls from the warm sector, which is aloft (fig. 3.1B), through the cold continental air in front and there freezes to snow. Far different from this is the situation when biting easterly winds blow over East Anglia bringing light snow flurries from the North Sea. The snow falls in powdery form and drifts almost like smoke, quite unlike the heavy, wet snow produced by warm fronts.

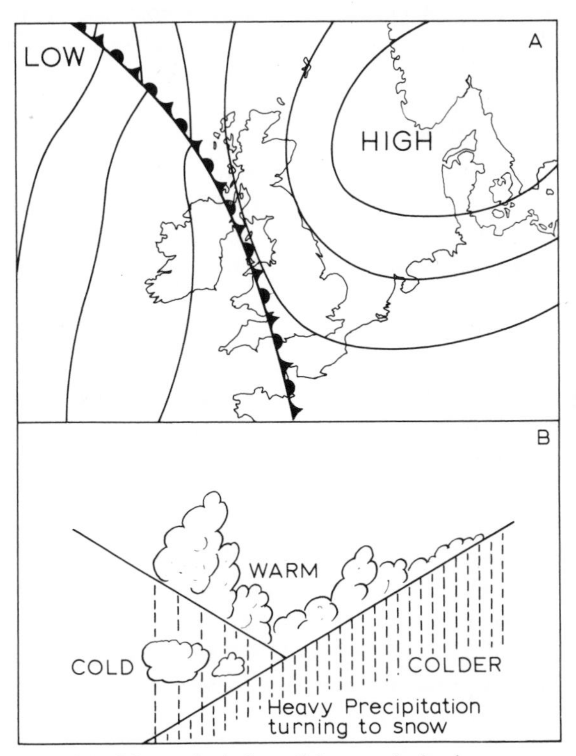

FIG. 3.1 Meteorological conditions producing heavy snow in Britain in winter.

Once snow falls the process by which it is converted into ice begins. Pressure at the points of contact between the grains results in local melting there and refreezing in the interstices between the grains. This leads to a general process of compaction and the degree to which it operates varies enormously with the temperature and the type of snow. Any small boy knows this. The heavy, soggy snow precipitated ahead of a warm front only needs a couple of quick presses with a pair of gloved hands to produce a dangerous ball of ice, whereas the dry, cold, powdery snow is virtually useless as ammunition, for, even when pressed together, it hardly coheres. The process of compaction is also clearly visible in the late stages of thawing after a heavy snowfall, when the fallen snow almost consolidates to granular ice before it finally changes to water.

This may help us to understand the varying rates of compaction of snow into ice. In areas such as the Alps the transformation of snow into ice, because of the comparatively elevated temperatures, can be very rapid. The same may well have been true of Britain in the Ice Age, if the cause of the latter was excessive snow rather than excessive cold. On the other hand, in Antarctica and comparable areas the fine powdery snow is only very slowly converted to ice because of the very low temperatures.

Pressure induced by the weight of the overlying snow effects the transformation into ice, which involves a very great increase in relative density. Fresh snow has a relative density of roughly 0·08 – hence the saying that a foot of snow roughly equals an inch of rain (this is one of those calculations which, unfortunately, will not go easily into metric units). After a year or so the density may be increased to about 0·5 as the result of the pressure of the overlying snow. At this stage the granular ice aggregate is usually known as firn or névé. The precise meaning of these terms is open to discussion, but irrespective of whatever original meaning they may have had, they are usually now used to designate the half-way stage between snow and ice. With age this material is metamorphosed – and the word is correctly used because recrystallization takes place – into true glacier ice with a relative density of about 0·9.

The time taken for the conversion into ice varies greatly with the temperature of the region. Embleton and King (1968) quote figures of about 25–40 years for conversion into ice in parts of the Alps and of 150–200 years in Greenland. This is one of the main differences between what are termed temperate and cold glaciers. Whatever the temperature conditions of the snow or firn field, it ultimately becomes converted to ice and then starts to flow.

Ice flow

Our primary object is to understand the landforms moulded by ice. In order to be able to decide on the rationality of our ideas we must know whether they square with what is known or surmised about the properties and flow characteristics of glacier ice. Although much of the measurement and observation of ice movement in the field has been made by geologists and geographers, theories of the nature and movement of ice have been mostly advanced by physicists not primarily interested in landforms. Partly because of this and partly because of incomplete knowledge, the links between ice movement and glacial landforms are in many cases still obscure.

The observation of ice flow is by no means easy, especially the observation of flow in depth. Surface movements are usually measured by surveying accurately the positions of stakes driven into the ice. Apart from the fact that such stakes tend to be thawed out by the melting of ice in summer – and to be of much use they should remain in as long as possible – the survey has to be of a high standard of accuracy, especially when it is being repeated at short intervals. Refraction effects due to very steep temperature gradients over ice can be a serious hindrance. The residual errors in a survey of a small Norwegian cirque glacier, Veslgjuv-Breen, made by Jackson and Thomas, were stated to have a possible maximum of 12 cm with the longest sight being of the order of 500 m. Greater accuracies than this have been claimed in other areas.

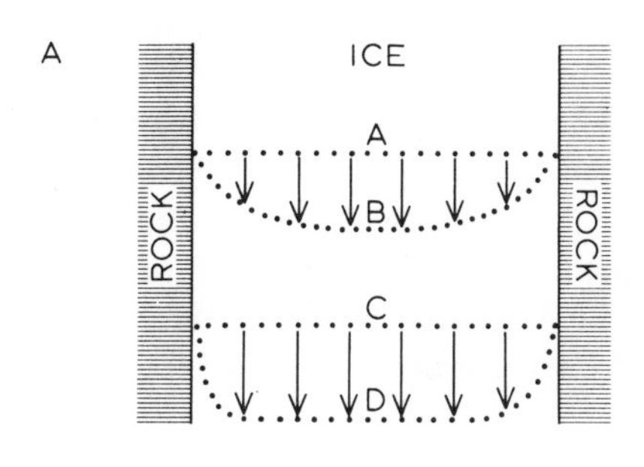

FIG. 3.2 Glacier flow. A, in plan; B and C, in vertical section.

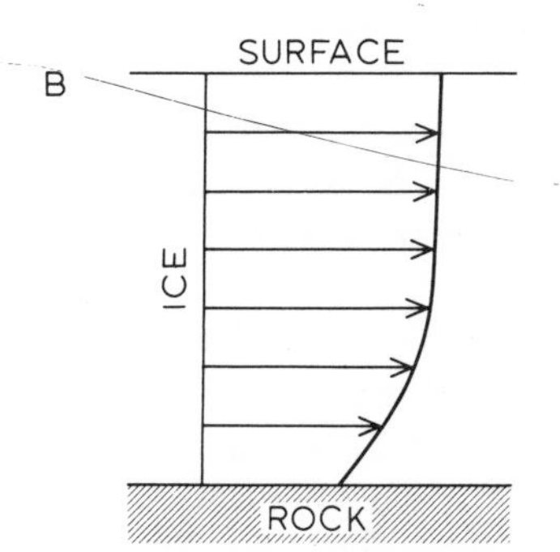

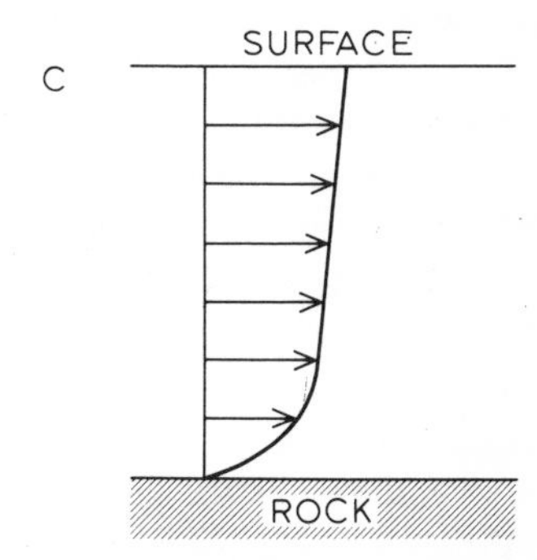

The observed cross-glacier surface velocity distribution is usually of the type indicated by the movement of the row of stakes from A to B in fig. 3.2A. Sometimes, however, a velocity distribution of the type shown by the movement from C to D occurs. It is called Blockschollen or plug flow and by allowing flow very near valley sides may contribute appreciably to erosion there.

The measurement of the velocity profile in depth is even more difficult because it involves drilling a hole in a glacier by some heat method and the insertion of a metal or plastic tube. Alternatively tunnels may be dug. Both methods are expensive either in terms of labour or of materials. Tubes have been inserted to depths of almost 500 m and their deformation measured by inclinometer. Although few results are available because of the difficulties of performing the experiment, those that we have show a vertical velocity profile of the type shown in fig. 3.2B. The rate of decrease of velocity with depth may be somewhat more than that shown, but no cases are known where velocity increases with depth. Again, how much the velocity drops off at the base depends on how much sliding over the bed contributes to the total glacier movement.

The two main components of glacier flow are sliding over the bed and deformation within the ice. The former is much aided by meltwater and so probably contributes greatly to the movement of temperate glaciers, which should have a velocity profile of the type shown in fig. 3.2B. On the other hand, cold glaciers, of which there have been no vertical flow measurements, certainly move more slowly and may be firmly frozen to their beds, so that shear does not take place between the ice and the bed but within the lowest layers of the ice (fig. 3.2C). It is reasonable to infer that erosion will be at a maximum in temperate glaciers with high sliding velocities and with plug flow, because these lead to the highest ice velocities at the base and sides of the glacier.

The complete absence of any increase of velocity with depth rules out the hypothesis of extrusion flow which was once so popular. This advocated greater velocities of ice flow with depth and hence greater glacial erosion at depth but does not seem to agree with the facts as far as they are known or, indeed, seem very plausible as one would expect the surface ice to be carried forward on faster-moving lower ice unless there was strong frictional drag against the sides of a narrow valley.

Variations in velocity also occur along a glacier. In the névé area, where the main process is the compaction and transformation of snow into ice, velocities are low. They are also low near the snouts of glaciers and consequently must be at a maximum in the middle reaches of glaciers.

Further, there are seasonal variations in glacier velocities. These probably affect temperate glaciers more than cold glaciers, but data from the latter are inadequate for reaching a final conclusion at present. It seems, however, that little seasonal variation is observable in Antarctica. The more rapid summer movements which characterize the main part of a glacier are probably due to the lubricating powers of the greater volume of meltwater available then. The meltwater also increases the extent of basal sliding. In cold glaciers the meltwater would not be available and the ice would remain firmly frozen to its bed. The possibility of meltwater increasing the rate of glacier flow is borne out by the correlation which seems to exist in some areas between temporary increases in the speed of glacier flow and the occurrence of heavy rainfall.

With all these variables it is hardly surprising that there are considerable variations in the rate of glacier movement. The normal velocity of valley glaciers is measurable in cm per day and a speed of 100 cm per day (cm/day) is rarely exceeded. The highest velocities observed, according to Embleton and King (1968), occur where gradients are steep and the accumulation zones very extensive, for example on those glaciers descending to the sea at the edge of Greenland, where velocities may be of the order of 20–30 m/day. Under certain conditions the whole glacier may get out of balance and a wave of high velocity, a kinematic wave or surge, may traverse the glacier. In such conditions velocities of two to six times the normal are observed and the whole process, which leads to a rapid transfer of ice down-glacier, is usually accompanied by a rapid but temporary advance of the glacier snout. Speeds somewhat in excess of 100 m/day have been observed in surges, though the reliability of such observations has been questioned.

Apart from the movement of the glacier over its bed, there is also the creep within the glacier that contributes to the general movement of ice. Ice behaves approximately in a plastic manner, i.e. it resists stress to a certain threshold value and above that deforms at a constant rate with time. The substance undergoing plastic deformation must undergo a permanent rearrangement of its particles without losing cohesion. If it loses cohesion it ruptures. The rearrangement of particles in glacier ice is probably brought about by more than one process. The ice crystals themselves probably move in relation to one another; there is evidence of some recrystallization, possibly some of it caused by pressure melting at the contacts between grains and refreezing in intergranular spaces of lower pressure; finally, there seems to be some shearing of individual crystals parallel to their basal planes. The processes are very similar to those experienced in metamorphic rocks when the stress is great.

Of these physical properties of ice probably the largest effects on landforms are caused by sliding between the ice and the sides and beds of glaciers, and also by the melting and refreezing of the ice. One would expect flow velocity to be important just as it is in river erosion. It is also true that at very low temperatures ice resists crushing more than it does near freezing point, so that it could probably transmit more pressure to its bed at very low temperatures. However, as we shall see, much remains to be learned of the relations between ice movement and landforms.

Aspects of glaciation in Britain

The general succession of conditions in relation to glaciation has been dealt with in Chapter 1. We do not really know whether ice sheets in Britain were cold or temperate, though the latter is probably more likely except just possibly at the maxima of glaciations in the centres of ice caps. Indeed, many of the landforms seem to demand temperate glaciers. Although the various facets of glaciation probably merge into each other we can probably distinguish the following with some justification:

(1) Areas above the ice in the mountains and hence subject to freeze–thaw as the main destructive process. Their size will have varied with the waxing and waning of the ice sheets but large parts of the mountain blocks of Scotland, the Lake District and Wales must have been in this condition for lengthy periods. Typical features are corries, arêtes and screes.

(2) Areas in the mountains overrun by ice. With these must be included low-lying rocky areas adjacent to mountains, for example the lower parts of the west coast of Scotland and adjacent islands. In these areas, scouring, grinding and plucking by the ice will have been mainly operative, together with whatever freeze–thaw action may have taken place beneath the ice. Typical features are the whole complex of forms associated with glacial troughs: truncated spurs, hanging valleys, steps, roches moutonnées, fjords etc.

(3) Lowland areas near the mountains where the ice sheets were centred. There may be some erosion here, but the moulding of glacial deposits is characteristic. Many glacial deposition features are preserved in this zone, because, being nearest to the mountains, it was obviously most affected by the minor readvances of the ice sheets in the waning phases of the last glaciation.

(4) Lowland Britain in general. In this zone one is not only considering more slowly moving and also possibly more temperate ice, but also very much softer rocks. There must have been a fair measure of erosion, but deposition and meltwater action were much more significant. Older, more subdued and more amorphous glacial deposition features are characteristic.

(5) Glacial interference with drainage. Although the main forms of interference vary from zone to zone there is overlap so that it is probably better to consider the various types of possible diversions together.

SCREES, CORRIES AND ARÊTES

Freeze–thaw and screes

Above the level of the ice, usually above valley glaciers because in the centre of glaciated highlands the ice cap probably often covers everything, the principal destructive force is freeze–thaw action. At the present time the number of freeze–thaw oscillations is greatest in the high alpine mountains of the world and not in the tundra, and the same was probably true of the Ice Age. In spring and autumn especially there must have been a great frequency of oscillations of temperature across the freezing point. Bare rocks, especially where steep south-facing slopes occur, would have heated greatly by day and cooled greatly by night. This is due mainly to the fact that rock is such a poor conductor of heat that all the daytime insolation is concentrated in the surface layer, which therefore heats considerably, while at night there is no flow of heat from the interior of the rock to compensate for the heat being radiated from the surface layers. The meltwater provided as a result of the diurnal heating gets into cracks in the rocks and freezes there at night, so creating stresses which literally prise the rocks apart. This is the simple account. It assumes that cracks are already present in rocks and that the full pressure of the expansions when water is converted into ice can be exerted against the confining rock.

It is probably true that there are nearly always pre-existing cracks or joints in rocks. Any structural pressures in the earth's crust exert stresses which result in the formation of two sets of fractures each at an angle of 45° to the principal stress. Such conjugate shear joints are even formed by very slight folding involving rocks in tilting of the order of a degree or two. In addition, rocks which are cooled and solidified from a molten state often develop shrinkage joints, and so, too, do sedimentary rocks when they dry out. Finally, many rocks are either formed deep in the earth's crust or have been buried there by mountain building and so have been subjected to enormous pressures. As the load of rocks above them is removed by denudation, the pressure is reduced and pressure release (or relief) joints are formed parallel to the earth's surface. For all these reasons rocks at the surface of the earth are fractured virtually without exception.

Were they not fractured the efficacy of the freeze–thaw process would be greatly reduced, as it would only be possible in a very much smaller number of rock types. For example, in the southern French Alps there are certain types of cellular dolomitic limestones characterized by ramifying and interconnected voids within the rock. Obviously such a texture is ideal for the entry of water into the rock which, given a location at the right elevation, can lead to great freeze–thaw activity and the production of very extensive screes, for example, east of the Col d'Izoard. Apart from such rather rare examples, water penetration and freeze–thaw would be greatly reduced in scale and confined to the tiny cracks weathered out between mineral grains. It cannot be too strongly emphasized that freeze–thaw does not usually crack rocks: it only forces rocks apart by using cracks already present.

It is not true that ice can always exert all its force against the rock. It can only do this if several conditions are fulfilled. Water reaches its highest density at 4°C and then there is a slight expansion with cooling to 0°C, at which temperature water is converted to ice with a sudden marked expansion of about 10 per cent. For maximum effect the freezing must take place in an enclosed space completely filled with water. Obviously water cannot get into a completely enclosed space and the next best thing is a deep crack with a very small surface area compared with the volume of water contained in it. One then assumes that, if the top freezes solidly first, pressure when the water below freezes will not be entirely dissipated in forcing up the ice plug. This seems less likely in open jointed rocks than in the cellular dolomite described above. For rapid freezing – and it may not be true that rapid freezing always or usually produces a greater disruptive effect than slow freezing – the latent heat given off should be quickly dissipated. But the poor conductivity of most rocks will militate against rapid heat dispersal. Finally, any dissolved gases in the water will form bubbles within the ice and this means that the pressure resulting from freezing can be relieved in two ways: either by forcing the rocks apart or by compression of the gas bubbles. It is patently obvious that the latter effect will take place first, so reducing the total ice pressure against the rock.

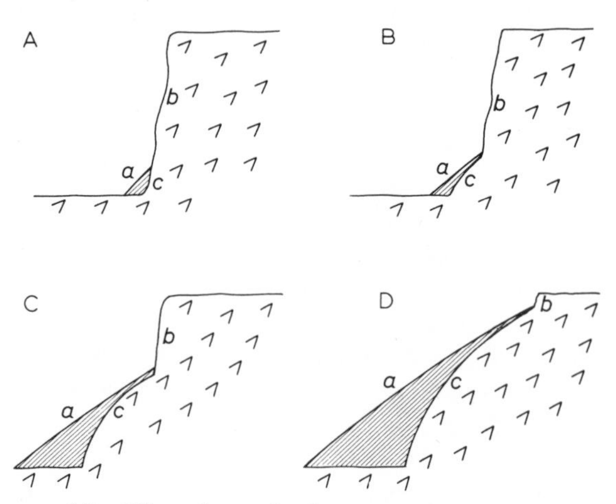

FIG. 3.3 Effect of scree development on buried rock profile.

Nevertheless the presence of screes (pl. 1) over much of the glaciated region of Britain is a clear indication of the wide effects of freeze–thaw processes. Scree development leads to the development of a constant slope (*a* on fig. 3.3) at the expense of the cliff, or free face, *b*, with the production of a convex buried face, *c*. The reason for the convexity is that as the scree, *a*, increases in area it will be fed by the cliff, *b*, which is decreasing in area, so that the rise of the scree will be slowed down. The result is that successive minute units on the cliff face are exposed to progressively longer periods of weathering and hence retreat progressively farther before being buried beneath the rising scree. The result is the convexity. The argument above assumes an equal density for rock and scree, which is not true because it will vary with the volume of voids in the scree. The greater the volume of voids the faster the rate of upward growth of the scree and hence the smaller the convexity developed on the rock below.

In practice many screes are funnelled from chutes in rock faces, follow existing hollows and minor valleys and so on. Many of the British screes are fossil. This can be clearly seen in valleys such as Dovedale in Derbyshire, where slope forms approximating to fig. 3.3C occur with the scree slope vegetated. Elsewhere the essentially fossil character may be less obvious. In many parts of Wales coarse, blocky scree has vegetation growing between the boulders, which are mostly firm and do not give when walked over. Turf and moss may even obscure the boulders but the presence of what is probably overgrown scree can sometimes be detected when one hears water flowing underground. Less obviously fossil are the examples where the scree is still bare, but lichens cover the upper surfaces of practically all the boulders. The Rhinog Mountains of the Harlech dome provide examples of most of these variants. A really active scree should have very much the same type of appearance, degree of stability and vegetation cover as the waste tips from such activities as slate quarrying. These are virtually man-made screes.

Of course, many natural processes are intermittent and in periods of inactivity their forms may appear to be fossil. The accumulation of stress in the material on a slope may take place over years until some trigger mechanism, for example an earth tremor, saturation by very heavy rainfall or deforestation sets a local catastrophic mass movement into operation. Screes, however, are probably not of this type. If they are still being added to, fresh stones should be visible on their surfaces. If they are active, they should be very tiring to climb because of movement, for example the Foxes Path up the north side of Cader Idris. On the whole, except possibly in the most mountainous areas, many of our screes and similar frost-shattered aprons of waste are probably inactive relics of the Ice Age.

Corries

Corries are essentially amphitheatric hollows in mountains (pl. 2), in which freeze–thaw, coupled with the action of the small glaciers that eventually come to occupy them, play an important role. There is no doubt whatever that they are Ice Age relic forms in Britain, for we have no examples of corrie glaciers and yet we have corries by the score in the main mountain areas. On the one-inch maps of Scotland they are usually termed ‘coire’ and on the maps of Wales ‘cwm’, though it cannot be guaranteed that every hollow so named is a corrie in the geomorphological sense. The same is true in France, where ‘cirque’ is the equivalent term.

In Great Britain corries are best developed or best preserved or both in mountain areas of resistant rocks, especially igneous rocks, though exceptions may be found to the latter generalization, for example in the resistant

Torridonian Sandstone of north-west Scotland. Quite typical is the preservation of corries on the hard volcanic and other igneous rocks of Snowdonia and Cader Idris (pl. 3) and their much greater rarity in the areas of Palaeozoic sedimentary rocks in Wales, for example the Berwyn Range. Although there are possibly simulating landforms to be considered below, the old nineteenth-century arguments of whether or not corries are glacial landforms are not worth raising for a number of reasons. Corries are intimately connected with present or past glaciated highlands. In the former they are often still occupied by corrie glaciers: the morphology of these glaciers has been most studied in the Jotunheim block of the Norwegian plateau, where a number of projects largely inspired by W. V. Lewis (1960) have been carried out.

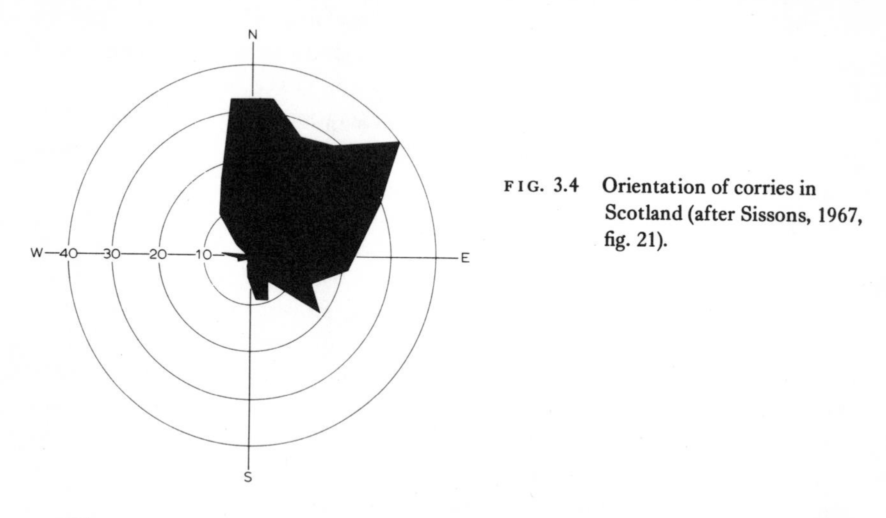

FIG. 3.4 Orientation of corries in Scotland (after Sissons, 1967, fig. 21).

Where corries occur they are nearly always orientated between north-west and south-east as Sissons (1967) has clearly shown for Scottish corries (fig. 3.4). There are two reasons for this. North-facing as opposed to south-facing means less insolation and the better preservation of small glaciers. The east-facing element is less explicable in terms of insolation, but it is generally believed that the main snow-bearing winds in the Pleistocene were from the west and eddies would ensure that snowbanks were preserved on the lee slopes, i.e. those facing east. It seems likely that corries develop in the first place from such snowbanks. Further, once corries had started to erode into the mountains, increased eddying would ensure that they received more than their fair share of the regional precipitation – almost an example of positive feedback. Indeed, in the present Norwegian corrie glaciers the thickness of the annual accumulation layers of ice suggests that the corries receive an annual precipitation above the regional average. This again confirms that

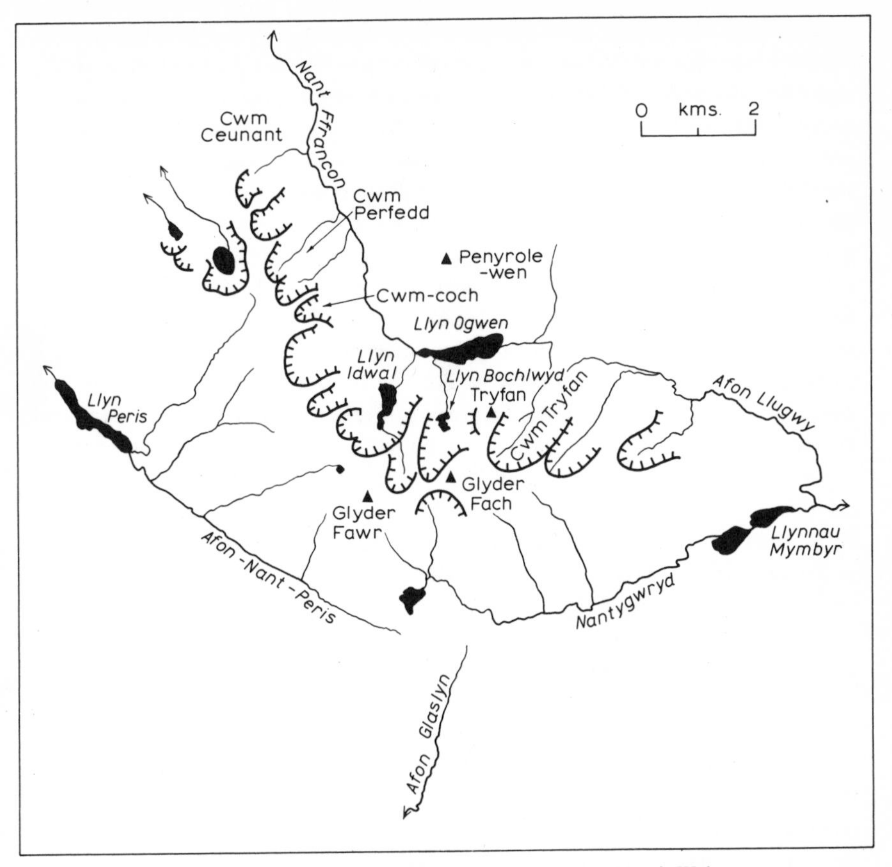

FIG. 3.5 Corrie distribution in the Glyder range, North Wales.

eddying is a main contributor to the snow accumulation total. General corrie orientation can be very well illustrated from the block of upland in North Wales containing Tryfan and the Glyders and lying between the Llyn Ogwen–Nant Ffrancon valley and the Llanberis Pass valley (fig. 3.5).

Yet, although correlation with glaciation is so firmly established as to need no statistics, many of the details of corrie formation are far from clear.

The ultimate origin of corries is uncertain. W. V. Lewis was a firm advocate of the idea that corries developed from snow patches lying in the slightest of preglacial hollows. This view of the formation of corries involved a snow patch in a hollow activating marginal freeze–thaw with the ensuing debris being removed by solifluxion until the hollow was large enough to be occupied by a small corrie glacier, after which stage the development of the corrie would be assured. We face here a risk which has often to be taken by those who wish to arrange landforms, concurrently existing in a landscape, into a time or development sequence. The danger lies in our ignorance of whether the various forms are arrested stages of development in a time-sequence or merely different landforms in a stage of finality. Between the snow patch hollow and the largest known corries the difference in scale is enormous. In the highest parts of the Himalayas and in Antarctica widths of corries range up to 15 km or so and heights of backwalls to 3000 m. This means that in the case of granite, which has a specific gravity of about 2·7, something of the order of one million million metric tons of rock have been removed, i.e. a true British billion tons, even though these are metric. It is difficult, if not impossible, to visualize such a quantity. Fortunately the corries of Britain are not of this order of magnitude. The largest are approximately 1 km across and 500 m deep. For the formation of such a feature the removal of a mere 700 million metric tons of granite would suffice. A train of standard 21-ton railway wagons to transport this would encircle the equator approximately six times or would stretch a little over half-way from the earth to the moon. These calculations are not precise; even the arithmetic is approximate. It has been assumed that corries are rectangular, that they are as broad as they are long, that their walls are vertical and that they have been cut into a constant-angle slope. If one halved the amount of material removed to allow for these assumptions, a formidable tonnage is still involved. Of course, how much of the corries was already there, formed by preglacial processes, is unknown.

It is very likely that many corries took advantage of pre-existing valley heads and that much of the material had already been removed. Nevertheless, the general form shows a very great modification by ice, and various ideas have been put forward about operative processes. An ideal corrie is shown in fig. 3.6. The backwalls of corries are shattered rock so that freeze–thaw must be regarded as the main process operating there. There can be little doubt that freeze–thaw is more active when the wall is not covered by the corrie glacier, as the greatest and most frequent temperature variations can then affect it. This would mean either that much of the shatter was accomplished before the glacier formed or that during the glaciation most shatter took place above the level of the glacier. It has also been alleged that freeze–thaw shatter takes place at the point where the bergschrund reaches the rock floor or within the randkluft. Both of these are major, more or less continuous crevasses found in some corrie glaciers (fig. 3.6). Various modifications have been made to these ideas, but several basic difficulties remain:

(a) Not all corrie glaciers have bergschrunds and not all bergschrunds penetrate to the rock floor. In fact some corrie glaciers must have been so large that it is very unlikely that such cracks could have remained open because of the plastic yielding of the ice at depth.

(b) Although meltwater may stream in from the top, especially down the randkluft, it is difficult to provide a mechanism to give frequent alternations between freezing and thawing. Observations made beneath the ice have not revealed fluctuations of the desired amplitude and frequency. Nor is it easy to see from first principles how they might develop. Any cold air formed by radiation cooling on and around the glacier will either flow off as a katabatic wind, a process which does not concern us here, or sink to the bottoms of the crevasses and so set up an absolutely stable stratification. Only direct solar heating or strong turbulence, neither of which is likely to occur in deep narrow cracks, could disturb it.

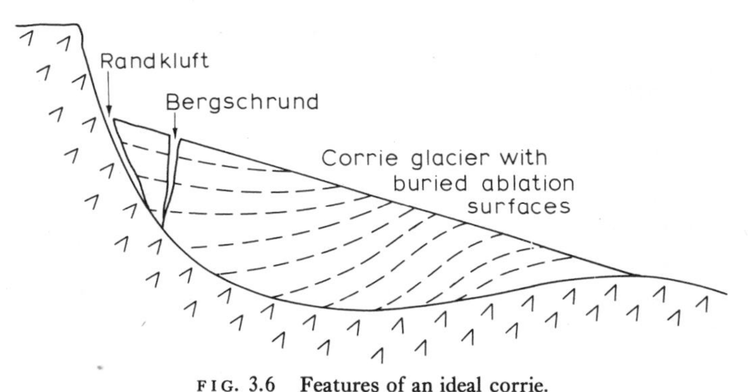

FIG. 3.6 Features of an ideal corrie.

(c) In this situation of small temperature changes, the problem of getting rid of the latent heat given off during freezing is awkward.

It may be that the ice merely drags away from the backwall blocks which have been loosened by earlier freeze–thaw. Yet there are limits to this process because of the low tensile strength of ice. If the forces holding the blocks in place exceed the tensile strength, the ice will merely break away from the rock, or within itself, and will not pull blocks of rock away.

However, by whatever process they are removed from the backwall, blocks of rock fall into the glacier down crevasses and are dragged by the ice across the floor of the corrie, grinding and striating it. McCall (in Lewis, 1960) calculated the likely effects of the movement of a metre cube of rock with a specific gravity of 3·0, i.e. something like a dolerite or a gabbro. Provided that the glacier was at least 22 m thick it could crush anything it was pushed against that had a bearing area of less than 18 cm^2, shear any irregularity with a cross-sectional area of up to 160 cm^2, or gouge out anything with a bearing area of less than 16 cm^2. Thus, the load in a corrie glacier could effectively smooth off and striate the rock bed beneath it. The fact that such actions took place can be very easily confirmed by observation.

Small corrie glaciers probably move over their beds mostly by sliding. The basic cause of this seems to be that ablation takes place down-glacier and accumulation up-glacier leading to a rotational movement to restore the balance. At one stage it was suggested that the annual ablation surfaces, shown on fig. 3.6, were in the nature of shear planes along which the ice rotated in discrete segments, but more detailed work has shown that movement is primarily a rotation of the glacier as a whole over its bed. Such a movement would help to explain the frequency of reversed slopes at the lip of the corrie by providing a mechanism giving a strong upward component to ice at the lip of the corrie. There is also an element of creep in the movement of a corrie glacier and a pronounced control exerted by rock features on the reversed slope at the lip. Haynes (1968) has discovered in corries in the Torridonian Sandstone areas of Scotland that reversed slopes are found where the rocks dip gently inwards into the corries and that they do not occur where the rocks dip gently outwards. At intermediate angles the effects of dip are less perceptible.

In some ways the corrie is a self-perpetuating system, i.e. yet another example of positive feedback. The form encourages rotational sliding, rotational sliding accentuates the form and so on. But there is even more to it than this. It has been suggested that pressure release joints will form parallel

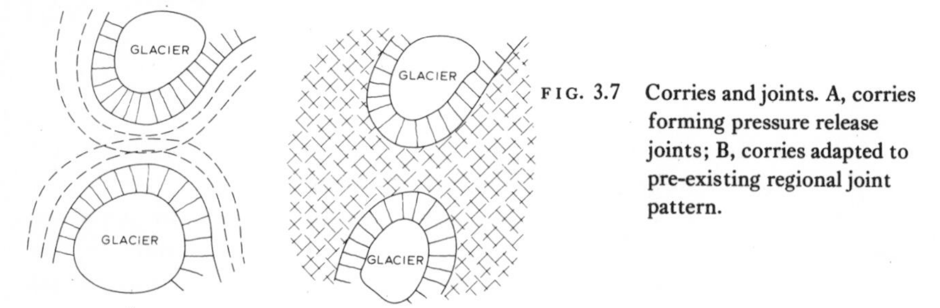

FIG. 3.7 Corries and joints. A, corries forming pressure release joints; B, corries adapted to pre-existing regional joint pattern.

to the corrie wall, as shown in fig. 3.7A, and so help perpetuate the form in yet another way. Observations by Battey in Norway (Lewis, 1960) tend to confirm this suggestion. However, it is always difficult to discover which caused what and the relations shown in fig. 3.7B, with an assumed regional joint pattern, caused by tectonic forces, controlling corrie form must also be considered. It must be remembered that the observation of joint patterns in the field is by no means as easy as in our diagrams, so that there is usually room for different interpretations.

Although there is still controversy over details, these essentially glacial features, left over from the Ice Age, can be seen to have a dominant effect on landforms in such areas as the Cairngorms, the Applecross area of Torridonian rocks, the eastern part of the Arran granite, the Cuillin Hills of Skye and Snowdonia.

Whether they affected the British landscape farther away from the real glaciated areas is much more difficult to decide. All the examples quoted are in resistant rocks. What would a corrie have looked like in soft rocks? Would it ever have had a bare rock wall? Even if it had, would it have been preserved? Might not a soft rock corrie have become merely a smoothly-rounded amphitheatre? A rock arrangement with near-horizontal beds and

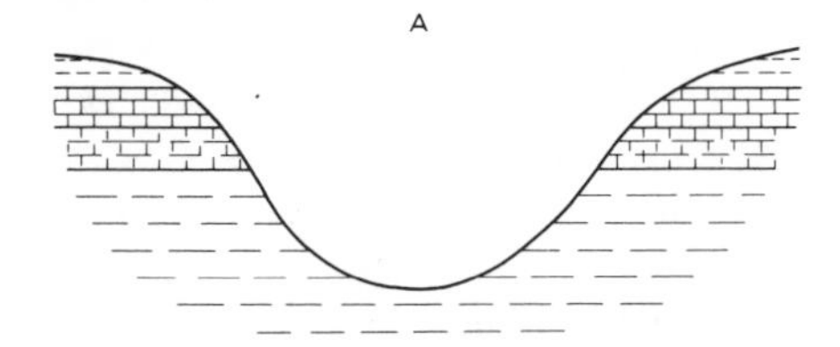

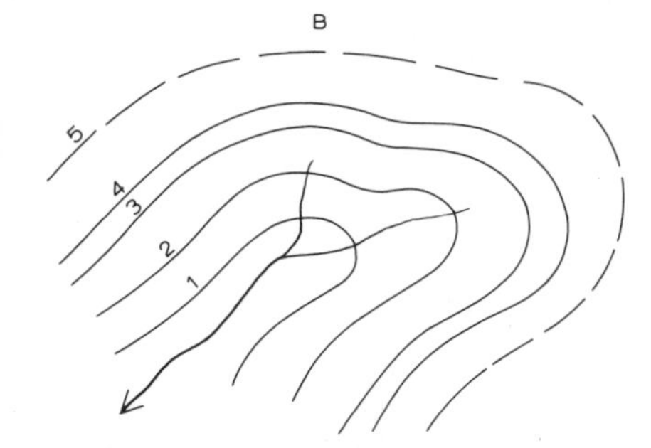

FIG. 3.8 Landforms in unglaciated horizontal rocks. A, U-shaped profile; B, rounded, corrie-like valley head.

harder beds towards the top of the valley-side slopes is very conducive to the development of U-shaped cross-profiles, typically a glacial feature (see below), and also to rounded heads to valleys (fig. 3.8B). There are plenty of areas in Britain where rock arrangements of this type are common, for example in the Pennines and in the various ranges of Old Red Sandstone mountains on the northern side of the South Wales coalfield. Are the forms there to be attributed to the effects of normal erosion on a special lithology or should they be regarded as subdued glacial features? The Black Mountains between Hay-on-Wye and Abergavenny provide ideal examples of such features and different interpretations are possible.

The same is true of Chalk lands. The highly permeable nature of both rock and waste mantle ensures the dominance of creep on the upper slopes, which results in very pronounced convexo-concave cross-profiles. The valleys are U-shaped and this has occasionally tempted people to call them glacial, even in the case of the South Downs which were never glaciated. In addition, armchair-shaped hollows in the escarpment face of the South Downs were interpreted by Bull (1940) as nivation hollows, though others would regard them as spring-heads or spring-heads modified by nivation, or nivation hollows modified by spring sapping. Undoubtedly solifluxion did occur on the South Downs and the presence of an ambivalent landform with an ambivalent deposit precludes certainty about the interpretation of such features.

In studying landforms one must always be prepared for the eventuality that very closely similar landforms may have been produced by different processes. There are usually tests which can be designed to separate them but the tests may be so arduous and expensive as not to be worth undertaking. A feature which might also simulate a corrie is the rotational type of landslip. This is not surprising as they are both to a large extent rotational slip features. Such features are fairly common on some slopes and probably date back to periglacial conditions in the Pleistocene in many areas (see Chapter 4). In the pristine state shown in fig. 3.9A they are hardly likely to be mistaken for corries, but if the slumped mass which has slipped from A to A′ has become thoroughly mixed up, the whole feature could look very much like an incipient corrie and moraine (fig. 3.9B). Violent arguments have been

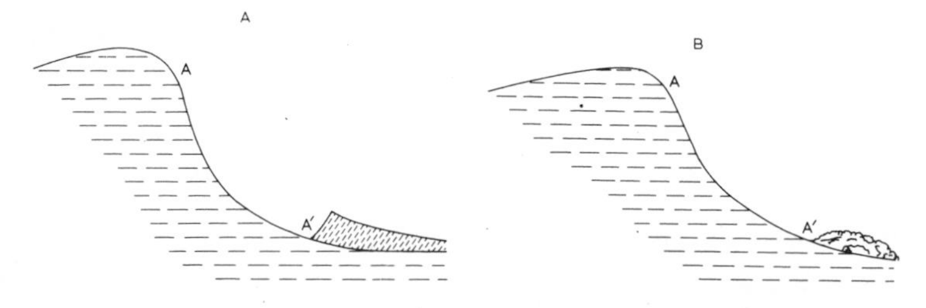

FIG. 3.9 Slipped mass, showing possible confusion with an incipient corrie.

known to arise in field parties on the Brecon Beacons over features such as these. However, spring-heads and the scars of landslips would provide ideal starting features for the development of corries, so that different processes may have supplemented each other in places.

Arêtes and pyramidal peaks

To return to our hard rock corries: as they bite back into the highlands, the ridges between them are reduced to knife-edge ridges known as arêtes (fig. 3.10), the profiles of which may be governed by pressure release joints. In fig. 3.10 are shown two intersections of groups of four corries; arêtes are forming between the corries and give rise to pyramidal peaks at x and y. Relief of this sort, which involves a fairly advanced state of corrie dissection, can be seen in north-eastern Arran (pl. 4), where Cir Mhòr is a good example of a pyramidal peak; in the Cuillins in Skye, the Ben Nevis group and the Mamore Forest according to Sissons (1967); in the Snowdon block where Crib Goch and Y Lliwedd are fine examples of arêtes; and in Striding Edge in the Lake District. Classic examples are found in high heavily-glaciated mountains such as the Alps and the Himalayas. One word of warning: although arêtes may be produced by pressure release joints parallel to corrie backwalls, our guess would be that the majority of British arêtes would not show any such features. Pressure release joints only occur markedly parallel to the present relief when no other joint system was present. Where other joints are present pressure is relieved largely through movement within the existing joint pattern. For example, on the north ridge of Goat Fell, down to the saddle between Glen Sannox and Glen Rosa in Arran, the joints are beautifully

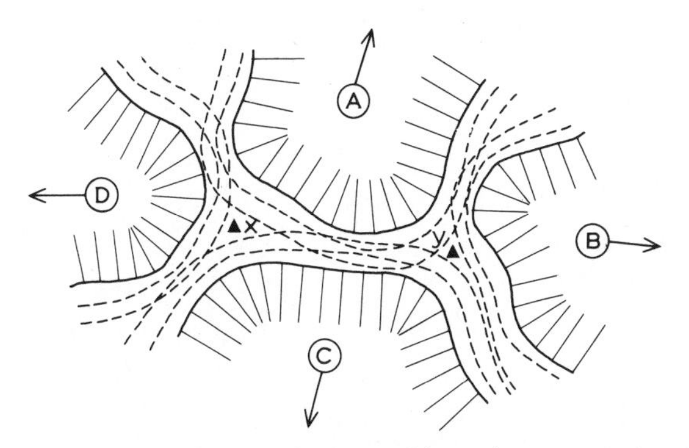

FIG. 3.10 Formation of arêtes and horns between corries.

parallel to the backwall of Coire nam Fuaran on the north, but, where they can be seen, they follow the same direction on the south side of the ridge instead of dipping down into Glen Rosa.

RELIEF DEVELOPED BENEATH ICE SHEETS AND GLACIERS

As ice sheets develop in mountains as a rule and extend beyond them on to the surrounding lowland still moving rapidly, some of the forms of erosion in the two zones must be similar. Thus, in Europe, Finland was mainly a glacially eroded area adjacent to the mountain block of Scandinavia, in North America the Laurentian Shield was overrun by an ice sheet, while in Britain ice flow from the main highland centres (fig. 3.11) spread outwards and was still rapidly moving over such rocky lowlands as the Lewisian Gneiss districts of far north-western Scotland and the Outer Hebrides.

If one visualizes an ice sheet building up to a great thickness and then moving out radially, grinding, plucking and tearing the rocks beneath, one is bound to envisage a tremendous amount of erosion. In Scotland the ice cap must have been at least 1000 m thick at times, and probably very much more in the opinion of some people. Yet the amount and location of erosion performed by such an ice cap is still a matter for discussion and argument.

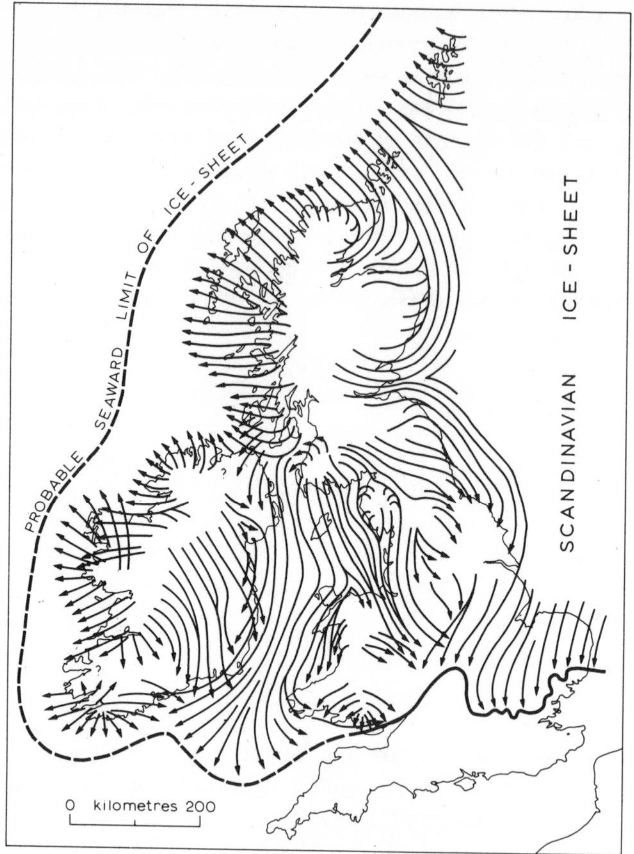

FIG. 3.11 Centres of ice formation and directions of movement in the British Isles (after W. B. Wright, *The Quaternary Ice Age*, 2nd ed., London, Macmillan, 1937, fig. 35).

Probably the main reason for this is yet another example of the circular form of reasoning which it is almost impossible entirely to escape in reconstructing the geography of the past. One assumes that an ice sheet must mark the landscape in varying degrees by smoothing and roughening the rocks in different places. Then, by plotting evidence of this type in the field, one uses it to establish the upper limits of the ice sheet.

Right at the start we may learn something from the physical properties of ice concerning its potential erosive power. Any substance has a certain degree of strength and if a force is exerted exceeding the strength of the material it will yield: usually the compressive strength is greater than the shear strength which, in turn, is greater than the tensile strength. For example, average values for these three strengths of granite are 1500, 200 and 40 kg/cm². Ice is very much less strong than granite and yields to much lower stresses, about 2 kg/cm² in compression and 1 kg/cm² in shear at 0°C. When these forces are exceeded, it deforms by creep – it does not fracture as the granite does at the critical forces quoted above. It is easy enough to calculate the thickness of ice needed to build up a force of 2 kg/cm². The specific gravity of ice is about 0·9 so that one cm³ of ice weighs 0·9 g. The length of column required to produce this force is equal to $\frac{2000}{0 \cdot 9}$ cm, i.e. 22 m approximately. So that once that depth is exceeded at 0°C, the ice deforms and cannot transmit greater downward pressures onto the boulders in its bed. Therefore, great ice thicknesses do not imply that there must be a much greater erosive force exerted through the boulders on the bed of the glacier.

The ice moving over its bed will exert two main actions. The debris carried along in its base, the ground moraine, will grind, scratch and groove the rocks over which it passes. Soils will be removed and incorporated in the moraine. The scratches and grooves will vary with the size of the fragments being carried. The general tendency to round underlying relief and to reduce the minor irregularities is probably due to the concentration of erosive action on the projections of the surface below. In addition to the striations, other minor features may be formed: these are mostly minor crescentic and conchoidal fractures and gouges, the exact mechanisms of the formation of which are not clear.

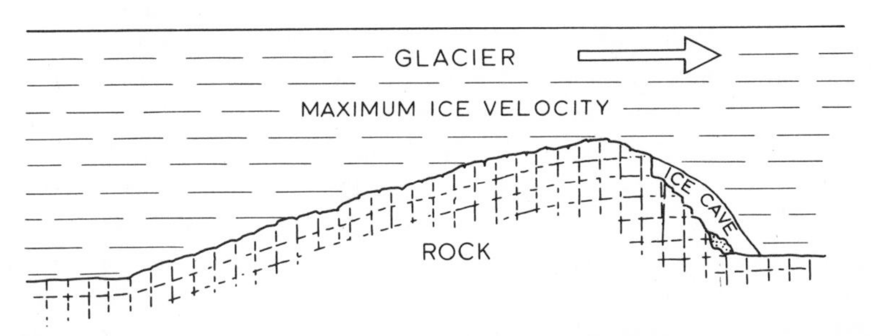

FIG. 3.12 Formation of a roche moutonnée.

The other main action is a plucking action exerted on the down-glacier, lee side of any projection in its path, whether this be an isolated rock or a step in the long profile of the valley. In a glacier in Switzerland Carol (1947) measured a maximum ice velocity near the top of a projecting rock, which suggests that maximum grinding may well occur there (fig. 3.12). He also suggested that freeze–thaw action might help glacial plucking in the lee of such upstanding rocks by invoking the following mechanism: on the upstream side of the rock the increased pressure of the ice against the rock lowered the melting point below the prevailing temperature of the ice and so meltwater was formed; in the cavity below the ice in the lee of the rock, where the pressure is greatly reduced, the melting point would be above the prevailing temperature of the ice so that the meltwater would refreeze there in the cracks in the rock. Some confirmation of this idea has come from later work, although the number of freeze–thaw cycles must be limited just as it is at the foot of the backwall of corries. Rocks which are smoothed on the up-glacier side and plucked on the down-glacier side are known as roches moutonnées (pls. 5 and 6). Similar effects are observed on the steps in glaciated valleys.

The plucking action on the lee side would normally be limited by the tensile strength of the ice, which is very much below that of the rock. When ice sticks to rock, it can only pull the rock away when the forces holding the rock in place are less than the tensile strength of ice. Otherwise the ice ruptures. This is why some preliminary shattering mechanism, such as the freeze–thaw action mentioned above, is vital. But the sheer drag of the ice on the blocks on the brow of the rock must also help. If this is pronounced, one would expect that the form would ultimately become more streamlined and might end as a rock drumlin, a smoothed striated form comparable in general shape with a true depositional drumlin (see below).

In many parts there are signs which might be interpreted as evidence of pronounced erosion: some of these are explicable by recourse to non-glacial factors, but others are not. The Lewisian Gneiss areas of north-western Scotland mentioned above look as though they had been ground almost out of existence by heavy glacial erosion. The landscape, sometimes known as 'knock and lochan', consists of a jumble of hills and mounds of roche moutonnée form alternating with an absolute maze of small lakes and peat-filled depressions where once lakes stood (pl. 7). The Lewisian Gneiss outcrops of western Sutherland and the Outer Hebrides, especially North Uist (fig. 3.13) and Benbecula, show this astonishingly well. But care must be taken not to attribute all this to glacial erosion in the Ice Age. If it had been due entirely to glacial scour and that scour had been very prolonged, one might have expected the rocky projections to have been largely eliminated. There are probably lithological factors involved here. The rock mostly affected, the Lewisian Gneiss, is very compartmented by faults, joints and dykes and these may well have contributed greatly to the chaotic landscape. Sissons (1967) also points to structural effects in Argyllshire round Loch Sween, where a linear structural pattern is strongly reflected in the glacial landforms. Much of the weathering may have been preglacial and penetrated more deeply along the fracture lines, so that the ice merely scoured out weathered material and trimmed up the rocky knolls between as it exhumed them.

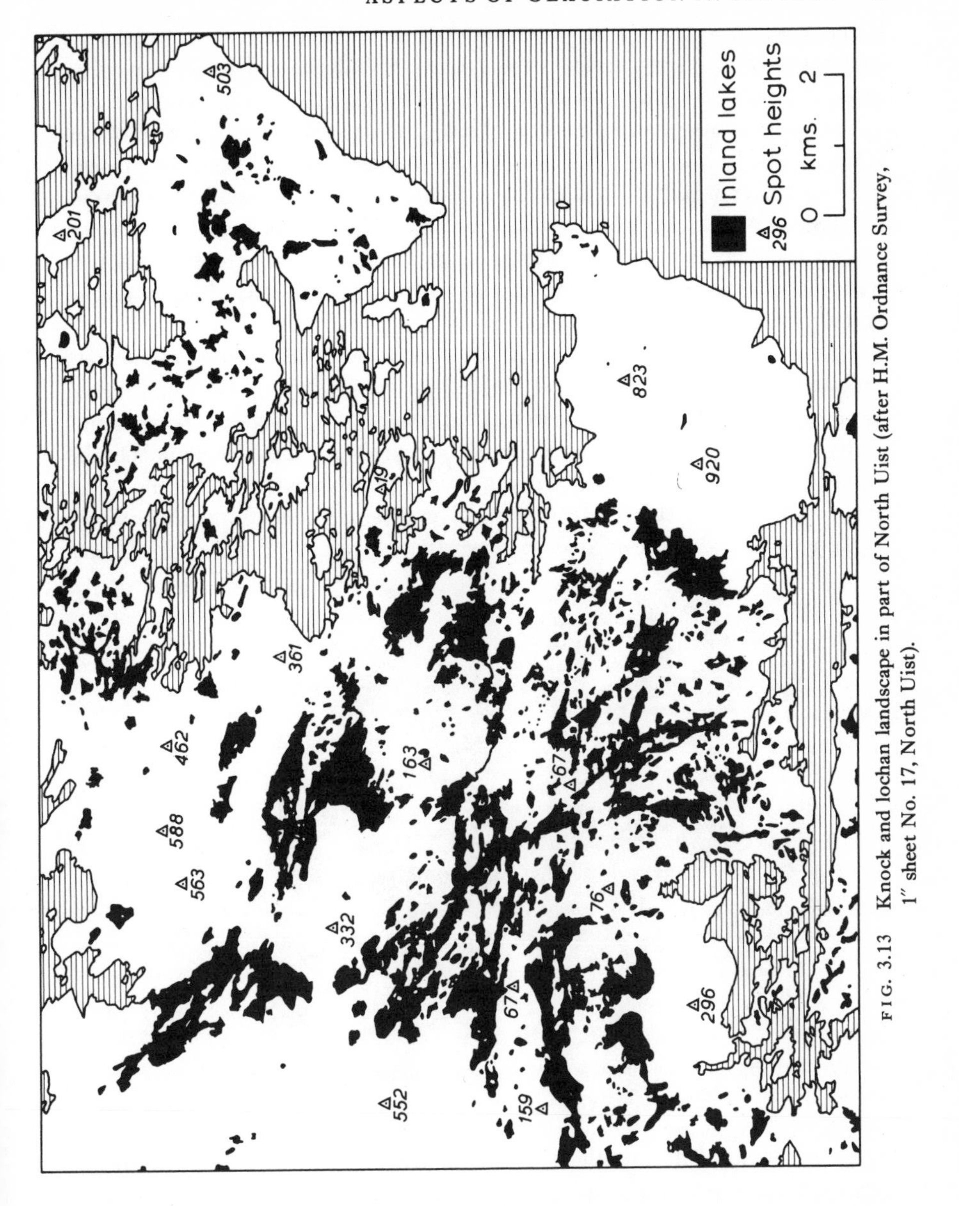

FIG. 3.13 Knock and lochan landscape in part of North Uist (after H.M. Ordnance Survey, 1″ sheet No. 17, North Uist).

There is, in fact, a general tendency today to suggest that ice erosion may be very selective. In a way, this is a partial reversion to some ideas expressed earlier this century that ice was relatively a protective agent, less erosive than running water. These views in their extreme form were untenable, as an attempt was made to interpret everything in terms of highly effective stream erosion and ice protection. Today the general view is that where ice is moving slowly it may perform little erosion, but that where it is moving rapidly it may do a very great deal of erosion.

Typical of this point of view is the recent treatment of the relief of the Cairngorm Mountains by Sugden (1968). In general terms the area consists of a high plateau surface of slight relief, which is deeply bitten into by glacial troughs and corries, while glacial through-valleys seem also to indicate strong erosion. In fact, there has been controversy as to whether this area was ever covered by an ice cap, because of the lack of signs of heavy glacial erosion on the surface of the plateau.

Some of the features of the plateau must have been derived from its preglacial history. On the broad scale, these include the rolling surfaces of faint relief, which seem best interpreted as surviving parts of the series of high plateaus, characteristic of much of the relief of highland Scotland. On the detailed scale, deeply-weathered rocks have been reported from this and other parts of Scotland: this means chemical weathering and chemical weathering is usually mainly associated with warm conditions, which are more likely

FIG. 3.14 Pressure release joint pattern truncated by corrie backwall (after Sugden, 1968, pl. II).

to be preglacial than interglacial. Tor-like features have been weathered out from the granite and still stand, though one or two have been smoothed on the up-glacier side and plucked on the down-glacier side. Sugden reports many cases of pseudo-bedding, presumably due to pressure release, parallel to the contours of the preglacial surface and truncated by glacial features such as the backwalls of corries (fig. 3.14). All this can be used to argue that the Cairngorm plateau escaped glaciation.

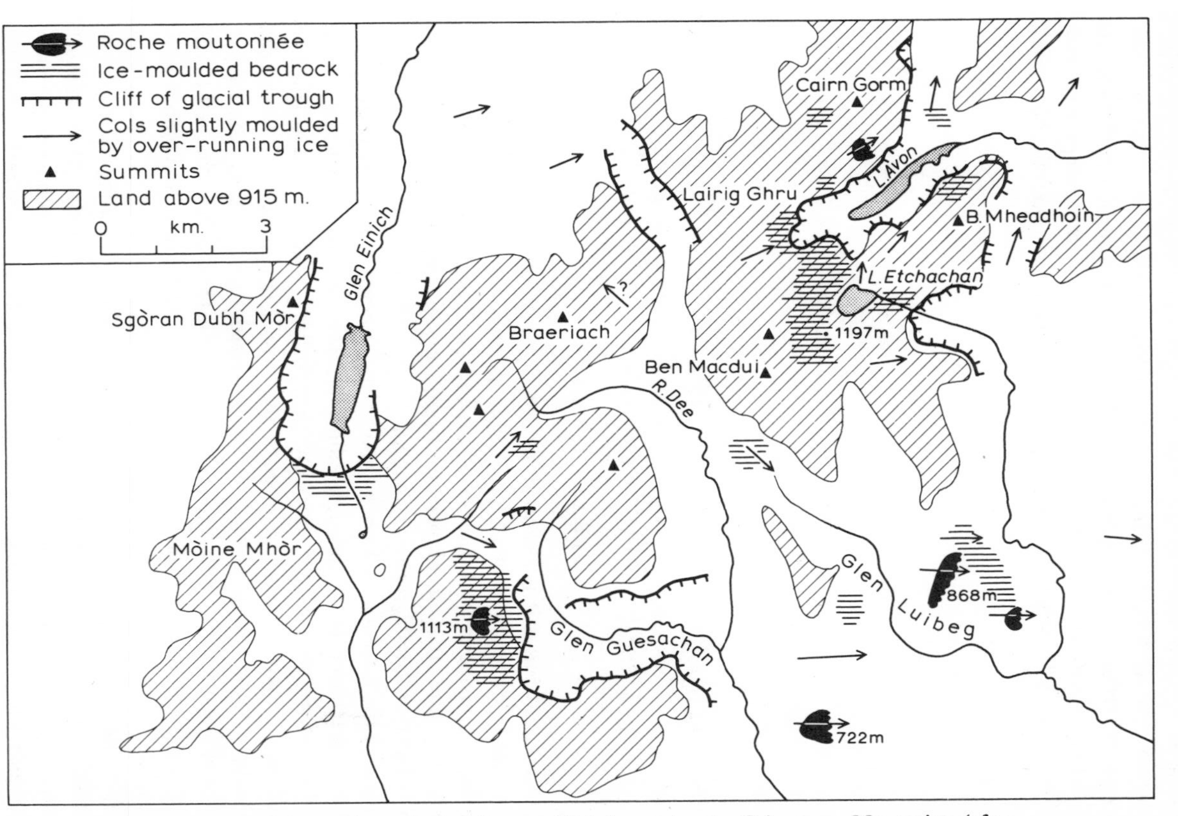

FIG. 3.15 Distribution of glacially moulded forms in the Cairngorm Mountains (after Sugden, 1968, fig. 3).

But such a conclusion would create difficulties and leave unexplained other features, if it were assumed that the Cairngorms merely suffered a valley glaciation with nothing more than perhaps a névé on the plateau. Many of the main glacial troughs have signs of considerable ice smoothing at their heads (figs. 3.15 and 3.16) and it is difficult to reconcile such forms (compare the two profiles on fig. 3.16) with the expected angularities of the backwalls of corries.

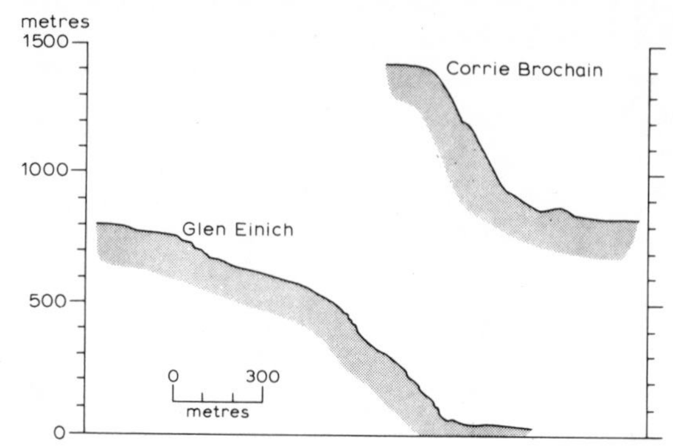

FIG. 3.16 Forms of trough head (Glen Einich) and corrie (Corrie Brochain) in the Cairngorm Mountains (after Sugden, 1968, fig. 4).

On the other hand if it is assumed, in order to explain the smoothing of the valley heads, that a powerful ice stream poured into the valley from the plateau, a considerable thickness of ice cap on the plateau is involved. An alternative idea that local plateau basins contained ice, which effected this, is hardly a satisfactory explanation.

Some of the main glacial breaches, i.e. places where the ice has overtopped a divide and scoured out a through-valley, are not consonant with the simple overflow of a valley glacier into the next valley. Indeed, the Lairig Ghru breaches the main Cairngorm watershed where it is estimated to have been originally 1130–1160 m high. If it is suggested that the area was subject only to valley glaciation and that these glaciers caused this breach, they would be required to effect a major breach at about the level of their collecting grounds, a very unlikely event.

Again, had there merely been local valley to valley breaching in an area of radial valleys the breaches should not have a definite directional pattern, whereas the main troughs of Glen Einich, Glen Guesachan and Glen Avon trend between north and east. Simpler, high-level, breached cols tend to have a similar orientation: breached to the north in the north of the area and to the east in the east. A similar direction of ice movement is witnessed by the major roches moutonnées (fig. 3.15). Certain of the major breaches do not follow this trend, the Lairig Ghru being one of them, but these seem to have been caused by the scouring out of structurally weak zones. Scouring of such zones, irrespective of the general direction of ice movement, is known to be associated with the present Greenland ice cap.

Much of this relief evidence seems to demand a general overrunning of the Cairngorms by ice from the south-west. This hypothesis can only be reconciled with the non-removal of many of the features of preglacial denudation by assuming that the ice sheet erosion was very selective and that it was concentrated where threads of high velocity ice followed the troughs and valleys and breached the watersheds. A similar combination of subdued plateau relief and deep glacial troughs, some of them through-breaches, occurs at the edge of the Greenland ice cap where high velocity glaciers reach the sea. Similarly, local ice streams of high velocity are known in Antarctica and some can be shown to be associated with troughs beneath the ice by gravity and seismic surveys.

If we accept this hypothesis of highly-selective glacial erosion the common juxtaposition of intensively glaciated and seemingly unglaciated forms becomes explicable. Unmodified V-shaped valleys may fall into glacial troughs from little-affected plateaus. Even among the major valleys the apparent impact of glaciation may vary widely, depending on whether the valleys were so orientated as to be used as major routes by the ice.

We are now in a position to return to the features of a glaciated trough. It is likely that former river valleys are both widened and deepened when they are converted to the steep-sided, flat-floored glacial trough form. Steepening of the head, due possibly to corrie action, and general reduction of the gradient, with in places overdeepening to give rock basins, are quite typical. So, the many glaciated troughs of the Highlands of Scotland (pl. 8), the radiating valleys of the Lake District in which the lakes themselves are situated, the major valleys of Snowdonia such as Nant Ffrancon and Nant Gwynant, all show some of the features listed below: wide flat floors, often with sluggish streams because of the low gradients, interspersed with lakes, sometimes with peaty areas where lakes have been filled in with debris; the lakes sometimes split into two by deltas built into them by side streams – the Lake District example of Derwentwater and Bassenthwaite is usually quoted, but Sissons (1967) gives some Scottish examples including Lochan na h'Earba south of Loch Laggan in the Ben Nevis area and St Mary's Loch and the Loch of the Lowes in the Southern Uplands; irregular masses of drift, sometimes hummocky, often slumped down the valley sides, and clearer moraines often marking late recession phases of the glaciers (these are discussed in more detail in the next section); cliffed or very steep valley sides patently out of equilibrium with present day processes in many areas and surviving because of the hardness of the rock; the main valleys straightened, with spurs truncated, and the tributary valleys hanging (pl. 9), so that streams enter the main valleys down falls and rapids; valley-floor steps approximating to roches moutonnées in form; and between the valleys and sometimes at the heads of valleys low passes with indeterminate divides where ice has breached from one valley to another. These features can be explained with varying degrees of success.

The concentration of glacial erosion along certain lines is indicated by Sugden (1968) and it may well be that the alignment of pre-existing valleys with the direction of ice flow is very important. Various processes have been suggested to help in the excavation of glacial valleys. Valley bottoms would be more constantly saturated than the valley slopes, so that in the very cold phase before ice covered an area, there might well have been concentrated freeze–thaw there, perhaps aided by the greater frequency and intensity of frosts in valleys. But overdeepening of the order of 300 m would be required in some Scottish rock basins and, although one can visualize shatter possibly of a few tens of metres, the results seem to be out of all proportion to the suggested cause, which would in any case cease to operate when ice covered the valley floor to any depth.

It has also been suggested that when a glacial trough is cut into a plateau surface a column of rock of specific gravity 2·7–3·0 is replaced by a column of ice of a specific gravity of only 0·9 (fig. 3.17). This would cause a considerable unloading of the valley floor, especially when it is remembered that the depth involved is fairly frequently of the order of 500 m in the highlands.

FIG. 3.17 Possible unloading effect in deep glacier-cut valley.

The result could be the weakening of the valley floor by the formation of pressure release joints, so making it easier for the glacier to prise the rocks away. It is difficult to see how this process might start, as the river valleys occupied by ice would have been there presumably with the stresses already relieved by joints. However, once started this could again be a self-accelerating process. In fact the whole formation of glacial troughs might be self-accelerating, for ice would tend to cut down most where it flowed at high velocity between confining valley walls, and the deeper it cut the faster it would flow and so on.

The deepening of glacial valleys must be attributed, apart from the possible action of the processes mentioned above, mainly to abrasion. Widening could be caused by two processes: freeze–thaw shattering just above the glacier surface and abrasion. The former would ensure that an ample supply of debris was contributed to the sides of the glacier and, given Blockschollen movement (see p. 43), great shear, leading to abrasion, could take place at the margins of the glacier. The results of lateral shattering and abrasion would be the truncation of spurs and, if the main glaciers were cutting down faster than the tributary glaciers (assuming that the tributary valleys were occupied by glaciers), of hanging tributary valleys.

The tendency for irregularities to develop in glacial trough floors may indicate a tendency for valley glaciers to magnify existing irregularities. Imagine a glacier bed with initial variations of gradient, such as might be caused by former heads of rejuvenation, different load-discharge characteristics in different reaches of the former river, hard rock bands and so on.

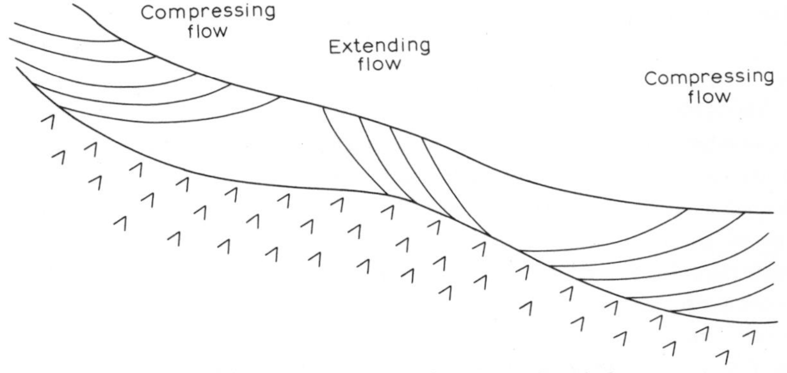

FIG. 3.18 Potential shear surfaces in a valley glacier.

Where the slope steepens the flow will accelerate and the ice will thin with the formation of a series of potential shear surfaces running down into the glacier bed (fig. 3.18): this state of affairs has been termed extending flow by Nye (1952). Where the slope slackens the ice will thicken and the potential shear surfaces will be orientated upwards, creating in Nye's term a zone of compressing flow. Embleton and King (1968) suggest that it is in these zones of compressing flow that the glacier will be most likely to carry debris up from its bed, thus further scouring and deepening pre-existing hollows. The general flattening of the valley gradient might be expected from pronounced deepening not far from the source of the glacier, where the change from the head slopes of the valley to the main valley gradient might well have caused a major zone of permanent compressing flow.

Although such processes might well account for minor rock basins in the floors of glacial troughs, overdeepening on a much greater scale has occurred in the fjords of such areas as Norway, British Columbia, southern Chile, New Zealand, Antarctica, and to a lesser degree in western Scotland. The maximum known depths of Greenland, Norwegian and Chilean fjords is of the order of 1300–1400 m: in Antarctica fjords almost 1000 m deeper are known. These great depths are usually separated from the open oceans by submerged sills, generally of solid rock, over which the depth is 200 m or so. Thus overdeepening of the order of 1000–1200 m, i.e. the height of Snowdon, must be considered even if the extreme conditions of Antarctica are ignored.

There is no doubt that glaciation must have been mainly responsible for these features. An earlier hypothesis, staunchly upheld by Gregory, that they were mainly slightly-modified, structurally-guided valleys, is not tenable for a number of reasons. They do not all follow weak structural lines; structural factors cannot explain the very deep basins; the fact that the deepest are usually the largest suggests that glacier size may have been very important; their general position is on high continental margins, usually on the west sides of continents where the gradients from ice caps fed by westerly winds were steepest. All these factors point indubitably to ice action. The general hypothesis that springs to mind, in view of what has already been said about sharply localized high velocities in the ice, is that something similar happened here and that the streams of high velocity were located by deep preglacial valleys. These, in turn, may have been originally located by structural features, but the latter cannot be regarded as the immediate cause of fjords. The element of positive feedback once again looms large: the steep gradient leads to high ice velocities, which lead to intense erosion, which accelerates velocity, which intensifies erosion, and so on. When the ice debouched from the valleys as, for example, on the coast of Norway at the edge of the Scandinavian block, it spread, thinned, retarded and so lost its erosive power. Flotation of the ice in the sea has also been suggested: it may have helped, but the ice in the Sogne fjord, for example, was much too thick ever to have floated. Again, it has been suggested that the depth of freeze–thaw shattering before the glaciation would have been much less near the sea than farther inland, so that a closed basin might be explained this way. But could it lead to basins 1000–1200 m deep? It seems very doubtful.

This in general terms is the fjord problem. It arises in a diminished form in the scenery of the west coast of Scotland. The distribution of rock basins in Scotland has been plotted by Sissons (fig. 3.19). Some of them are inland. Many form the sea lochs of the west coast (pl. 10). Some occur drowned beneath the sea. Many of their features are those of fjords on a reduced scale. The deepest loch is Loch Morar (A), now separated from the sea by a low sill, in a basin about 310 m deep; next are Lochs Ness (B), about 235 m deep; Lomond (C), about 190 m deep; Lochy (D), about 160 m deep; and Ericht (E), Tay (F) and Katrine (G), in that order and each about 155 m deep. It is

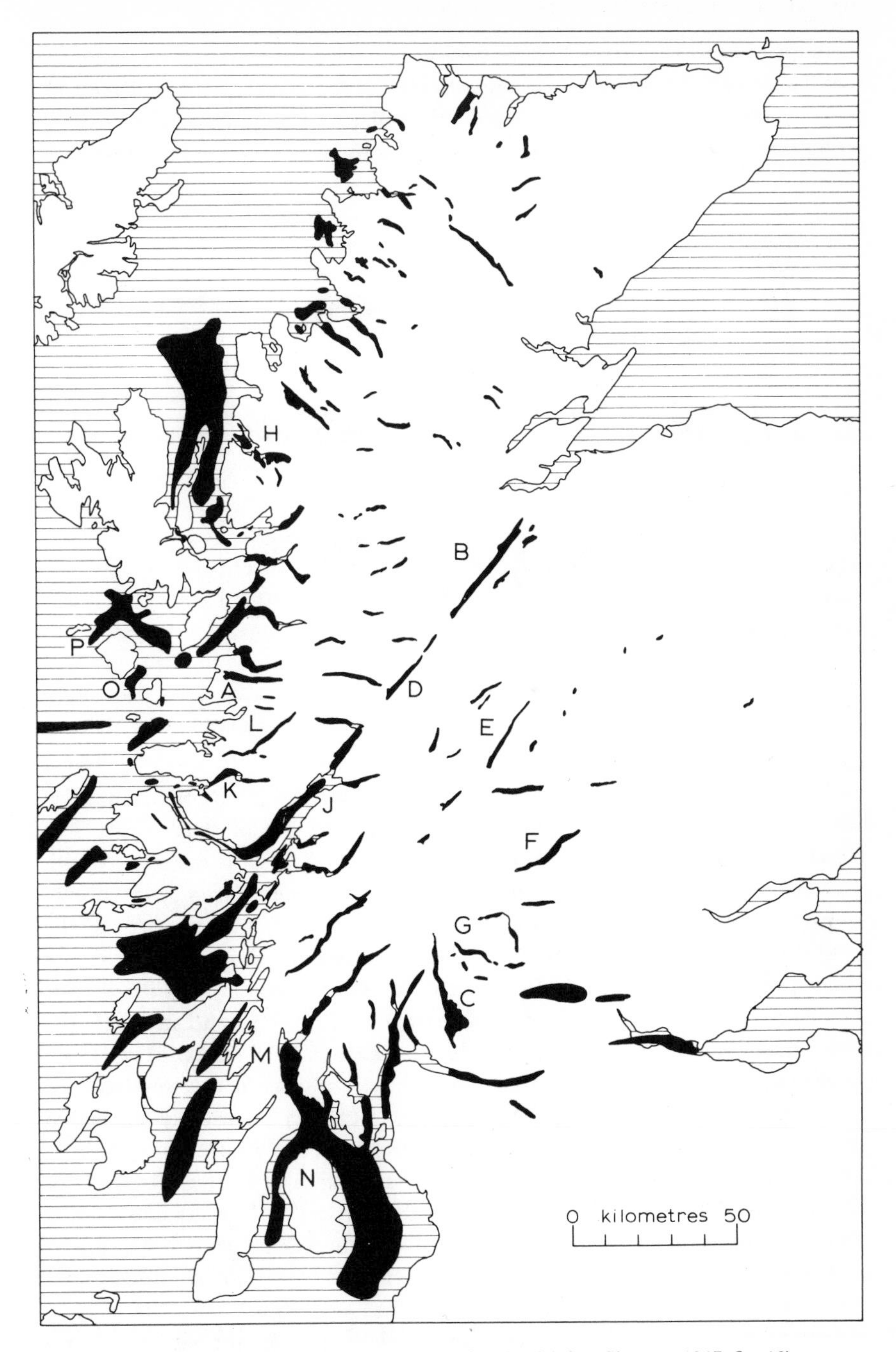

FIG. 3.19 Overdeepened rock basins in Scotland (after Sissons, 1967, fig. 18).

curious how the majority of these very deep rock basins are not on the west coast, although the greatest concentration of rock basins is there. All are true rock basins (fig. 3.20) of varying depth and profile; irregularities in long profile may usually be related to variations in the lithology of the underlying rock.

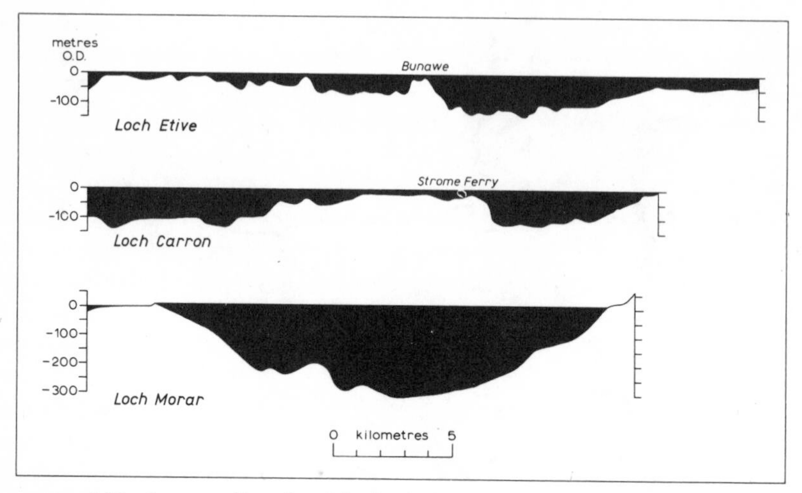

FIG. 3.20 Long profiles of rock basins in Scotland (after Sissons, 1967, fig. 17).

The greatest frequency of rock basins on the west coast is consistent with rapid flow down the steep gradient of the western side of the ice cap. The radial pattern of alignment of the basins on the west coast is consistent with the general flow pattern of the ice (fig. 3.11), north-westwards in the north, westwards in the centre and south-westwards in the southern parts of the west coast. Yet structural effects complicate the general pattern. North of Loch Torridon (H on fig. 3.19) the structural trend coincides with the ice flow trend and the lochs are almost all parallel. South of Loch Linnhe (J) the structural lines again coincide with the generally parallel lochs. From Loch Torridon to Loch Sunart (K) the general ice trend was to the west and the basins are generally aligned in the same way, but they are mostly smaller in size than in the northern and southern stretches of the west coast and there are important exceptions, e.g. Loch Shiel (L) which is fault-controlled. In this section of the coast lies Loch Morar, the deepest basin of all and aligned almost at right angles to the known rock structures. The general alignment of major basins along the Great Glen fault (Lochs Ness, Lochy and Linnhe) again shows structural control cutting across and dominating the general direction of ice flow, as also does Loch Ericht. In fact the situation on a vastly greater scale is similar to that shown in the Cairngorms, where major troughs also seemed to follow the general direction of ice flow except where pronounced structural weakness intervened.

Where the ice spread out onto the lower ground, usually at the coast, the basins cease to occur. Loch Lomond (C) is the prize example of this: in the north, where it is in a narrow trough, it is almost 200 m deep; in the south, where it spreads out in an area of far less resistant Old Red Sandstone, it is no more than 30 m deep. Sissons suggests that the lack of true basins at the northern end of Kintyre (M), where the structural trend coincided with the direction of ice flow, is due to the relatively low land there allowing the ice to spread, retard and become ineffective as an eroding agent.

Deep basins inland are usually associated with structural weakness (Ness, Lochy and Ericht) or with the confluence of glaciers (Tay).

Sissons also argues that certain broader basins on the sea floor may be due to larger-scale concentration of ice flow between major blocks of highland and islands, especially where local ice caps may have added to and deflected regional flow patterns, for example round northern Arran (N on fig. 3.19), between the island and Kintyre on one side, and between the island and the mainland on the other. Other examples are the deep trenches between Rhum and Eigg (O) and between Rhum and Canna (P). Perhaps some of this sea-bed excavation was helped by a distribution of fairly readily eroded rocks there, either Old Red or New Red Sandstone, and by the presence of structural weakness. Little is yet known about this.

These are the main features of subglacial erosion with the important exception of the glacial breaching of watersheds, a process which often results in the permanent modification of drainage patterns, which will be discussed later in this chapter.

GLACIAL FEATURES OF LOWLANDS NEAR MOUNTAINS

There may be erosive features where the ice flowed from highlands to lowlands, such as those on the Lewisian Gneiss, but these have already been discussed. Depositional forms predominate and may be of two main types:

(a) Forms produced by ice-moulding consequent on the rapid movement of ice in these situations.

(b) Depositional forms produced by melting or retreating ice and best preserved in these localities. In Britain the last ice sheet had a more limited extent than the previous two ice sheets (fig. 5.18) and generally left the Midlands and the English scarplands unglaciated. This was its limit at the maximum of glaciation. Its general retreat was punctuated by local re-advances which went on as late as zone III of the postglacial period (see Chapter 5). Each successive readvance normally reached less far than its predecessors, so that forms of glacial deposition become increasingly fresh as the highlands are approached and the valleys entered. Scotland and Ireland, because of their larger share of the later episodes of the Ice Age, show much better constructional forms than are usual in England.

Drumlins and associated forms

In the areas covered by the last glaciation a variety of ice-smoothed forms occurs, reflecting the tendency for moving ice to produce streamlined forms over which flow is smoothest. The general forms of such features are elongated hills with a rounded blunter end towards the ice and a longer, tapering end in the down-ice direction. These forms approximate to a half egg shape, the egg being sliced through in the plane of its longest axis. Thus drumlin fields, for they usually occur in swarms (pl. 11), were formerly known as 'basket of eggs topography'. The dimensions vary widely but Flint (1957) gives the following ranges for ideal drumlins: length 1000–2000 m, width 400–600 m and height 15–30 m. They may become much more elongated and it is usually held that greater elongation results from greater pressure consequent on more rapid ice movement. Chorley (1959) has pointed to the fact that other streamlined shapes vary similarly. The aerofoil section of the wings of high-speed aircraft is more elongate than that of slower aircraft. Similarly, birds that lay large eggs for their size – and are hence presumably subject to greater pressures – lay more elongated eggs!

The general distribution of true drumlins is shown in fig. 3.21. It is very difficult to know what a true drumlin is. Ideally, it should be a hill of glacial drift falling within certain ranges of shape and dimension, such as those quoted from Flint above, but unless there are breaks in the series of forms at either end of the range of quoted shapes and dimensions, there seems to be little point in trying to separate part of a series which may range from small circular mounds to very elongated, almost rib-like features. Again, 'drumlin' features may vary from pure drift forms, through features with rock cores to features formed largely of rock, and finally to streamlined forms moulded out of relatively non-resistant rock. As it is very difficult to differentiate these forms in the field without tremendous labour, the nature of 'drumlins' in any area must be largely a matter for conjecture for it is unlikely that many of them will be conveniently sectioned either by nature or by man.

In fact, Sissons (1967) in his description of ice-moulded forms in Scotland includes the gamut from forms moulded largely in solid Devonian and Carboniferous rocks to drift drumlins. The solid rock forms occur around the Firths of Forth and Tay, in Caithness, and especially in the lower Tweed basin where a N.E.–S.W. structural trend was moulded by ice moving towards the north-east into a series of discontinuous ridges and hollows (fig. 3.22). Another very good area is south and south-west of Falkirk: here there is a superficial layer of till up to 6 m thick but the solid rock beneath has been

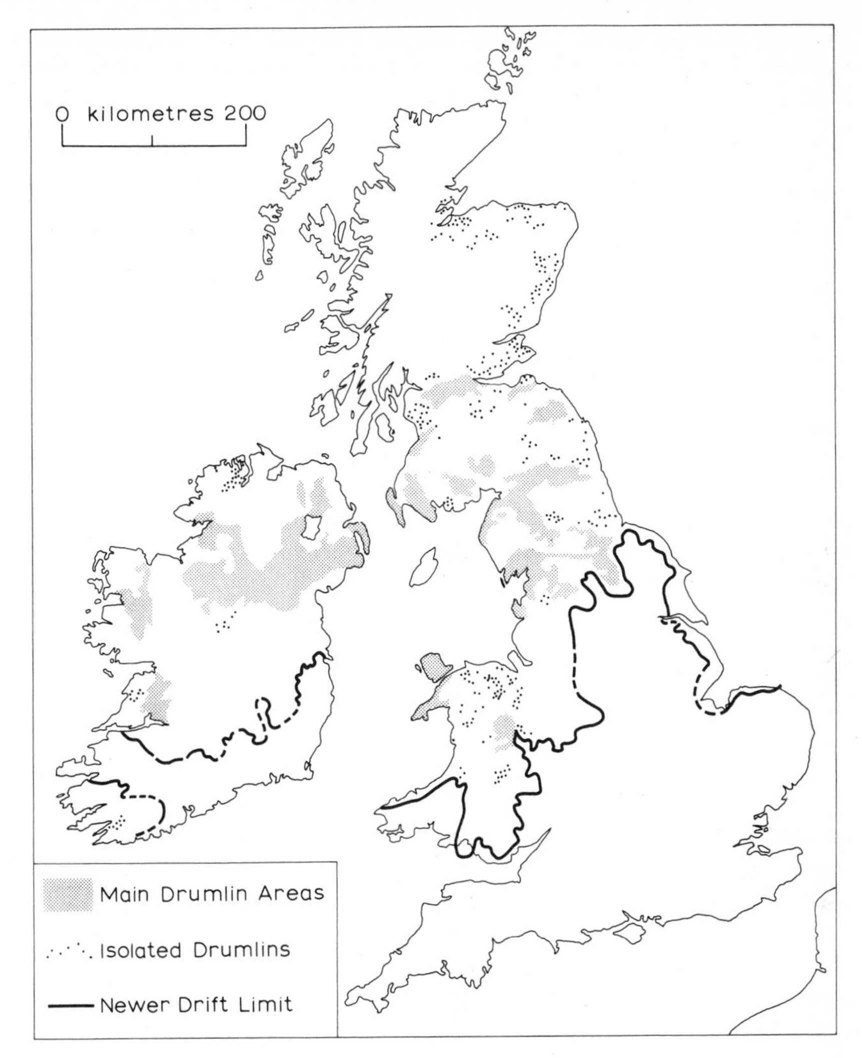

FIG. 3.21 Distribution of drumlins in the British Isles (after Embleton and King, 1968, fig. 14.4).

moulded by the ice into a series of tapering ridges almost as smooth as the till surface.

At the other end of the scale predominantly drift drumlins are very well developed south of the Southern Uplands around Castle Douglas and Newton Stewart, again in the Eden valley, and are widespread in Ireland (fig. 3.21) where they are probably as well-developed as anywhere in the world.

Sissons would put crag and tail features (fig. 3.23) in the same general class

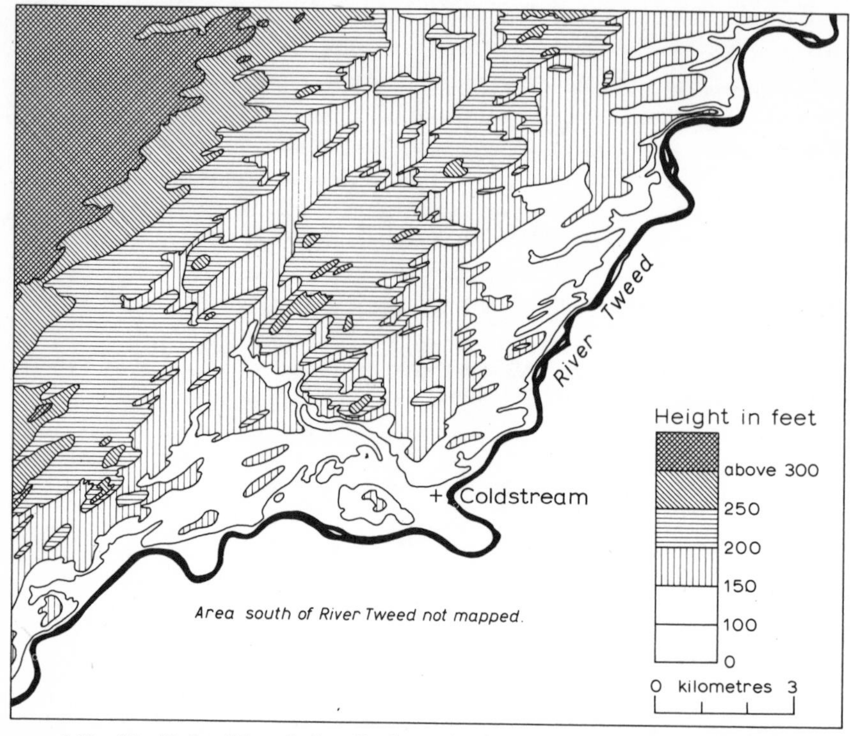

FIG. 3.22 Moulded solid rock drumlin forms in the Tweed valley (after H.M. Ordnance Survey, 1″ sheet No. 64, Berwick).

of glacially streamlined forms. There seems to be agreement that resistant rocks, usually igneous, form the crags and that a long tapering tail is developed in the lee, but there is difference of opinion about the composition of the tail. Many writers, including Flint, show the tail to be mainly drift (fig. 3.23C) but Sissons shows convincingly for some of the Scottish examples that it is readily-moulded, soft sediment with only minor patches of drift (fig. 3.23B). Some of the best-known Scottish examples of these features are in and around Edinburgh: Castle Rock with the Royal Mile following the tail is the classic example but Calton Hill forms another though less spectacular case.

There have been a number of arguments concerning the origin of drumlins: whether they were originally deposited in that form as streamlined accumulations of unstratified drift, or whether they represent the readvance of an ice sheet over an already existing sheet of drift. The second idea arose from the fact that the material in some drumlins has been found to be stratified. The argument seems relatively pointless because, if there is a whole range from drumlins moulded from solid rocks right through to drift forms, drumlins moulded from pre-existing drift sheets are merely a special case of drumlins moulded from very soft rock. It has also been suggested that till might have been squeezed into hollows beneath ice sheets where the ice rode over projections in its bed. This might be possible in the case of very thin, stagnant ice sheets, but drumlins seem to have been formed by thick moving ice under which cavities are not likely to have remained open.

Summing up, it might be said that drumlins occur where ice from adjacent centres is still moving rapidly, that they were caused by some tendency inherent in an ice sheet to mould either the bed rock or its own deposits into streamlined shapes. This is really an 'it's in the nature of the beast' explanation, like that which says that meanders are an inherent characteristic of many rivers. But that is all that can be done at present.

The regional pattern of drumlins means that they have large-scale effects on the landscape, nowhere better exemplified than in Ireland. A clear brief description of these effects has been given by Charlesworth (1953). In the

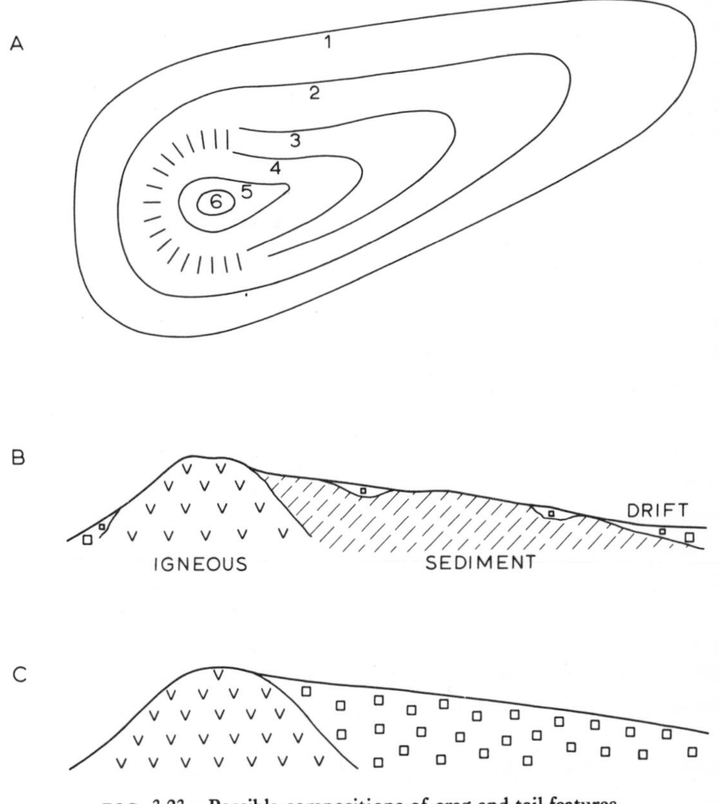

FIG. 3.23 Possible compositions of crag and tail features.

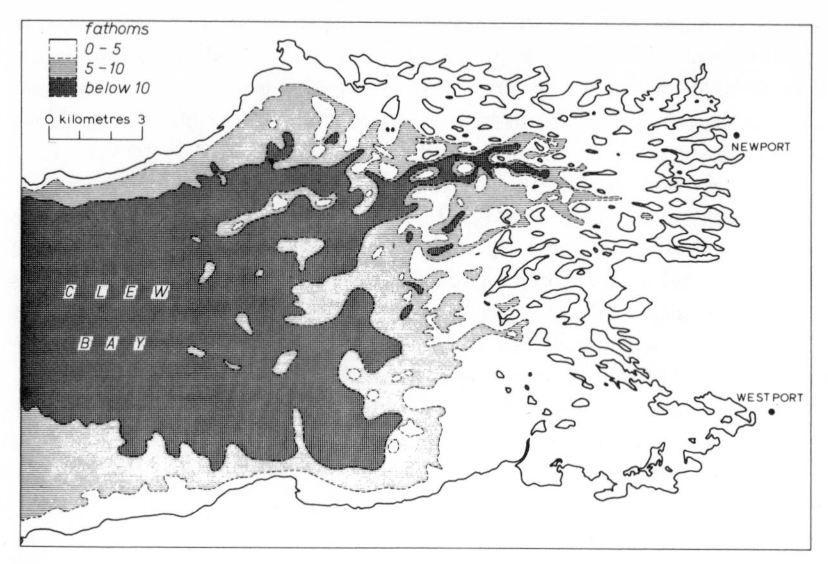

FIG. 3.24 Submerged drumlins, Clew Bay, Co. Mayo, Eire (after Charlesworth, 1953, fig. 90).

drumlin fields north and west of Donegal Bay and between Belfast and Downpatrick the smooth, rounded hillocks are separated by small lakes, peat bogs and marshes linked together by intricately-winding streams. Where they are partly submerged, e.g. Strangford Loch or Clew Bay, from which area the term was originally derived, archipelagos of low, flattish islands result (fig. 3.24).

Moraines

When ice melts it dumps the material it is carrying. Ice can melt at all its margins. It can be thawed underneath by heat flowing out from the earth and by pressure lowering the melting point. It can be thawed at the surface by solar radiation. It can be thawed at the sides and the front by a combination of solar radiation and heat reflected from adjacent rocks. Wherever it thaws there results a natural concentration of the debris which it has been carrying. This is moraine.

Moraine formed at the bottom of an ice sheet is known as ground moraine, and as sheets of till or boulder clay it is mostly characteristic of lowlands well away from the highlands where the ice sheets originated. It is therefore described in the next section. It is often unsorted and in places very clayey.

Moraine caused by the thawing of the surface of an ice sheet is known as ablation moraine. Because of its mode of formation the finer constituents tend to be washed out of it by meltwater and what is left is sand and gravel, which may be let down irregularly on to the ground moraine when the ice sheet finally melts. In the case of the melting of stagnant ice even the ground moraine may be largely re-sorted by meltwater streams within and under the decaying ice.

Moraines formed by melting of the sides and fronts of glaciers and ice sheets are known respectively as lateral and end moraines. For any thickness of these to form, the ice front must remain more or less stationary. Such a state implies an active glacier still flowing, but with flow only just balancing the marginal thaw rate. A steadily retreating glacial margin will not have a marked moraine though halts in retreat may be marked by recessional moraines.

The nature, form and succession of deposits capable of being formed by the thawing of an ice sheet are so varied that it is probably better to be aware of the processes which may be going on rather than to attempt a rigid classification with all the attendant difficulties of trying to fit newly discovered forms into it.

A fundamental distinction might be made between ice, the front of which remains stationary because the wastage rate equals the delivery rate, and ice in which there is virtually no movement and which melts away irregularly as a stagnant sheet. In the first instance movement ensures that any potential cavities in the ice remain closed. Upwardly directed thrusts at the snout, where compressing flow occurs, ensure a continuous supply of material for the growing end moraine while ground moraine, either in the form of lodgement till (see below) or moulded as drumlins, is deposited below the ice sheet. An ablation moraine will probably be forming at the surface as shown in fig. 3.25A.

If for some reason, such as the cutting off of the supply of ice because of the emergence of a relief barrier (fig. 3.25B), the ice stagnates, it will develop different features. Melting will induce in the ice a form very much like the karst landforms that develop on limestone. Local melting of the ice will produce holes of various sizes and depths, flow will occur in tunnels and steep-sided chasms probably developed from crevasses, just as limestones are often pitted with solution holes, riddled with caves and passages and traversed by steep-walled gorges. In these holes, tunnels and chasms resorted glacial deposits will be laid down: these are really fluvioglacial or, as Sissons calls them, glacial river deposits.

Finally, the ice will thaw leaving the ablation moraine to be deposited on the collapsed intra-glacial deposits (fig. 3.25C), after which further resorting by meltwater may take place. Lumps of ice remaining in the moraine will eventually melt and lead to the formation of small enclosed hollows (kettle-holes): such a moraine is often described as a kettle moraine (pl. 12).

The result, whether one looks at the surface or at a section of the deposits,

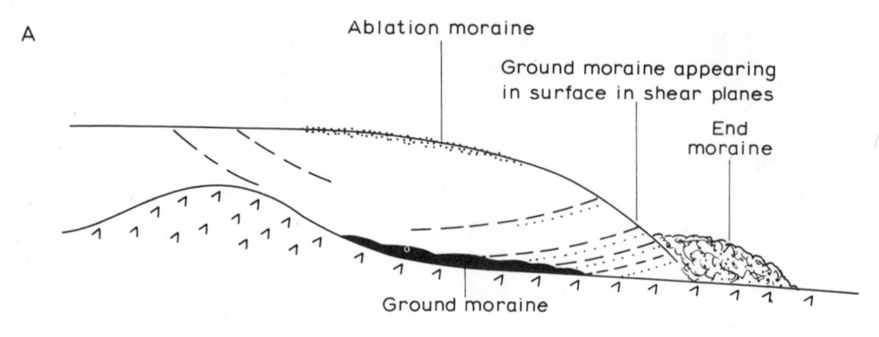

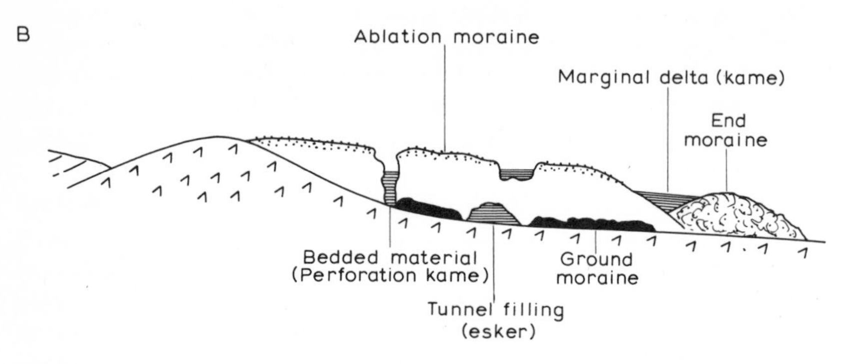

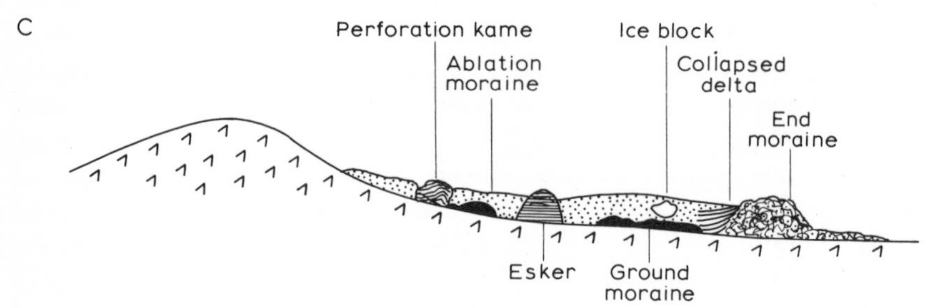

FIG. 3.25 Formation of a moraine complex.

can be a bewildering confusion of bedded and unbedded deposits. If the deposits have been further disturbed by periglacial action, as they well might, and perhaps also overridden by a readvance of the ice, the interpretation of the moraine complex (pl. 13), as it is perhaps better to call it, is best left to expert geomorphologists and glacial geologists.

But among the many forms certain salient ones may be discerned. End moraines are probably the most conspicuous of these. Recessional moraines must be grouped with them because the form is the same. Such features occur both on the lowlands and within the highland valleys, though they are usually better shown within the valleys. Flint (1957) states that end moraines rarely exceed 30–50 m in height in lowlands, but may go up to 300 m in mountain valleys because of the much greater flow and hence supply of debris in valleys. They may appear to be even higher where they are underlain by rock. None in Britain reaches this size. The form of the moraine and the character of the sediments will vary with the place and mode of deposition. If deposited in a lake the form will be different from that of a moraine deposited on land. Where meltwater plays a greater part in deposition there will be more bedded material than where moraine sludges off the melting front of the ice.

Moraine complexes occur at the edges of the last glaciation in England, for example the York and Escrick moraines in Yorkshire, or the Woore moraine, which extends from near Whitchurch in a curve eastwards to near Woore in Cheshire. The latter probably represents a retreat stage of the ice sheet. Another example along the Norfolk coast inland from Cromer and Sheringham is described below.

Sharper moraine features mark the various readvance stages in the valleys of the Highlands of Scotland, though they are by no means all clear arcuate ramparts such as might be expected from an ice lobe. Many are little more than roughly-aligned irregular dumps of gravel. One of the best examples is the arcuate moraine along the southern end of Loch Lomond marking the limit of the Loch Lomond glacier, where it spread out from its narrow valley onto lower ground. Other smaller but good examples are to be found in the western part of the Highlands of Scotland and mark the last readvance of local valley and corrie glaciers. The age of these features is little more than 10,000 years, or approximately one third or one quarter of that of the deposits marking the limit of the last glaciation in England. For this reason alone one would expect them to be a lot better preserved. Lateral moraines are often associated with these end moraines.

In the valleys above the end moraines moraine complexes, consisting of an apparently chaotic distribution of mounds, probably represent in general terms the state shown in fig. 3.25C, i.e. a mixture of ground moraine with ablation moraine, the whole probably to a considerable extent resorted by meltwater streams.

Apart from forming ridges and mounds in the landscape moraines may occasionally impound lakes, while irregular hummocky moraine may contain kettle lakes or small stretches of water or peat in the hollows between the hummocks.

Kames and eskers

These deposits, which have been mentioned above, owe their existence to the action of glacial rivers. The essential difference between the two, though the

A

B

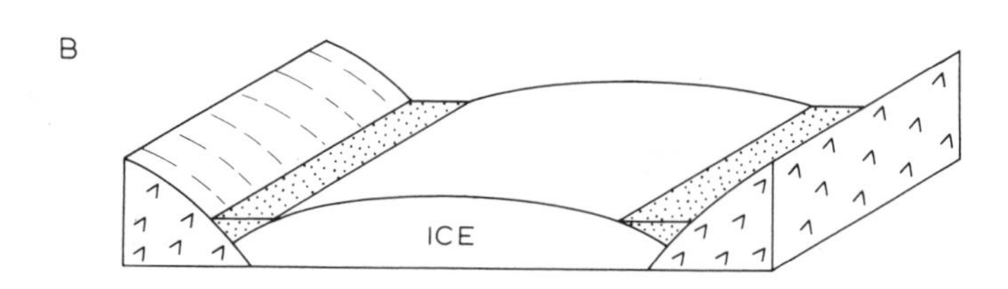

C

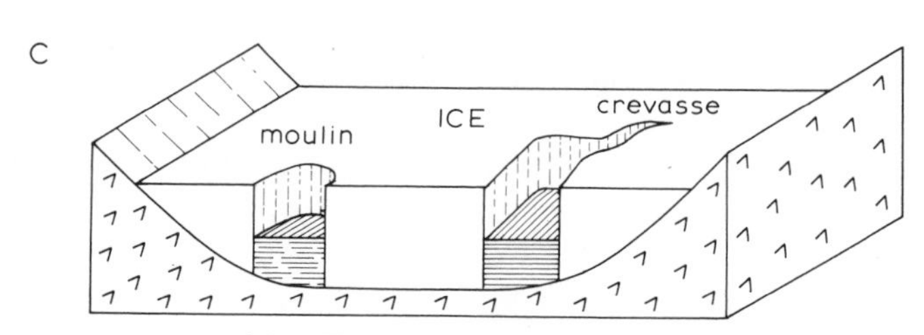

FIG. 3.26 Types of kame and their formation.

terms have been used interchangeably by some authors, is that kames are features produced at the margins of the ice and that eskers are primarily forms developed beneath the ice, though one of the hypotheses of esker formation, which are discussed below, effectively bridges the gap between the two.

The margins of the ice include not only the exterior margins, but also, in the case of fairly advanced stagnation of the ice, any open holes or enlarged crevasses within the ice. Because of the freely-connected water-table in a stagnating mass of ice, all kame features associated with a given body of ice at the same time should reach approximately the same elevation. They may be divided into four main types but variants are obviously possible.

(a) Ice front kames usually take the form of small deltaic deposits produced at the edge of the ice (fig. 3.26A). The mound will become more symmetrical when the ice melts and the ice-contact slope slumps.

(b) Kame terraces are formed by streams flowing along the edges of the ice where it abuts against higher ground. When the ice melts the edges of the terrace will slump and any included lumps of ice melt to give kettle holes (fig. 3.26B).

(c) Perforation or moulin kames are formed in hollows and perforations through the ice (fig. 3.26C) and on collapse give rise to isolated mounds.

(d) Crevasse fillings or kames are little more than elongated versions of (c) above. They may assume angular dog-legged forms when they represent the infilling of a stream following the elements of a rectangular crevasse pattern.

Kame features are common at the edges of the Southern Uplands of Scotland and are usually associated with intricate patterns of meltwater channels, as indeed they should be. The Nith valley, a few kilometres north of Dumfries in the vicinity of Duncow, provides a very good example of such an association, but many more examples can be found along the northern edge of the Southern Uplands. In the Highlands kames and kame complexes are widespread although not always easy to distinguish from the generally hummocky drift. The two in fact merge into each other in places.

Eskers (pls. 14 and 15) are long, more or less winding ridges composed of sands and gravels which are normally bedded. They are usually thought to be the casts of streams formed either on the ice, or beneath the ice. They wind upslope and downslope unlike any subaerial river. There are two possible explanations for this. If they are the beds of streams formed on the surface of the ice, such stream beds may have been let down on to the land below irrespective of its relief, when the ice melted. However, it is very unlikely that they would have been let down without great disturbance and without being mostly washed away by meltwater forming during the whole time the process was taking place. Further, although eskers run uphill and downhill, they usually cross ridges at their lowest points, a feature which would not have been found had they been haphazardly superimposed from the ice. There is, of course, no reason why a stream conditioned by the form of the ice surface should in any way reflect the underlying relief in its course. On the other hand, if eskers are the casts of streams running in tunnels at the base of the ice, they would have flowed under hydrostatic pressure and hence have been capable of flowing uphill. They might also be expected to cross ridges at their lowest points, for this would seem to be a reasonable advantage even to a subglacial stream.

A variant is the beaded esker, which has local broadenings, the beads, at intervals along its course. These have been explained as deltaic, kame-like bulges formed by a stream leaving an ice front in periods of standstill during the general retreat of the ice front. It is this hypothesis which makes it very difficult to separate absolutely kames from eskers.

Eskers are extremely common in the central lowlands of Ireland, where they form ridges of different vegetation, often followed for long distances by routes, standing above the peat bogs developed in the slacks between. In Scotland eskers are found associated with the network of kame and meltwater channel features on the northern margins of the Southern Uplands, and again on the lowlands on the sides of the Moray Firth, for example around Dornoch, Golspie and Brora. They commonly occur where glacial channels and glacial river deposits occur and must merge into kames, or, even if they do not merge, short eskers so closely resemble long kames that it is very difficult to tell the two apart. The ideal criterion, that the kame should be flat-topped, is very difficult to apply in practice when settlement of the deposit and subsequent erosion have affected the picture. Usually eskers reflect in their orientation the last movements of the ice sheet concerned with their formation.

GLACIAL FEATURES OF LOWLAND BRITAIN

Most of the lowlands of England differ from the areas near the highlands in that the glacial features are much more amorphous and subdued. This could obviously be due to initial differences or to differences caused by denudation since the deposits were laid down. Even the youngest of the lowland landforms are three or four times as old as the moraine and glacial river features described above, while the majority of those south of Yorkshire and Cheshire date mostly from the two glaciations before the last, known chronologically as the Lowestoft and Gipping glaciations. There are occasional fresh-looking forms associated with these, but for the most part the glacial deposits form monotonously smooth till sheets.

It is often said that ice sheets in lowlands are agents of deposition. It is undoubtedly true that deposition is the main effect, but the deposition of large quantities of material implies its erosion somewhere or other. The main product in lowland Britain is ground moraine and the raw material for this has not been derived from the highlands where most moraines are largely sand and gravel. In the lowlands they are predominantly clays derived from no great distance. This can be seen in the way in which the nature of the till usually varies with the rock on which it lies, though obviously its character cannot change immediately one passes from one outcrop to another. Thus, all through the east Midlands and even on the Chalk to the south-east of this region, till is a slaty-blue, stony clay, weathering buff (pls. 16 and 17). Individual erratics may have been derived from any outcrop down the east side of England or by two stages even from Norway, but the majority of the clay comes from the Lias, Oxford and Kimeridge Clays of Jurassic age and the Cretaceous Gault Clay. Similarly, some way east of the Chalk escarpment of west Norfolk, the till becomes predominantly chalky. It is difficult to make an estimate of how much material has been eroded and redeposited because the thickness of till is very variable, and one does not know how much till has been resorted by meltwater streams and is therefore no longer recognizable as such. In places near Cambridge, such as the plateau between that town and St Neots, there are till thicknesses of up to 60 m (fig. 3.27) and an average till thickness of possibly

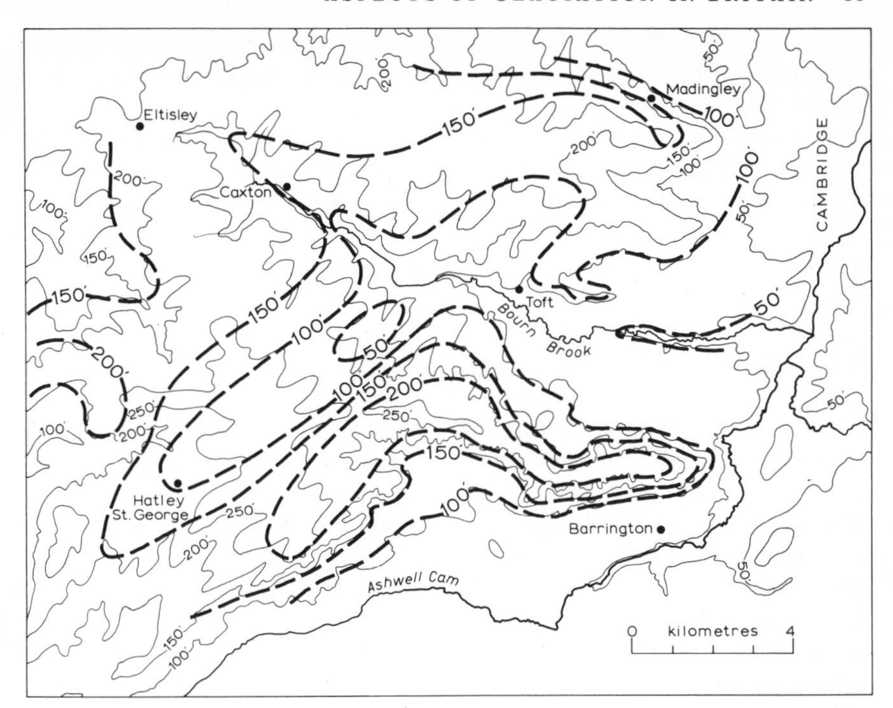

FIG. 3.27 Surface contours (continuous thin lines) and sub-drift contours (heavy pecked lines) west of Cambridge (after Sparks and West in J. A. Steers, ed., *The Cambridge Region*, British Association for the Advancement of Science, 1965, fig. 13).

10–15 m. On the other hand, if one takes into account the large areas with no till or very thin till, the average till thickness would probably be of the order of 2–5 m in the East Midlands and East Anglia. Allowing for the conversion of some till into bedded, meltwater deposits this must mean considerable erosion, probably mainly of clays already disintegrated by freeze–thaw under periglacial conditions.

There is evidence of spectacular, massive movements of material, but they are all very local. Embleton and King (1968) list a number of large erratics up to 800 m long, but usually with a maximum of about 300 m, which have been moved short distances. Many are smaller than this but are still surprisingly large in relation to the generally slow rate of movement of continental ice sheets. A famous erratic near Ely of Chalk, Gault and Lower Greensand measured about 400 m by 50 m. Its weight must have been of the order of a million tons or perhaps even two million. In the cliffs at West Runton in Norfolk, masses of chalk, the largest about 600 m long, have been sliced up by the ice presumably from offshore exposures. The larger erratics still have undisturbed bedding and flint layers, but the smaller ones have often been milled out into lens-shaped masses. The distance that these erratics have been moved is most likely quite small.

However, the general effect in lowland Britain is for the formation of till plateaus by lodgement. It has been suggested that lodgement till, the boulder clay of older geological terminology, has been plastered onto the landscape in a semi-dry state as the glaciers advanced and pressure melting took place at the base. As it is a stony clay it is much less susceptible to meltwater erosion than sandy moraine, because clays are much more coherent and require much higher stream velocities for erosion than does fine sand. One can appreciate lodgement against relief features lying in the path of the ice sheet as the pressure of the ice increases there. By analogy with drumlins one can also appreciate a tendency for an ice sheet to convert somewhat irregular relief into broad, smooth forms over which it could ride easily. One cannot understand why it seems to have done this in some places and not in others. However, there is no doubt that in most places the effect of the deposition of lodgement till has been to simplify the relief, to increase the absolute height but often to reduce the relative relief, and to form broad, smooth, badly-drained plateaus. Such features may also have been formed by the settlement of saturated till after deposition: this material is known as flow-till.

The amount of meltwater available at the margins of a continental ice sheet must have been enormous and its effects spread over a very wide area in the course of time as the ice sheet advanced and retreated. Yet evidence of meltwater activity, which is very common for example in the north European lowland, is not all that widespread. Most of the till plateaus have been formed in interfluvial areas and till is rare in most of the major valleys of eastern England, so rare in fact in some of them that it has even been suggested that they were formed after the deposition of the till. Yet within the valleys there are often great sequences of terraced sands and gravels, for example in the Thames valley and in the valley of the Bedfordshire Ouse and its tributaries. Many of these terraces are interglacial, but much of their raw material, to judge from its content of erratics, was derived from glacial deposits, either till sheets or from outwash deposited in the valleys and later resorted. This may account for part of the story, but the absence of ablation till from the plateaus is also a problem unless it is suggested that this too was washed down into the valleys by later denudation.

In a few places dead ice outwash (pl. 18) complexes are known to exist. One of the best of these, and unique in Britain because it seems to date from the glaciation before the last, is the series of morainic forms around Holt in Norfolk. This consists of two clear outwash plains marking two successive positions of the ice front (fig. 3.28) and a whole series of kames, kame terraces and one very good compound crevasse filling, formerly known as the

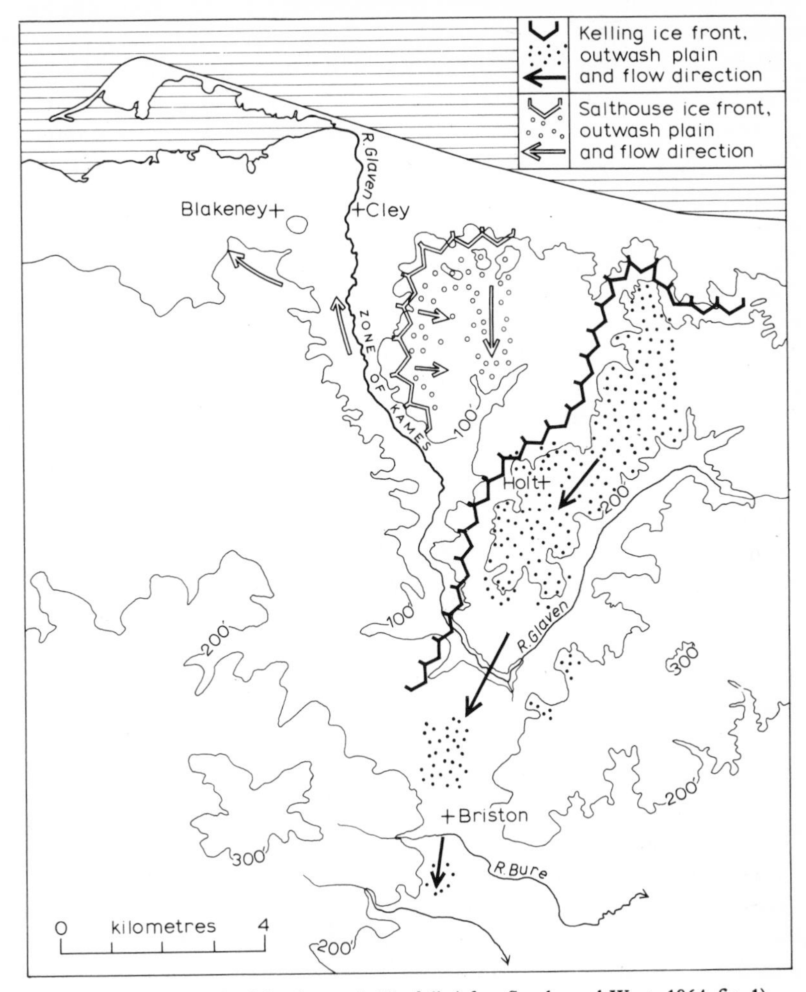

FIG. 3.28 Outwash plains in north Norfolk (after Sparks and West, 1964, fig. 1).

Blakeney esker, in the lower Glaven valley (fig. 3.29). The preservation of this fine series of features over such a long period has probably been due to the extremely coarse gravel giving very high permeability and hence near-immunity from dissection.

Outwash also occurs within many of the highland valleys and features corresponding to outwash plains there are known as valley trains. They are often terraced later by rivers and have to be carefully distinguished from both river terraces and kame terraces.

The southernmost limits of these older glaciations in Britain are not marked by any well-defined end moraine features. It is more likely that such

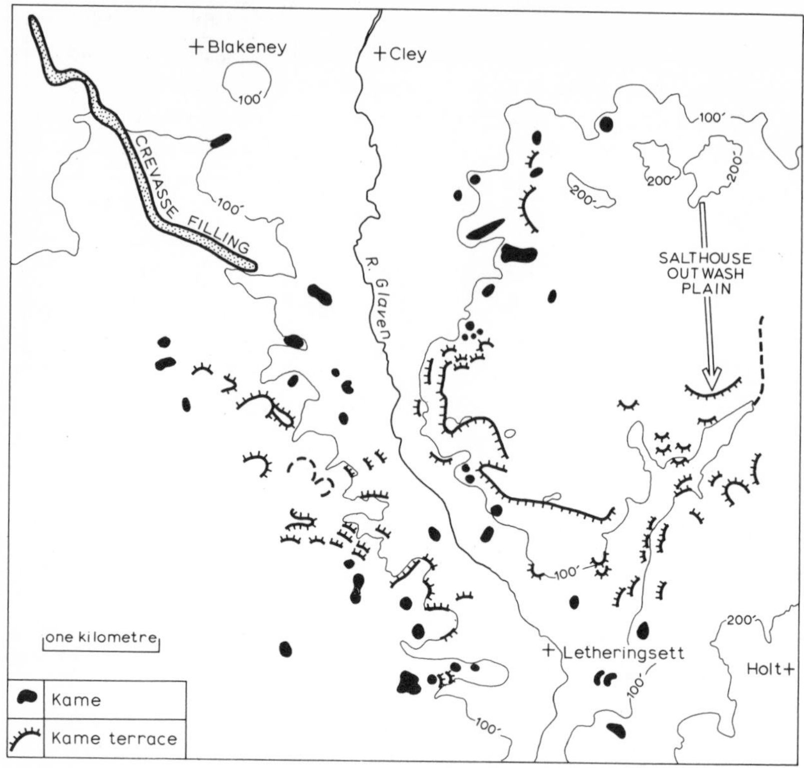

FIG. 3.29 Kame features in the Glaven valley, north Norfolk (after Sparks and West, 1964, fig. 3).

features never existed than that they have been subsequently destroyed. Hundreds of kilometres from the centres of ice dispersal it was probably much more difficult for a balance to be achieved between flow and melting, which has to occur for the formation of a well-developed end moraine. If the ice advanced to its farthest limits and there stagnated one would not have expected a good end moraine to be formed. It must be remembered, however, that in Germany and Poland end moraines do occur, so that one cannot make a generalization that end moraines do not occur far from the centres of dispersal of ice sheets.

In, or rather beneath some of the river valleys of East Anglia, especially those of Suffolk, are features which may be comparable with eskers in their origins. These are buried channels now filled with combinations of glacial, fluvioglacial and fluvial deposits. They might just conceivably be due to intense erosion consequent upon rejuvenation during a period of low glacial sea level, but this explanation is unlikely for a number of reasons. In the first place, the long profiles of the channels are not continuous but possess sections of reversed gradient, a feature much more likely to occur in a stream under hydrostatic pressure beneath an ice sheet than in a normal subaerial stream. Secondly, the channel following the river Stour reaches a depth of about −105 m below present sea level at Glemsford, between Clare and Long Melford in Suffolk, which would seem to be far too low for it to have been graded to some former glacial shoreline somewhere in the North Sea, as it is unlikely that this sea level would have been appreciably lower than the bottom of the channel at Glemsford. Thirdly, the channel, which approximately follows the course of the Cam towards Saffron Walden but in the opposite direction, seems to possess valley side slopes of the order of 45° or more in one place. This is far too steep for a subaerial valley in Chalk, the maximum slopes of which are about 30°. It is generally thought, therefore, that these buried channels were carved by subglacial streams under hydrostatic pressure.

GLACIAL MODIFICATIONS OF DRAINAGE

Lakes of various sizes are plentiful in glaciated regions. Many types have been mentioned incidentally in connection with other features in the preceding pages. They include corrie and valley lakes in rock basins, moraine-dammed lakes in corries and on lowlands, irregular mazes of lakes in knock and lochan country and also in drumlin or hummocky drift country, kettle lakes and so on. In fact the general proliferation of lakes is a characteristic feature of glacial scenery except in older till sheet areas such as southern Britain. All the types of lakes in Britain can be seen on a larger scale elsewhere in glaciated areas: the knock and lochan landscape is magnified in the glaciated Laurentian and Fennoscandian shields; the valley lakes of the Highlands of Scotland and the Lake District can be paralleled and surpassed in the Alps and Scandinavia; the moraine-dammed Loch Lomond at the edges of the Highlands is matched by the north Italian lakes.

Apart from lakes, which will not be further discussed, glaciation can seriously interfere with drainage patterns. Wherever complexes of deposits formed by melting of stagnant ice occur, so too may systems of glacial drainage channels, for example on the margins of the Highlands and Southern Uplands of Scotland. They are readily cut in incoherent glacial deposits and even in quite hard rocks, because of the high velocities and loads of the streams concerned. In solid rocks they often possess features entirely uncharacteristic of present river channels. Surface meltwater channels followed the relief of the time when they were formed. That relief was formed partly of rock and partly of ice. Since then half of the relief has melted away so that the channels often bear little relation to present relief. They may cut across spurs of rock, completely at variance with the regional pattern (pl. 19). They often cross watersheds. They may have been formed by valleys with rock on one side and ice on the other, so that when the ice melted they remained as ledges on the hillside. Channels may have been formed under the ice, within the ice and on the surface of the ice and eventually let down onto the rock beneath. The results of these conditions are series of usually shallow channels, up to 10 m or so deep, diverging from each other, converging, starting nowhere and ending nowhere and generally causing an intricate patterning of the landscape.

Apart from such channels glaciation can cause major and permanent modifications of the general drainage pattern in the following principal ways:

(1) By almost total disintegration of the drainage pattern in heavily glaciated lowlands.
(2) By glacial breaching of watersheds.
(3) By the formation and overflow of proglacial lakes.
(4) By modifications caused by englacial and subglacial streams.

The first example needs little further elaboration. The formation of knock and lochan topography (fig. 3.13) obviously disintegrates the drainage pattern and the mainly slight gradients militate against its reintegration. Conditions in drumlin and hummocky drift country are basically similar though the rocks are less resistant, so that a disintegrated pattern will probably not survive as long.

Glacial breaching of watersheds

When ice cannot get away down a valley fast enough, either because its valley is blocked lower down by other ice or because it is building up faster than the valley can evacuate it, it will overflow at the lowest available point. The process is known as glacial diffluence. When a whole series of diffluent ice streams overflows every col in a watershed so that it becomes a regional feature, the term glacial transfluence is sometimes used. The effect is the erosion of a col or gap through the watershed; hence, the self-explanatory term glacial breaching of watersheds (pl. 8). The phenomenon was recognized by A. Penck in the Alps at the beginning of the century, but was not really studied in Britain until Linton (1949, 1951) greatly improved the explanation of certain drainage anomalies in Scotland by substituting the hypothesis of glacial breaching for that of river capture. Since then it has been favoured by other geomorphologists and is now widely applied to the relief of Scotland, the Lake District and Wales, but especially to the first.

It is really impossible on physical grounds to explain many of the Scottish examples in terms of river capture. In order to maintain that river capture is likely to have taken place one must be able to attribute definite advantages to the alleged capturing stream, such as a steeper gradient to the sea giving more erosive power, or a greater discharge giving a similar advantage, or a course over much more easily eroded rocks. In some of the Scottish examples there are no such advantages. Imagine trying to explain the present course of the Callop river west of Fort William (fig. 3.30) in terms of its diversion by a tributary of Loch Shiel through the watershed east of Loch Shiel. Loch Eil is directly connected to the sea via Loch Linnhe, so that why any stream from Loch Shiel, which is a few metres above sea level, should

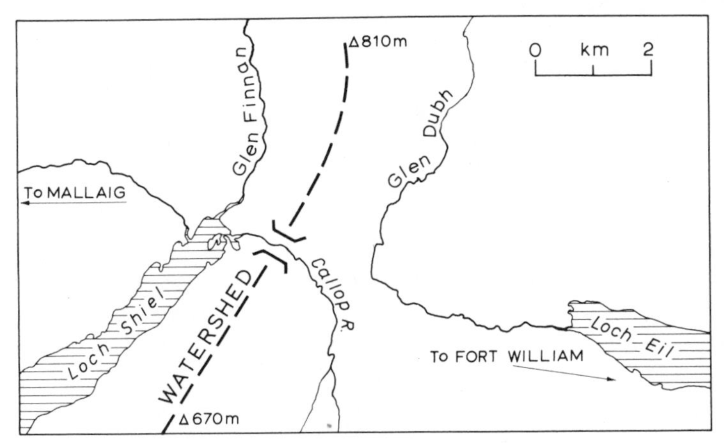

FIG. 3.30 The diversion of the Callop river near Fort William.

be able to cut through a divide, which Dury (1953), who wrote about this example, estimated to have been about 45–60 m high here, and then capture the upper Callop, is completely incomprehensible. Dury, instead, visualized the gap scoured out by an overflow stream from a lake ponded up in the upper end of the Loch Eil trough during a temporary ice retreat and then the gorge scoured out by the ice readvancing through it westwards. This is really a combination hypothesis of overflow channel and glacial breaching, but it illustrates very well the implausibility of river capture.

In the western part of the Highlands near the centre of the Scottish ice cap the pattern of glacial breaches is very complex. Not all breaches were reoccupied by streams after the Ice Age, because some are only high-level cols which were never cut down to valley-floor level. Sissons (1967) has an effective diagram (fig. 3.31) of the area north-west of Fort William showing the main east–west troughs, probably of Tertiary origin, scoured out and linked by high-level cols formed by diffluence. The whole pattern is so complex that it is difficult to know which valleys have been modified and which initiated by glacial diffluence.

North of Mull on the western side of Scotland the main ice divide was east

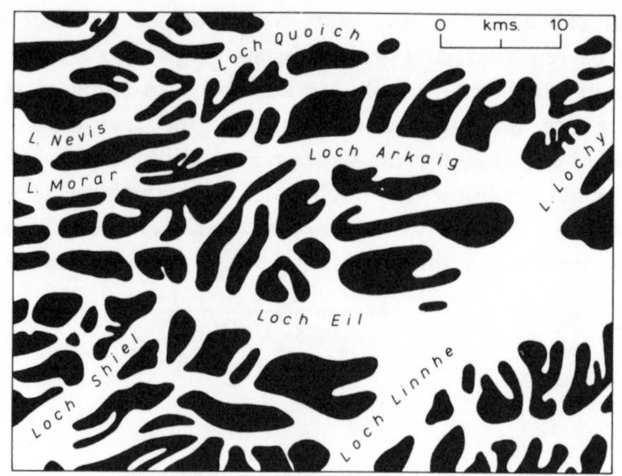

FIG. 3.31 Fragmentation of highland by glacial breaching north-west of Fort William (after Sissons, 1967, fig. 14).

of the watershed, so that the westward-flowing ice caused wholesale transfluence through the main watershed giving some thirty major breaches and many minor ones (fig. 3.32). Major breaching also took place through the main Grampian and Cairngorm divides.

In North Wales breaches occur but not on the wholesale scale found in Scotland: the breach between the Tryfan-Glyder block and the Pen-yr-Oleu-Wen block to the north of Llyn Ogwen (fig. 3.5) at the head of the Nant Ffrancon valley is one of the best examples and was due to westward-flowing ice.

Proglacial lakes and overflow channels

The classical idea of the proglacial lake and overflow channel can be summarized very briefly. An ice sheet holds up meltwater between itself and rising ground. The meltwater lake rises and overflows at the lowest point either into another lake dammed up by the ice front or away from the ice front (fig. 3.33A). The overflow channels could cut across ridges or occur between the rock and the ice giving rise to channels which will have, after the ice has melted, the forms shown in fig. 3.33B. Finally, in the theoretical example, lake C is shown draining away from the ice front. It is this type of channel which is most likely to cause permanent diversion.

It has been alleged that there should ideally be several lines of evidence to support such a hypothesis. There should be the overflow channel itself, lacustrine deposits in the former lakes, and benches or terraces marking the former lake margins. Too often the only piece of evidence is the channel, from which the lake is inferred, and then by means of the inferred lake the overflow channel is explained, a form of circular argument we have met several times already. Yet it is questionable whether one should expect much in the way of lakeside terraces. Presumably if the hypothesis works the lake level will rise, steadily or spasmodically, until it overflows. Then, if the overflow channel has the erosive power we are led to believe it should have, the height of the overflow col will be steadily lowered. There seems to be no reason why lake level should stand at overflow col level for any length of time and therefore no real reason why a marked shoreline should have been

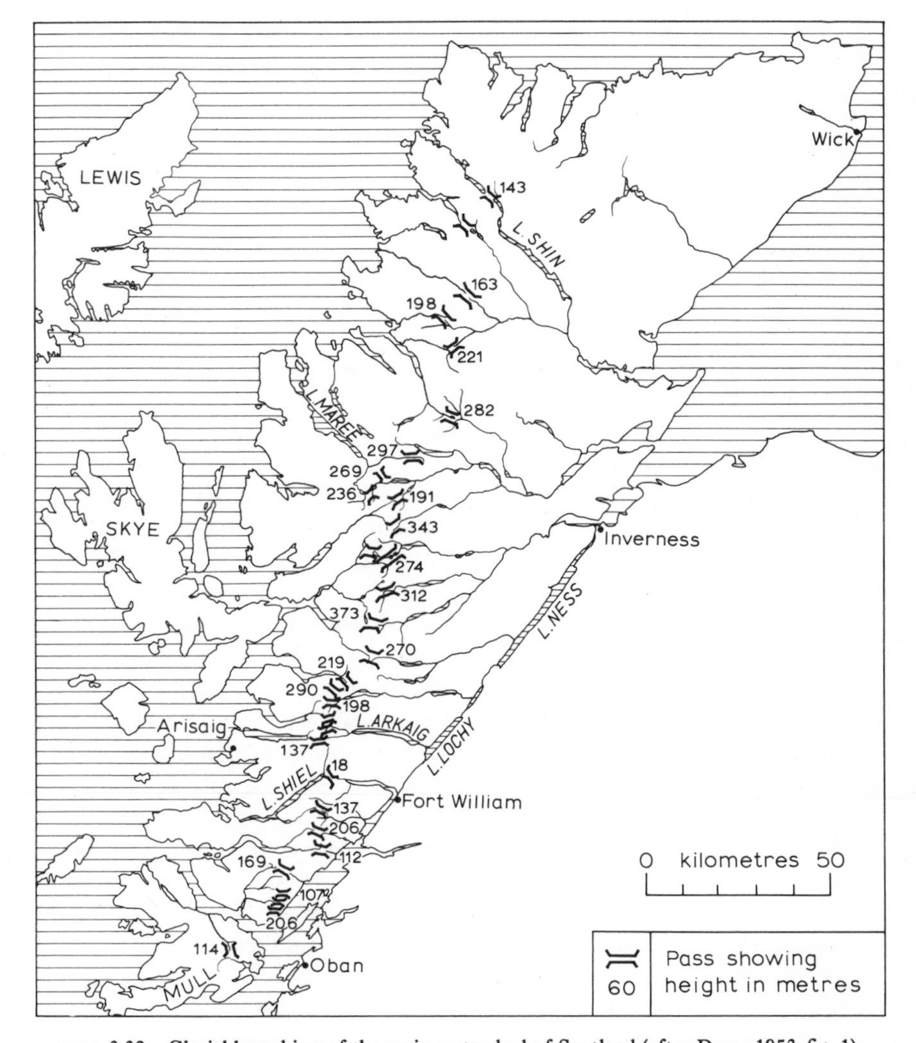

FIG. 3.32 Glacial breaching of the main watershed of Scotland (after Dury, 1953, fig. 1).

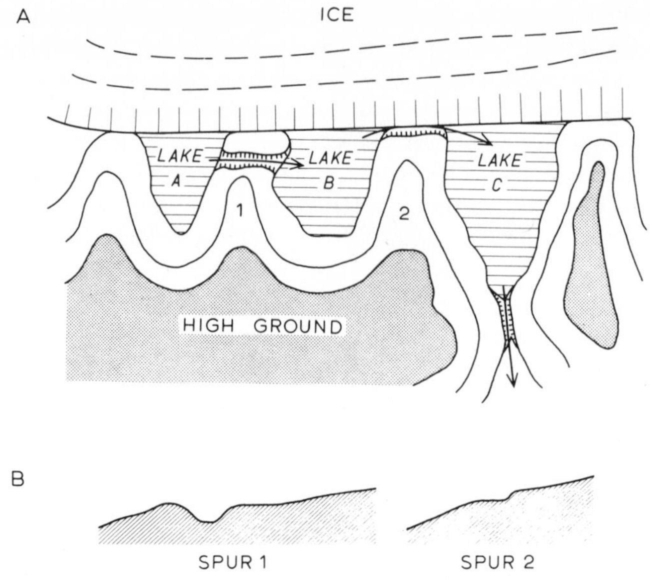

FIG. 3.33 Ice-dammed lakes and overflow channels.

formed at this level. Lake deposits we might expect, but in the case of short-lived lakes these would probably be in the form of ice-margin deltas and some thin fine silts on the former lake floors rather than any great thickness of bedded lacustrine deposits.

The proglacial lake and overflow channel idea held sway for the first half of the twentieth century in Britain, the classic example being the Cleveland Hills drainage described by P. F. Kendall (1902). Ice virtually wrapped itself round the uplands of the East Riding but did not override them. Eskdale was blocked at its eastern end and Kildale at its western end, and a combined Kildale–Eskdale lake came into existence (fig. 3.34) when water rose over the divide between them. The lake gradually extended and finally overflowed across the North York Moors through Newtondale at the lowest point of the Cleveland anticline. Newtondale is a typical, steep-sided, flat-floored overflow channel. As Newtondale was lowered Lake Eskdale split up into the lakes shown on fig. 3.34 with ridges reappearing above its surface. The evidence for Lake Eskdale has been reconsidered by Gregory (1965), who thought that much of it would be more consistent if interpreted as the result of subglacial channels and stagnant ice (see below). If this is correct, it would greatly affect the size of the suggested lakes, but not alter the main fact of the southwards escape via Newtondale. The overflow through Newtondale helped to swell Lake Pickering, where there are deltaic and lacustrine deposits, which was formed when the course of the river Derwent to the

North Sea was blocked by ice. Lake Pickering finally overflowed through the Kirkham Abbey gorge into yet another proglacial lake, Lake Humber, but that takes us beyond the limits of the present area. This series of ice-dammed lakes resulted in a permanent diversion of the drainage, not through Newtondale which was abandoned, but through the Kirkham Abbey gorge which is still followed by the Derwent as this river never reoccupied its direct course to the North Sea.

There are many other examples of proglacial lakes and overflow channels, some with more strong evidence than others. A famous example is the Ironbridge gorge in the middle course of the river Severn, which is interpreted as the result of overflow from a lake ponded up in the Cheshire Plain by the Irish Sea ice.

The drainage pattern of the Welsh Borderland around Ludlow is most anomalous: unoccupied gaps and also the way in which some of the rivers occupy tortuous gorge-like courses instead of the obvious open valleys are both indicative of glacial diversions. The upper Teme provides a good example. Originally (fig. 3.35) it flowed southwards from Leintwardine and on through Aymestrey. The advance of the Wye glacier blocked the southern

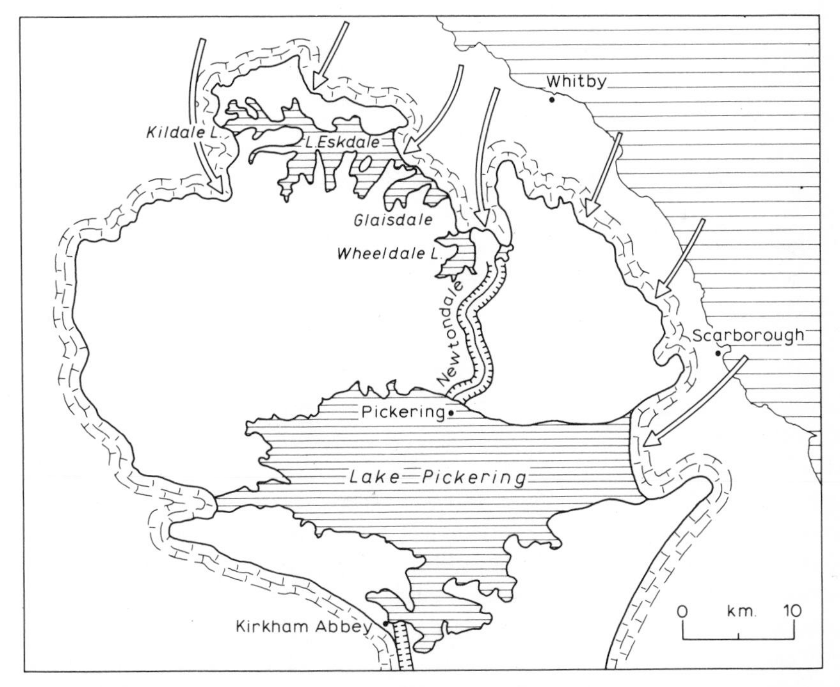

FIG. 3.34 Glacial lakes in the North York Moors (after P. F. Kendall, in *Quarterly Journal of the Geological Society*, 1902, p. 571).

outlet through Aymestrey, held up a lake in the Vale of Wigmore which overflowed and formed the gorge between Burrington and Downton Castle. This it continued to occupy after the Ice Age. Evidence for the proglacial lake in the Vale of Wigmore is provided by a gravel delta washed out from the ice front. Farther east, the north–south course of the Teme south of Ludlow originally led into the river Lugg, but this too was blocked and the Teme was diverted eastwards through Tenbury Wells.

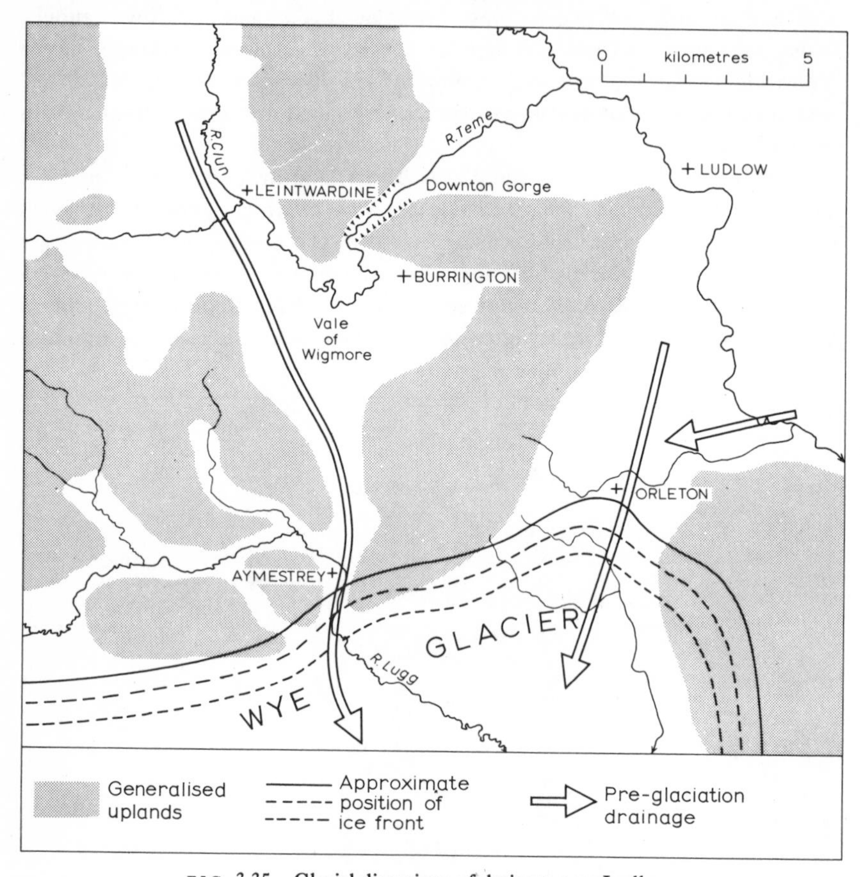

FIG. 3.35 Glacial diversions of drainage near Ludlow.

The Lake Pickering and Lake Humber examples quoted above are part of a whole series of lakes down the east side of England, which some geomorphologists believe ended in a lake in the Fenland. This finally overflowed eastwards across the ill-defined Little Ouse–Waveney watershed – or perhaps not finally because it may have overflowed into a great proglacial lake in the southern North Sea, which ultimately overflowed the ridge on the northern side of the Weald connecting the English and French Chalk outcrops and so initiated the Straits of Dover.

Even the river Thames owes its present position largely to several progressive displacements around the ice front from its original course northeastwards along the Vale of St Albans to a former outlet somewhere on the Suffolk or Essex coast.

Some of the postulated lakes must have been enormous features and one is entitled to become a little sceptical about their existence, especially about the size, for example, of a lake blocked up in the Fenland and finally overflowing eastwards through the Little Ouse–Waveney depression. There is very little evidence for such a feature, but it is becoming apparent that at some stage late in the Pleistocene there was an extensive lake in the area. Gravel sheets, which can be shown by their fossils and general condition to be associated with the last glaciation, were spread by rivers such as the Great Ouse, the Cam and the Wissey where these streams emerged on to the margins of the present Fens. Those of the Wissey are particularly impressive. They are all at about the same elevation and it is very tempting, although it has not yet been definitely established, to regard them as deposited in a vast lake held up by the ice front of the last glaciation across the mouth of the Wash. They are far too low for the lake to have overflowed anywhere other than around the northern coast of Norfolk when the ice pressure was relieved.

Subglacial channels

The ice-dammed lake hypothesis began to fall into disfavour as the sole cause of glacial river channels just before 1950, largely as the result of some very careful surveying of two Northumbrian spillways by Peel (1949). Two types of objection were raised to ice-dammed lakes: one that the form of the spillways or overflow channels was not consistent with the hypothesis, the other that it was difficult to believe in such lakes when one thought fully about what was involved.

First the evidence: if one thinks of an overflowing lake it is obvious that the gradient of the overflow channel should be one continuous slope away from the site of the alleged lake. Peel showed that the long profiles of some alleged overflow channels were humped. Under normal conditions water cannot flow uphill so that the up and down form of the channel profile is difficult to explain. On the other hand it could be explained by subglacial flow under hydrostatic pressure. One could always produce *ad hoc* arguments and special pleading to explain away the channel forms, but the boldest course was to admit that they might not be subaerial overflow channels from proglacial lakes and to see where this admission led one.

The objections based on credibility were concerned with doubts whether ice could provide a sufficiently impervious barrier to hold up a large volume of water. This was least likely in the case of stagnant ice which would probably be riddled with freely-intercommunicating channels. Sissons visualizes the actual ice-dammed lakes as being very small and consisting of a marginal lake plus a subglacial water table below which any passages in the ice are full of water (fig. 3.36). The whole of this would become quickly filled with sediment. Deeper in the glacier the still solid ice would form the effective dam. During ice advances, or in phases when the ice front was stationary but the ice still moving, the front would be steeper and the ice much more solid so that ice-dammed lakes would be much more likely.

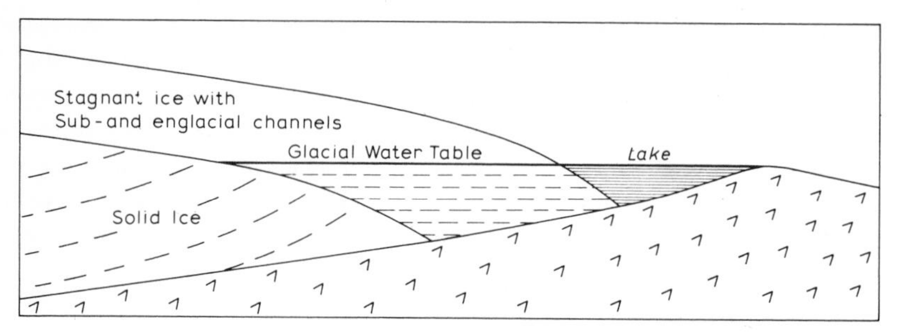

FIG. 3.36 The possible nature of a proglacial lake (after Sissons, 1967, fig. 54).

To a great degree these ideas accord with the facts. Ice-dammed lakes are known at the present day but they are usually very much smaller than those thought to have existed in the Ice Age. A famous one is the Märjelen See held up by the Aletsch glacier in the Alps, but others are known in glaciated mountain regions such as Alaska, Patagonia and the Himalayas. Temporary ice-dammed lakes are well-known features of some of the headwaters of the river Indus system, especially the Shayok river. Here, advancing glaciers from the side valleys from time to time block the main valley and impound a lake. At this stage they are probably solid ice, but with stagnation their characteristics change and the lake is rapidly drained, often with the result that there are disastrous floods lower down the valley.

An alternative idea for the release of water from glacially-dammed lakes is the flotation idea. When the depth of water in a lake reaches about 0·9 times the thickness of the ice barrier, the latter will begin to float and so allow the water to escape. A difficulty lies in the fact that, as soon as the water level falls, the ice dam should drop back into position and stop the escape, but this does not seem to happen. Perhaps as the ice lifts it fractures, and the volume of escaping lake water is able to enlarge the fissures as fast as resettlement tends to close them.

Yet the number of Ice Age glacial river channels in Britain is so large and their position and form so little consistent with overflow from ice-dammed lakes that there has been an increasing tendency to regard them as subglacial or englacial channels. Some of these channels are merely subglacial chutes, short channels down the steepest slope, which start and end quite suddenly. Other channels are closely connected with esker patterns, the one merging into the other, and this too reinforces the subglacial idea for eskers are essentially formed in this position. Some channels have subglacial chutes tributary to them and eskers in their floors, both of these relationships implying a subglacial origin.

It may well be that many channels, the evidence for which now suggests that they were subglacial, originated englacially. The advantages of the englacial idea over the subglacial idea is that one is not forced to envisage streams flowing beneath great thicknesses of ice where pressure should have resulted in the closing of their tunnels. On this hypothesis up and down gradients could be caused by streams becoming lowered as the ice melted and superimposed on to the rock beneath. When they first met rock their lowering would be retarded but would continue on either side as the ice continued to melt. Flow would continue under hydrostatic conditions. Under such conditions it is possible to imagine the development of extremely complex drainage patterns as the englacial drainage became progressively transformed into subglacial drainage, as the ice sheet wasted and superimposed the drainage on to the rocks beneath.

However, the subglacial and englacial hypotheses have been mostly applied to the development of the multitude of small glacial channels related to the retreat stages of ice sheets. It is possible that some of the major permanent drainage diversions started in this way, but many of them were probably due to overflow channels from proglacial lakes. Where the sites of the lakes were not overrun by ice this idea cannot be challenged by the englacial and subglacial hypotheses.

But, let it be noted, the arguments are between different types of glacial modification and not between glacial and non-glacial ideas. There is no real question about the profound and lasting modification of drainage patterns as well as of other landforms by the Ice Age.

4

Periglacial Features in Britain

Beyond the ice margins stretched a belt of tundra climate; the term is used in a broad sense and with no implication that the climate was identical with that of the present northern tundras. This broadly is the region described as periglacial. It is not possible to define it satisfactorily in precise quantitative terms, but, ideally, it is a zone of permanently-frozen subsoil (permafrost), seasonally-thawed topsoil (active layer), frequent changes of temperature across the freezing point, and an incomplete vegetation cover of herbaceous plants and dwarf species of trees, for example dwarf birch and willow. This periglacial zone migrated back and forth with the waxing and waning of the ice sheets.

Its physical conditions encourage specific physical processes, so that one can speak not only of periglacial climate but also of periglacial processes. Yet one cannot say that there is one universal set of periglacial processes any more than one can say that there is one universal periglacial climate. In parts of North America forest came right up to the ice sheet: in parts of Siberia extremely low temperature tundra probably extended for very great distances from the ice margin. Climates must have varied from maritime periglacial to continental periglacial; latitudinal variations of insolation pattern, day length and sun's angle of incidence were briefly mentioned in Chapter 1; the width of the zone and the character of its climate would depend on whether the winds were predominantly onto or off the ice sheet. With the climatic variation there must have been process variation and with the process variation landform variation, so that features typical of one periglacial area may not be found in another equally periglacial area.

In general the forms found in Britain bear a family likeness to those occurring in present-day periglacial regions, but they are not all identical. The active processes are freeze–thaw, the development of ice accumulations in the soil and waste mantle, rapid mass movements of the mobile layer, wind action on unvegetated ground, and highly seasonal river régimes with consequent enormous variations both in capacity to move total load and competence to move large boulders.

Periglacial activity was recognized by many nineteenth-century geologists. Though they usually described what they saw in simple English, there is no doubt that they understood its potentially very great erosive powers. Clement Reid's (1887) torrents of enormous scouring and transporting powers which he considered to be the cause of the South Downs dry valleys were essentially periglacial. The study of periglacial phenomena was greatly intensified after 1945 by the Americans and Canadians in the tundra of Alaska and Canada, by the French and, above all, in Poland where Jan Dylik inspired a school of periglacial geomorphology and established an international journal of periglacial geomorphology, the *Biuletyn Peryglacjalny*. With this wholly admirable intensification of study has come a wholly regrettable spate of jargon, not all of which fortunately has found favour. Thus, most people still speak of frost shattering rather than congelifraction, active layer rather than mollisol, permafrost rather than pergelisol. Such classical coinages have been supplemented by the introduction of dialect terms for various features from the little known languages of the peoples inhabiting the far north. There is of course plenty of precedent for this sort of thing from other sections of glacial geomorphology, e.g. kame, esker, cirque, firn, roche moutonnée. In fact, if one is truthful, one accepts the familiar jargon and rails against the unfamiliar.

Broadly speaking, periglacial action has three major classes of effect:

(1) The formation of new deposits, such as screes, wind-deposited (aeolian) rocks, and solifluxion deposits of various types.

(2) The modification and the alteration of the structure of existing superficial deposits. The geomorphologists, with their main interest in landforms, have been known to speak of this slightingly as embroidery. Yet it is vitally important in revealing the nature of past climates and in some cases detailed studies of these features allow surprisingly complex reconstructions of sequences of earlier climates.

(3) The modification of landforms usually by rapid large-scale movement. It may be that a fair proportion of the landscape we see is in equilibrium with past periglacial processes and hence just as much a fossil landscape as the forms of glaciated mountains.

Periglacial deposits

FRAGMENTED ROCK DEPOSITS

Screes (pl. 1) are, of course, periglacial deposits but they are so closely bound up with the sphere of denudation responsible for corries that they have been included in the last chapter. They are confined mainly to resistant rocks which are only permeable along their fractures and not through the rock as a whole. If the rock as a whole is permeable, scree does not result. In Britain screes are confined to the highlands, because that is where the hard rocks are. In soft rocks such as chalk, the rock will be fragmented (pl. 17), the fragments heaved about by frost action (see below) and the end products will be a sort of conglomerate consisting of rounded chalk fragments set in a chalky paste: such deposits can be seen in areas where chalk has been subjected to intensive freeze–thaw, e.g. western Norfolk.

There are various other features which may be of periglacial origin: accumulations of boulders known as blockfields and linear arrangements of boulders usually called stone streams (pl. 20). Some of the trains of coarse boulders occupying hollows in the Welsh and Scottish mountains may be features of this type rather than true screes. They may have been formed in several possible ways. In present-day periglacial regions rock glaciers are sometimes encountered. On the surface these are rock streams, but at depth there is interstitial ice and it seems that the feature, which moves very slowly, behaves essentially as a highly-loaded glacier. Others may have originated as mud flows, which because of their high density are capable of transporting large boulders. Later surface washing may have removed the fines from the upper layers and allowed the included boulders to creep slowly downhill giving rise to a stone stream. In the Falkland Islands, where stone streams were first observed, it seems that some of the streams are not confined to valleys but follow individual outcrops and are composed almost entirely of rock from that outcrop. In these cases the stone streams probably represent the final stages of freeze–thaw degradation of the outcrop.

In present-day, high-latitude regions it is easy to exaggerate the importance of freeze–thaw. In fact the highly-seasonal climate minimizes the number of times the temperature crosses the freezing point both in summer and winter. On the other hand, high-altitude, periglacial climates have the maximum number of freeze–thaw alternations. It seems most likely that Britain more resembled the latter in the Ice Age, because our climate did not have the exaggerated seasons of the polar regions and the alternating days and nights throughout the year would have promoted diurnal freeze–thaw.

SOLIFLUXION DEPOSITS

Under intense periglacial climates the soil, subsoil and underlying rocks may become frozen to great depths. This frozen layer will thicken until the cooling from the earth's surface is balanced by the heat flow from the earth's interior. Equilibrium may be achieved only at considerable depth and solidly frozen (permafrost) layers 500–600 m thick are known in northern Siberia. The thickness of the permafrost layer will decrease away from the pole and the permafrost will become discontinuous and confined to favourable localities, before it eventually disappears.

In summer a layer of varying thickness will thaw at the surface of the earth and will be thoroughly saturated both by the melting of interstitial ice, which usually forms a high proportion of the deposit, and by the addition of meltwater from the overlying snow. This is the active layer in which movement takes place. Permafrost is not essential for the formation of an active layer and solifluxion. If winter freezing penetrates to some depth, the surface will be thawed before the ground at depth so that a phase of solifluxion may intervene before the ground is completely thawed in summer. This phenomenon, in a limited development, can be seen in Britain even now in very bad winters, such as that of 1946–7, when a soft mud layer forms in a thaw above the still-frozen, bone-hard underlying ground.

The permafrost beneath the active layer ensures that the ground is impermeable, so that the active processes are confined to the surface layers. It is sometimes said that meltwater lubricates the upper layer. It would be more accurate to say that the water in the soil reduces both the internal friction and the cohesion so that the shear strength is greatly reduced and the soil fails. A general lack of vegetation probably helps the process as no root systems are present to bind the soil together. The result is mass movement over low angles of slope. Exactly what these angles are is difficult to ascertain, though, if the interpretation of some British deposits as solifluxion is correct, solifluxion must have been initiated on slopes of about 2° in places and maintained on slopes of not more than about 0·5° for distances of several kilometres.

The type of solifluxion envisaged here is that approximating almost to a liquid mud flow. But the term is often used to include a variety of phenomena, ranging from very liquid movements to the downhill creep of material caused when the formation of frost heaves particles up at right angles to the slope angle which, on thawing, fall back perpendicularly under the action of gravity so maintaining a slow downhill movement. The very fluid type of solifluxion must merge with the addition of meltwater into an approximation to a meltwater stream. Indeed, one can visualize in the early stages of thawing a slow creep movement, which accelerates into a mud flow later in the summer and probably almost to a highly-loaded meltwater stream in the height of summer. The latter will occur in valleys and there wholly or partially resort sediments brought to it by normal solifluxion down the valley sides. The pattern of sediments deposited under periglacial conditions is thus complex, ranging from unbedded, unsorted material, difficult to distinguish from till, to fairly well-sorted and bedded material which is virtually a stream deposit (pls. 21 and 22). These varying facies may of course succeed each other in a deposit which in general terms is solifluxion throughout.

In Britain, deposits of this type have been recognized for over 100 years.

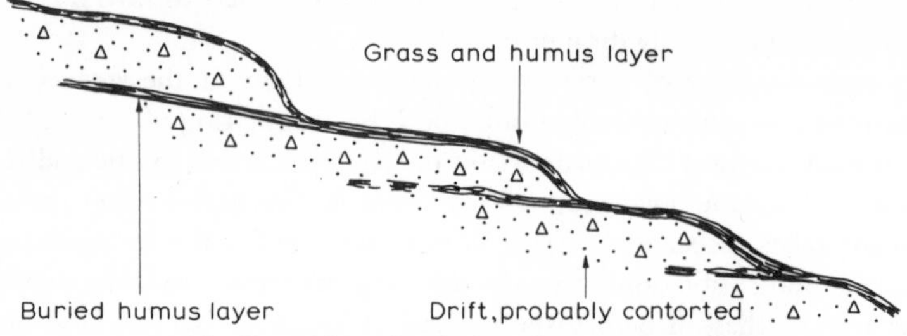

FIG. 4.1 Formation of turf-covered terrace by solifluxion.

They generally occur as fans and spreads at the foot of escarpments in lowland Britain and as narrow strips of deposition in the bottoms of dry valleys. In section they often have a convoluted pattern especially if they have moved in a fairly dry state. The names used to describe them are many: head, rubble drift, coombe rock (for a deposit derived from Chalk), trail, congeliturbate (fortunately not generally adopted as a term), sludge (a delightfully onomatopoeic term probably a little too homely for the specialist), tjäle or taele gravel, and solifluxion.

The deposits vary in composition with the rocks from which they have been derived, for they have naturally travelled only short distances. The coombe rock of the Sussex coastal plain consists of either chalky rubble, chalk and flinty rubble or flint and sand rubble, depending on the exact nature of the Chalk surface from which it has been derived. Limestone head is characteristic of the Jurassic areas, for example the Isle of Portland. The taele gravels of Cambridgeshire, which lie on the Middle and Lower Chalk in front of the Chalk escarpment, have been derived from the boulder clay capping the Chalk and so include a great variety of glacial erratics although flint is the dominant constituent.

Such deposits are usually spread as featureless sheets, but occasionally they are responsible for minor landforms. In Britain these take the form of low-angle waste fans at the mouths of valleys. Often these are only revealed when they are sectioned. Such fans occur at the mouths of some of the South Downs dry valleys, for example the Findon valley north of Worthing. They also occur in that branch of the river Cam which drains north from the Saffron Walden district towards Cambridge. They are difficult to distinguish here from subdued valley-side spurs of solid rock.

Another feature ascribable to solifluxion is the turf-covered terrace or lobe. These represent linear or lobate features where the constituent material has been held in one mass by the turf cover, although the latter may have been breached here and there. Features of this type have been described from the upland regions of both Wales and Scotland. If lobate flows have occurred one would ideally like to see an arrangement in section of the type shown in fig. 4.1.

AEOLIAN DEPOSITS

During and immediately after glacial periods there must have been vast spreads of unvegetated, unconsolidated, glacial deposits exposed both on the present lands and also on the floors of shallow seas, such as the North Sea, laid bare by the glacial fall in sea level. Periglacial action itself would add further to the total of bare deposits. On this expanse wind action was very effective, though presumably less effective in Britain than in the more continental parts of Europe, where vegetation cover may have taken longer to become established than in the maritime climate of Britain.

Wind action probably drove particles of sand grain size over the surface, polishing, faceting and grooving stones lying there. Such stones are known as ventifacts. This wind-blown sand, like that of present-day deserts, ultimately formed sand dunes. These features are less common in Britain than in parts of continental Europe such as Poland, though wind action must have contributed to the accumulation of sand in the Breckland on the margins of East Anglia and the Fens and to the common occurrence of polished stones therein. Even at the present, wind action attacks these areas very readily when they are exposed by ploughing. Sheets of sand can drift across roads in strong winds almost anywhere in the Breckland proper and on the cultivated sands at its margins, for example at Wretham Heath north of Thetford. We have known the visibility reduced to a hundred metres or less at Red Lodge near Barton Mills in severe conditions of blowing sand and silt.

It may seem a little strange to attribute such strong wind action to a phase when other considerations would lead to the belief that many of the deposits were saturated and, therefore, not capable of being attacked by the wind. Blowing may well have been mainly concentrated after the summer melt, as it is in some periglacial areas at present. But we strongly suspect that blowing might have been pronounced in summer as well. At this season there would have been no frost to bind the particles together and the possibility of stronger winds may have been greater than later on, especially if the cold winters were associated with anticyclones and either light winds or calms. There is probably a simple and valid analogy with coastal conditions here. On those beaches where vast stretches of sand are laid bare at low tide, for example along the north coast of Norfolk, the sands often dry out sufficiently at low tides for blowing to take place in spite of the fact that saturation must occur at approximately 12-hour intervals. Surely the same may well have occurred in periglacial conditions in summer, especially as some of the material was finer than beach sand and more readily moved.

Much more important than the movement of sand along the surface by the process of saltation (this is described in most general books on geomorphology, e.g. Sparks, 1960) is the lifting of silt-size particles and their movement suspended in the air. Even today dust clouds are characteristic of many periglacial and tundra regions in autumn. The material is deposited either when wind velocities slacken or when precipitation carries the dust down to earth. It forms the deposit known as loess. Loess extends as a belt of diminishing width and thickness from the Ukraine of the south-east of the U.S.S.R., through Poland and Germany, where it forms a zone of varying width at the foot of the Hercynian highlands, into north-eastern France, where it is usually called limon, and, in attenuated form, into Britain. This distribution probably reflects the fact that climate decreased in continentality westwards, and that vegetation growth was more vigorous to the west.

The British loess is usually called brickearth, although brickearth, a term which obviously refers to the suitability of the material for making bricks, includes things which are not loess. Brickearth, defined as the equivalent of loess, is spread widely on the Sussex coastal plain (pl. 23) west of Worthing. It occurs on many of the Thames terraces in and around London, or did until much of it was used up in the manufacture of London stock bricks. It is present as a thin cover over many parts of the Weald, where it is difficult to distinguish a thin Pleistocene silt, buff in colour (loess), from the buff, weathered, silty, Weald Clay beneath it.

The origin of loess presents some problems. It has been suggested above that it is a wind-borne deposit. Certain features support this suggestion. It is characteristically unbedded, whereas water-deposited material is characteristically bedded. But it does contain occasional flints which cannot have been airborne. Its mode of occurrence as a widespread and, in some areas, possibly unrecognized superficial deposit is indicative of an aeolian origin. Its fauna, mostly non-marine molluscs (snails and related bivalves), points to mixed dry and damp conditions. The majority of the species are dry-loving or dry-tolerant species (xerophiles), such as *Pupilla muscorum*, which occurs in vast numbers locally, but with these are found marsh snails, such as species of *Succinea* and *Agriolimax*, the latter a slug genus, and occasionally other species which can tolerate very poor freshwater environments, for example *Planorbis leucostoma*. From this type of fauna, which occurs both in British brickearth and in European loess, it seems that loess may have accumulated in an arid environment, but that the surface consisted of a mosaic of microhabitats including dry mounds and damper slacks between. It is also probably true that some deposits have suffered short secondary movements, perhaps merely a short distance downslope, so allowing the incorporation of a few flints in places.

RIVER DEPOSITS

Streams in periglacial conditions have tremendous variations in flow. They may remain frozen solid for eight or nine months of the year and build up to a high peak with summer melting for a very short period. The regime is very similar to that of outwash streams. When such enormous volumes of water are available, enormous loads can be carried over very gentle gradients. On this basis one might expect periglacial terraces to have gently-sloping profiles. But in periglacial areas in summer the load supplied from the land by solifluxion may also be very high, so that even the swollen periglacial streams need to maintain steep gradients to reach equilibrium with the available load. Thus, we might expect periglacial terraces, which were graded to low sea levels, to be below interglacial terraces in the lower parts of drainage basins but above them in the upper parts; i.e. the terrace profiles would cross. Examples of this sort of relationship are rare but the possibility must be borne in mind in interpreting sequences of terraces of mixed origins.

The contents of such terraces are partly distinctive, though very difficult to tell from outwash streams, a fact which is understandable as both are meltwater streams. They may contain chalk pebbles, which are unusual today in the streams of Chalk areas where most of the material is flint gravel. Such chalk pebbles were probably incorporated in a frozen state. The same explanation probably applies to the flattened lenticles of clay sometimes seen in periglacial and meltwater gravels.

Periglacial structures in superficial deposits

Under conditions of very rapid freezing, ice tends to be disseminated throughout the frozen ground. With slow freezing ice tends to segregate in the soil, using this word not in an accurate pedological sense but in a general sense to indicate the superficial layers. Segregation is much more effective in some materials than in others. Generally sands are too coarse for segregation to occur and clays are too fine, presumably because of the resistance that the small pore spaces oppose to water movement. Between these, the silts show the maximum development of segregation. It seems that the attraction of water upwards to the ice is more easily achieved under tension when the column of water is thin, i.e. in fine but not too fine sediments. Ice segregation, usually in the form of lenses, leads to the heaving-up of ground and to its collapse when the ice melts giving rise to a number of characteristic structures and minor relief features.

ICE MOUNDS

The outline evolution of these features is sketched in fig. 4.2. For some reason a lenticular ice segregation occurs and heaves up the ground surface (fig. 4.2A). Material sludges off the hillock so formed (fig. 4.2B). The hillock may be quite steep because modern ice mounds range up to about 300 m in diameter and 60 m in height. A hill of these dimensions would have a slope of

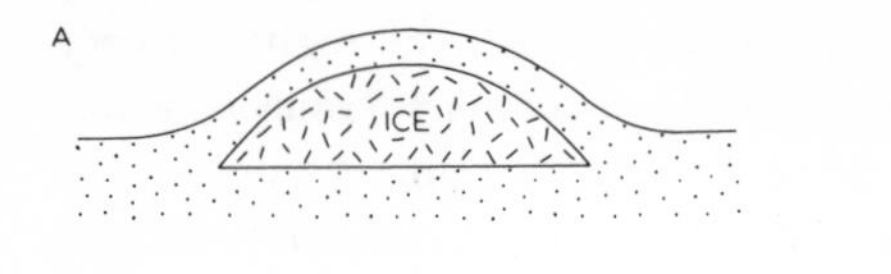

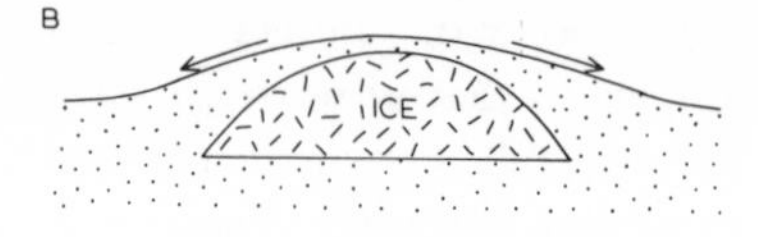

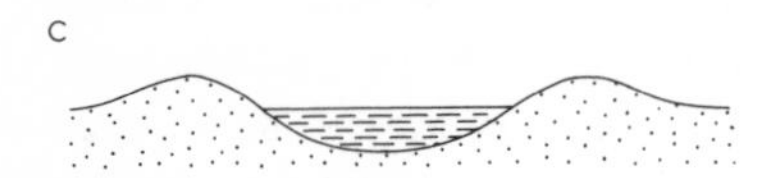

FIG. 4.2 Formation of a hollow by the thawing of an ice mound.

22° if it were conical, so that its actual maximum slope angle would probably be 25° or more. This results in a thickening of sediment on the sides of the mound. When the ice thaws the sediment is left as a rampart surrounding a hollow, which often contains a small pond (fig. 4.2C).

These and similar features have been described under a variety of names, pingo, hydrolaccolith (a term derived from an igneous analogy), bugor, pals, frost mound (a general term under which Maarleveld (1965) reviewed them all), but ice mound seems to represent most clearly the general idea and to be a descriptive term with no genetic implications and, therefore, desirable in the present state of knowledge. Genetic terminology must ideally be based upon modern, active forms, and it is very difficult to know how British fossil forms evolved, or sometimes even whether they are ice mound features at all.

At the present time two main types of pingo are recognized. The first of these is the Mackenzie delta type. This area has the greatest known concentration, some 1400, of pingos. The word itself is derived from the Eskimo dialect of the area. According to Mackay (1963) these pingos are practically all confined to old lake beds and they reflect the fact that beneath lakes the ground remains unfrozen longest. Their suggested evolution is shown in fig. 4.3. When the lake has been sufficiently shallowed by sedimentation it freezes to its bed and permafrost begins to develop below it. At the same time permafrost encroaches from the side and the pressure expels water from the sediment creating upward pressure on the thin permafrost beneath the lake. This is domed up and becomes the location of the ice core.

The other type, the East Greenland type, is explained in a different way. If progressive freezing produces a permafrost layer above an unfrozen layer, pressures may build up and lead to local doming and rupturing of the permafrost layer. This may even lead to the eruption of water and debris from this type of pingo. In this case the ice in the pingo is fed under artesian conditions with water from below, very much like the magma in the old-fashioned description of a laccolith.

In Britain ice mound features are obviously only present in the form shown in fig. 4.2C. They were first recognized by Pissart (1963), who pointed to some very good examples near Llangurig in central Wales. But the greatest

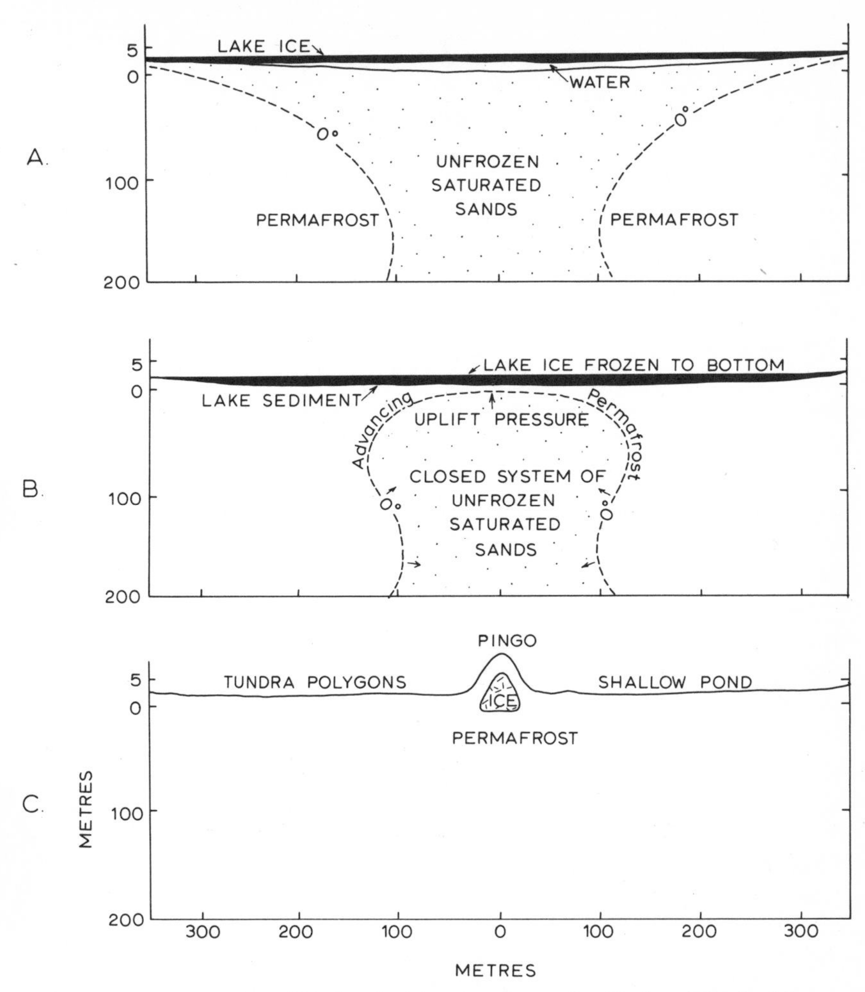

FIG. 4.3 Formation of Mackenzie delta type pingo (after Mackay, 1963, fig. 34).

concentration of these features, like many other periglacial features, is in East Anglia on the Fenland–Breckland margins and extending from this area both into the Breckland and south-eastwards along the Fen–Chalk margin. They seem to be of two ages, probably related to the latter part of the last glaciation and to the cold conditions of zone III of the Late Glacial, i.e. the very end of the last glaciation (see Chapter 6). The older series are shallow depressions, often degraded by cultivation but showing clearly as whitish circular markings on aerial photographs taken in the right conditions. These whitish marks are due to the Chalk being nearer the surface in the rims while sand and gravel accumulate in the depressions. A very good area to see them

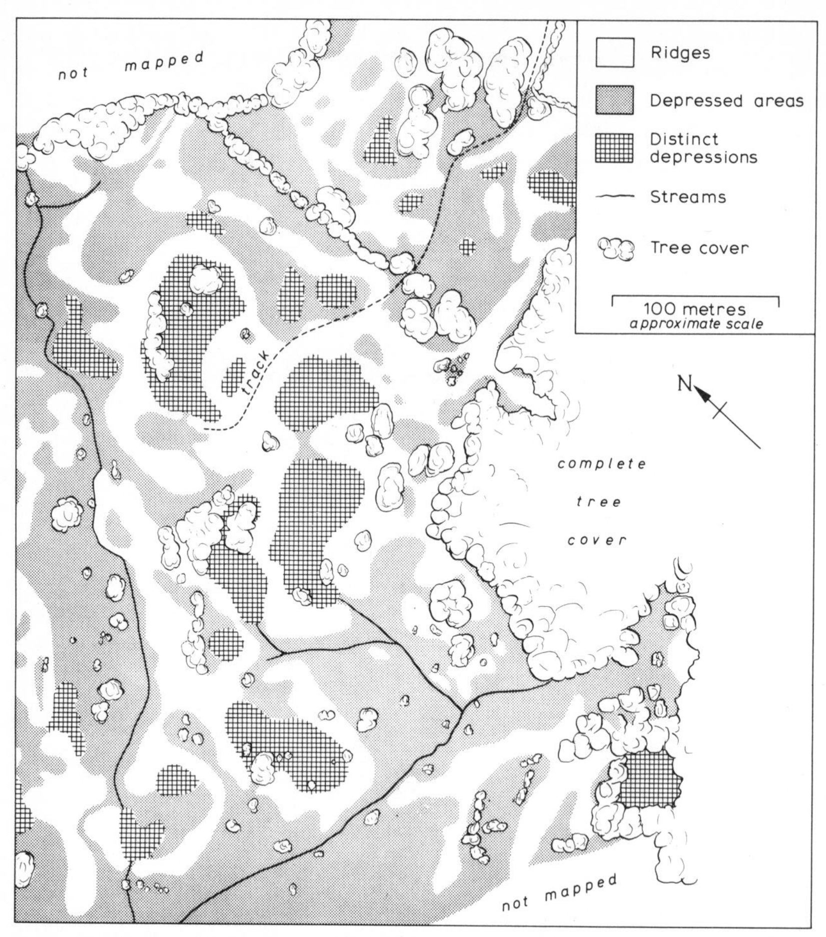

FIG. 4.4 Ice mound features on Walton Common, near King's Lynn, Norfolk (after B. W. Sparks, R. B. G. Williams and F. G. Davies, in *Proc. Roy. Soc.*, **A, 327**).

is east and south of Marham aerodrome in west Norfolk. The more recent ones are on lower, damper ground: Walton Common (pl. 24), in west Norfolk again, could be designated the type locality, but Foulden Common also has them very well-developed. A map of part of Walton Common (fig. 4.4) shows very clearly the pattern of ramparts and marshy ponds left by the melting of the ice mounds.

Whether the British forms are exactly comparable with any modern type is uncertain. Walton Common is an area of springs and rises (fig. 4.5) where water can escape from the Chalk: farther west the Chalk is sealed by the Nar Valley Clay, an impermeable interglacial deposit of Hoxnian age. Thus the conditions here seem to be partly artesian. It is certainly not true that the ice mounds have developed in former lake beds. Again there are more depressions per unit area in this part of East Anglia than there are pingos per unit area in the Mackenzie delta. They seem to have formed successively in slightly different places so that the pattern is very much composed of mutually-interfering ramparts and depressions, the latter of very variable size from a few metres up to about 100 m in diameter, and up to about 3 m in depth at a maximum.

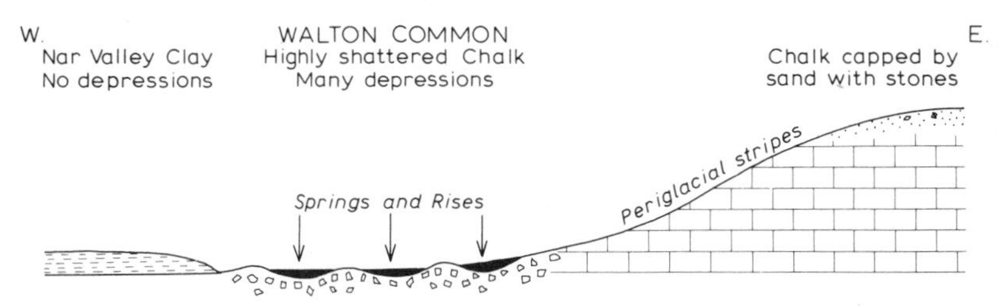

FIG. 4.5 The hydrological situation at Walton Common, near King's Lynn, Norfolk (after B. W. Sparks, R. B. G. Williams and F. G. Davies, in *Proc. Roy. Soc.*, **A, 327**).

On a smaller scale than the Walton Common features are basin-shaped deposits of fine sediment within the Weichselian sand and gravel deposits at Wretton, near Stoke Ferry in south-west Norfolk. These reach diameters of about 8 m and depths of nearly 2 m. They do not produce surface features but most likely represent some form of ice mound feature. It is known from a study of their mineralogy that they do not represent forms injected from below, but are most likely to have been formed by the development of ice lenses in silts.

THERMOKARST

Thermokarst is the name for a landscape of closed depressions formed under periglacial conditions. The various descriptions of it vary somewhat. In some the depressions are ragged, irregular hollows resembling the active solution forms of limestone regions: in others the depressions are much more smooth and rounded. It is alleged to be due to the irregular thawing of ground ice, but if nothing else was involved this could not produce depressions which would last after the whole of the ground ice thawed, when the original ground surface should be restored. It seems more likely that irregular development of ice segregations would lead to humps and hollows, that solifluxion would transfer material from humps to hollows, so that when the thicker segregations thawed under what had been the humps, hollows would replace them. Water bodies tend to thaw their own margins, hence the rounding of some of the hollows into thaw lakes, and, once a certain size had been exceeded, wind-generated waves might cause mechanical enlargement of the lakes. Thermokarst has been reported from New Jersey and the Paris Basin. It is quite possible that some comparable process played a large part in the formation of the East Anglian hollows described above and also possibly in the development of the Breckland meres.

INVOLUTIONS

Many Pleistocene deposits in section are seen to have an involuted structure (pl. 25) of the type shown in fig. 4.6A. Such features usually occur where there are lithological contrasts in the deposits. Very often either weathered bed rock or gravel will ascend as tongues or flames of orientated stones into a siltier deposit above, as suggested by fig. 4.6A. If one drew the third dimension these features would be seen to be pocket-shaped masses with no linear continuity: hence it is possible for the type of section shown in fig. 4.6B to occur.

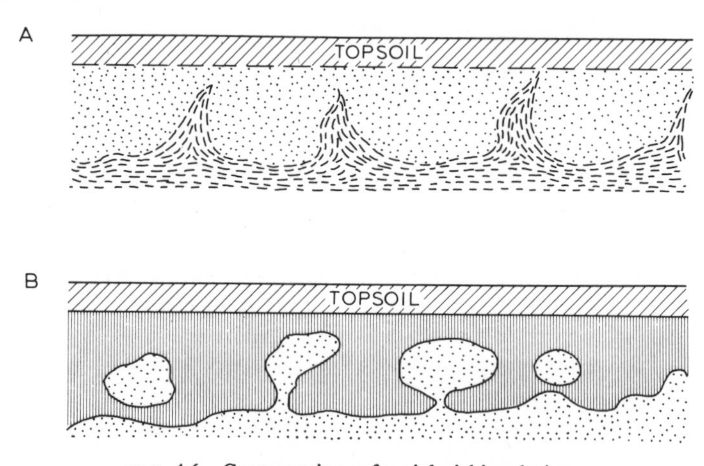

FIG. 4.6 Cross-sections of periglacial involutions.

Such features are common wherever periglacial action has disturbed deposits in Britain, but their exact mode of origin is uncertain. Two hypotheses are commonly advanced. One attributes them to movement in a mobile layer caught between the permafrost below and a rigid frozen layer reforming on the surface in autumn. It is a little difficult to visualize how this could result in the rough cellular pattern so commonly shown by such features. However, on the assumption that the hypothesis is correct it is often said that the depth of the involution indicates the depth of the seasonally-thawed active layer. The other hypothesis is that different susceptibility to freezing means that very susceptible materials, such as silts, freeze first, expand, and contort the coarser beds which remain in an unfrozen state longer. It would seem to be necessary to assume that such silts were either lenticular in the first place or that continuous beds froze irregularly.

ICE WEDGES

When thoroughly saturated superficial deposits are subjected to intense winter cold, i.e. temperatures of the order of −20° to −40°C, they contract. It is true that on first freezing expansion occurs, but with very low temperatures contraction ensues. This leads to the development of a network of cracks in which ice can form. Every winter the ground will re-open along the cracks, which represent lines of weakness, and so the cracks will be occupied by widening wedges of ice. Modern ice wedges in the tundra are known to reach 10 m in thickness at the top and it is calculable that 1000 years might be necessary for their formation. So runs the theory, and no one would deny the existence of ice wedges or that there are plenty of examples of ice-wedge structures (pl. 26) in British gravels to be explained (pls. 27 and 28). But certain points are not quite clear. If the cracks appear only at very low temperatures how does one get water available to fill them both at their initiation and when they are re-opened every year at very low temperatures? Does very shallow thawing of the surface take place while the cracks are open below, because, if the temperature rises generally, it would seem difficult to understand how the cracks were not closed by the accompanying expansion of the ground?

However, ice-wedge casts are common features and have been reported from localities all over Britain. They are very common, for example, in the Weichselian terraces which the influent rivers have deposited at the margins of the Fens, e.g. the Great Ouse at Earith (pl. 29), Huntingdonshire, and the Wissey at Wretton, Norfolk. Their use in deducing successions of climatic conditions is considered in Chapter 5.

PATTERNED GROUND

This topic follows automatically from the study of ice wedges, though it includes features not due to ice wedges. On flattish ground polygonal patterns of varying dimensions are common in periglacial conditions. On steeper slopes many of these are replaced by striped patterns, a fact which is often explained by the suggested transformation of polygons into stripes by movement.

It was said that ice wedges are caused by the cracking of ground when it contracts at very low temperatures. Now this contraction of a sheet of material is very similar to the contraction of a sheet of cooling lava which results in a polygonal network of joints: such joints are often well-displayed in basalts. Similarly the cracks associated with ice wedges form a polygonal pattern, but on a very much coarser scale than in basalts. The usual diameter of the polygons is of the order of 20–50 m. The polygonal patterns often show up in aerial photographs (pl. 29), but they are difficult to see in the field on the surface. The major characteristics of such polygons are their large diameter, the deep and narrow nature of the bounding ice-wedge features, and the essentially non-sorted character of the sediments. In sand and gravel pits the wedges show as casts (fig. 4.7A) of varying size (pls. 26 and 27): they are best preserved in sand and silt beds and are usually far less clear in gravels. When seen in the face of pits the polygonal pattern, revealed by aerial photography, may remain unsuspected.

On an altogether smaller scale are stone polygons and stripes. Where characteristically developed in the present-day high-latitude and high-altitude periglacial zones, these occur in great numbers. They consist of centres of finer material margined by coarser debris. The diameter of the polygons or the width of the stripes rarely exceeds 3 m, so making them distinguishable on mere size from ice-wedge polygons. Although these polygons are very common in high-altitude tundras of the present day, e.g. the Patagonian Andes, and can even reform after disturbance at high levels in the Alps and the French Pyrenees, they are not common in Britain. They have been reported at high levels in the Lake District and the Tinto Hills in southern Scotland, where they are capable of reforming when disturbed. Their formation does not require permafrost as long as there is severe winter freezing.

There have been many theories to account for the formation of stone polygons. They include theories which invoke convection: when water reaches 4°C it is at its maximum density and so sinks, being replaced by upward currents of warmer water from below. Such a process occurs in lakes but it is not now generally believed to be likely in soils and hence not a sound explanation of polygons.

The real problem is to account for the initial differentiation into coarser borders and finer centres for the polygons and stripes. Perhaps contraction cracking originally set up a network of fissures, which were used as minute channels by meltwater and so became the sites of concentrations of stones as the finer materials were washed away.

Once the sorting has started reasons can be suggested for the permanence of the polygonal pattern. Because of their better conductivity, the coarser, drier borders freeze more readily than the finer, wetter centres, which are thus laterally compressed. Both the pressure and later freezing tend to cause updoming, so that any surface stones, when thaw starts, will tend to move towards the polygon margins, thus perpetuating the sorting. Within the finer centres freezing starts from the top and proceeds downwards. Thus any stone is frozen into the soil first by its upper surface and will be carried up when the soil below freezes and expands. Thawing starts from the surface, so that the stone will then be held in position by the frozen ground at its base while the thawed layer collapses round it, thus aiding the relative movement

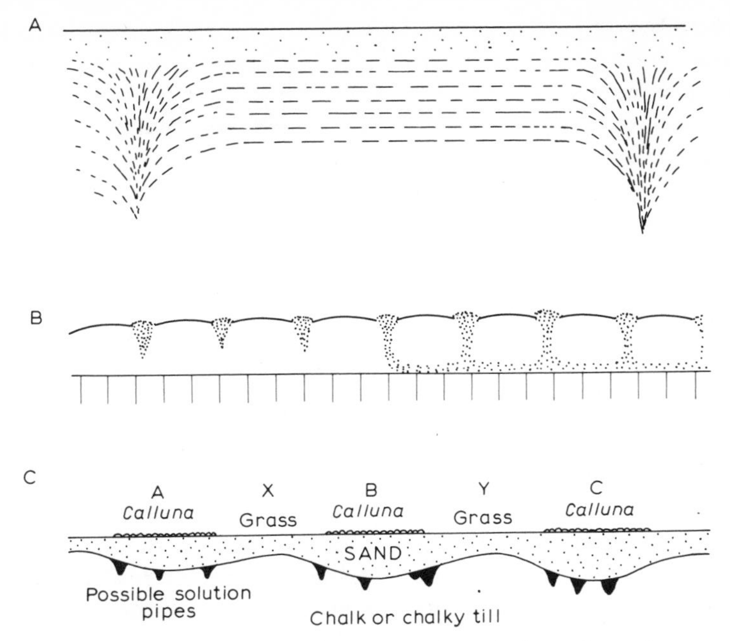

FIG. 4.7 Periglacial polygons in cross-section. A, ice-wedge polygons; B, stone polygons; C, Breckland polygons.

of the stone upwards through the soil. This may be the explanation of the upward transport of stones in soil to the surface of the polygon centres, whence they migrate as stated above towards the coarser borders.

It must be emphasized that the development of stone polygons and stripes is not really understood and that many suggestions have been made about their origin. Many of these features are shallow, some may be deeper-seated, as shown by the left- and right-hand sides of fig. 4.7B. On slopes of 3–7° the polygons become drawn out, and on slopes steeper than this stripes are formed according to Cailleux and Taylor (1954). Characteristically, the features are small, well-sorted and the margins of coarse material are relatively narrow and deep. They are not as common in Britain as ice-wedge polygons or the Breckland-type polygons which are described below.

The Breckland type of polygon is exceedingly common where it occurs on the exposed East Anglian Chalk and chalky till and to a far smaller degree in Thanet (fig. 4.8), but is virtually absent elsewhere, unlike the ice-wedge polygon which has a fairly wide distribution outside the limits of the last glaciation.

They are intermediate in size between ice-wedge and stone polygons. Williams (1964) quotes the average distance between polygon centres as 10·5 m and the average distance between stripes, which replace polygons on slopes exceeding 1–2°, as 7·5 m. Like stone polygons and unlike ice-wedge polygons, which persist as polygons on quite steep slopes, Breckland polygons are replaced on slopes by Breckland stripes. It seems that the stripes are not merely drawn-out polygons affected by solifluxion because as soon as the gradient flattens the stripes are replaced by polygons, and it is hardly likely that solifluxion would have stopped as soon as the slope flattened. These Breckland polygons are clearly visible on aerial photographs and, under the right conditions of ploughing, are also readily visible from the surface. Usually they can be seen north-east of the road from Elveden to Bury, especially along the Duke's Ride (pl. 31) between this road and the abandoned railway from Thetford to Bury. They show clearly on the gentle slopes of the area as stripes of alternate chalky and sandy material.

The general structure of these polygons in cross-section is shown in fig. 4.7C. It will be seen at once that not only are they intermediate in size between the other two types, but that some of their other characteristics are also different. They are unsorted polygons, the variation being in the depth of the sandy material overlying the Chalk or chalky till. In addition, the troughs of sandy material are very broad and shallow, not relatively deep and narrow as are the margins in the other two types of polygon. Not only do they show up as colour variations on cultivated land but also as vegetation patterns, with acid-tolerant plants such as *Calluna* where the sand is thick (A, B and C on fig. 4.7C) and more base-demanding grasses where it is thinner and more calcareous (X and Y on fig. 4.7C). They have been compared by Watt *et al.* (1966) with tussock–birch–heath polygons at present forming in Alaska, but the exact mechanism of origin is still a matter for speculation.

The distribution of the patterns also requires explanation (fig. 4.8). Chalk and chalky drift are among the most frost-susceptible materials exposed at the surface in Britain. Hence there is an obvious reason for their location on these materials, but no obvious lithological reason why they seem to be missing from such large areas of the Chalk outside East Anglia. The greater continentality of the climate of this region probably needs to be invoked as well, but this too raises difficulties as there are areas of Chalk in the eastern part of the Paris Basin, in Belgium and in Denmark, which should show these patterns, unless the lack of sand cover makes them difficult to spot or cultivation has destroyed them.

Periglacial landforms

When one sees the number of deposits and patterns which are either periglacial in origin or in modification, one begins to wonder how much of our landscape is a relic of periglacial conditions. The possibility is greatest in

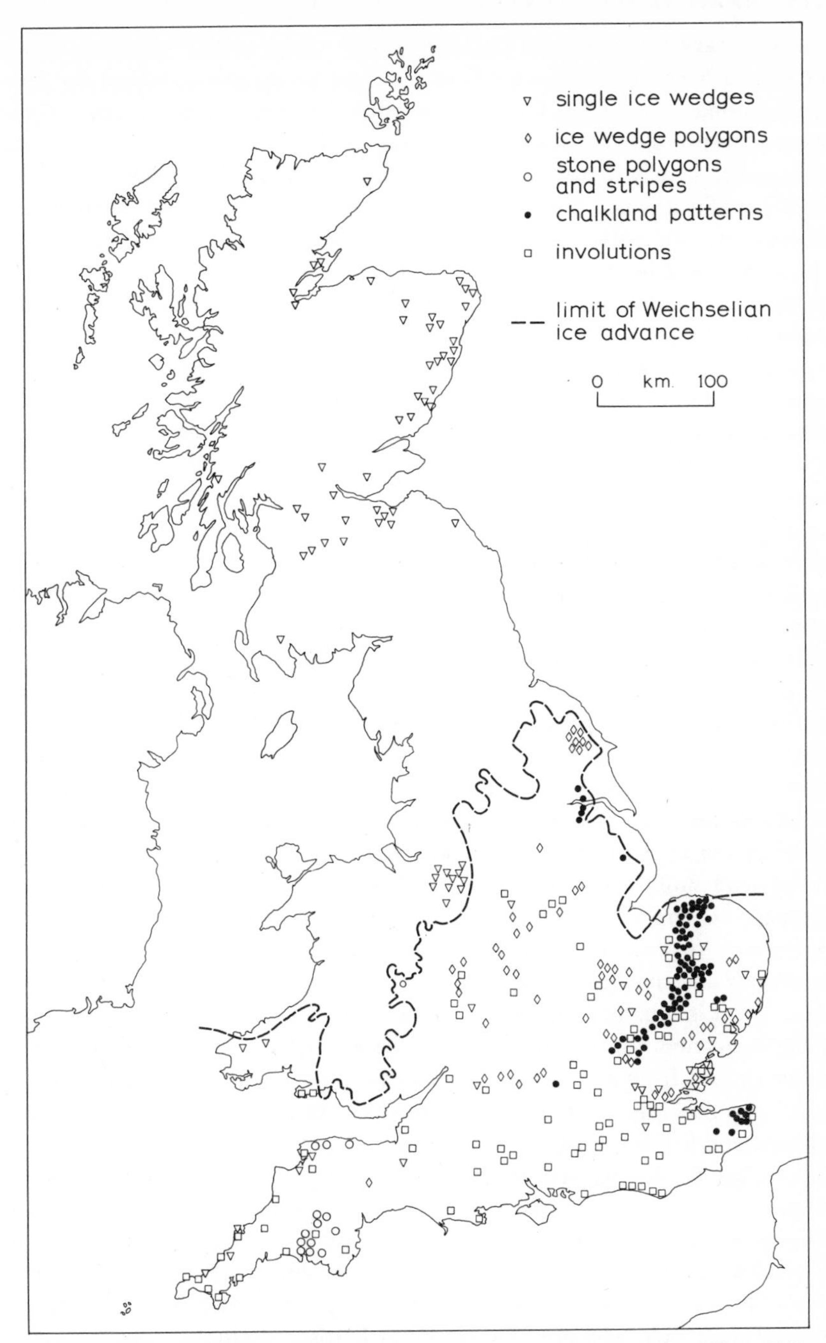

FIG. 4.8 Distribution of periglacial patterned ground in Britain (after West, 1968, fig. 12.21).

areas of very gentle slopes, because, where slopes are steeper, present-day processes have stood a better chance of modifying the legacy of the periglacial conditions of the Ice Age. Many of the phenomena described above have widespread distributions in East Anglia. Where patterns of polygons and stripes are widely preserved over large areas it seems that slopes have been little modified since their formation. One cannot deny all modification because we are not sure about the original depths of the patterns, so that there may have been some truncation without destruction of the patterns by slope movements after the Ice Age. Very much the same might be said of the widespread distribution of alleged ice mound features.

Again, if one approaches the question from a theoretical point of view, it could be argued that solifluxion can take place on slopes very much gentler than those required to promote present-day creep. The result could be that the very gentle slopes of much of Britain are still in equilibrium with Ice Age periglacial processes and are not capable of being affected by present processes. This general possibility should be carefully considered in the explanation of practically any landforms in Britain.

The origin of dry Chalk valleys is another problem in which periglacial processes have been called into account. Like many geomorphological controversies the whole question has been clouded by people taking extreme views – the valleys too often are alleged to be wholly periglacial or completely unaffected by periglacial processes. The basis of the periglacial idea is that the Chalk was rendered impermeable by deep freezing and that shallow summer thawing provided solifluxion and meltwater from the snowfields, the whole giving intense erosional power which scoured out the valleys. This view of the origin of dry valleys was strongly supported by Clement Reid (1887) and Bull (1940), largely on the basis of studies of the South Downs, almost the farthest Chalk area of Britain from the ice front of the last glaciation. In its extreme form it is an unlikely hypothesis for it is almost inconceivable that no valleys had ever existed on the Chalk before the Ice Age. Yet it is equally difficult to see how the Chalk valleys could have escaped periglacial modification: the sheets of periglacial Coombe Rock must have come in part from the valleys; some of the valleys are graded to base-levels lower than the present one which suggests erosion under conditions of cold climate and low sea level; the sharpened and undercut form of some concave dry valley bends suggests fairly recent modification by large volumes of water and meltwater seems to be the most likely explanation; finally, the known susceptibility of Chalk to freeze–thaw processes implies that modification was highly likely. In short, it seems that many Chalk dry valleys have probably been scoured out but not initiated by periglacial processes. If this is true of the Chalk, why not of every other British outcrop?

But in certain parts of Britain there are valleys which are probably periglacial valleys in the sense in which that term is understood by geomorphologists. These valleys are usually short and closely-spaced: their average distance apart is of the order of 400 m and their length is usually about the same, whereas the average valleys in most parts of lowland Britain are about 2 km apart at least and of very variable length. Such small, usually non-integrated sets of valleys have probably been initiated by meltwater runoff, and may have been extended by a combination of spring sapping and further meltwater action. If valleys are initiated by meltwater when the mass of a permeable rock is frozen, any springs that occur when the rock unfreezes will tend to be located at the heads of valleys, where their activity will promote the headward extension of the valleys. If snowfields later cover the area the meltwater from them will be guided by and help in the further excavation of the valleys.

Good examples of valleys of this type are located in the moraine country around Pretty Corner inland from Sheringham on the north coast of Norfolk. They have steep, often largely straight slopes, and a close spacing. Others probably of the same general type are those described by Kerney *et al.* (1964) from the North Downs escarpment at Brook in Kent. These seem to have been formed, or at least greatly modified, by solifluxion in zone III of the Late-Glacial at the end of the last glaciation. Similar valleys are also known from Royston Golf Course in Hertfordshire. The close spacing of such valleys is probably due to their development time being so short that they have not been integrated into a set of simpler major valleys.

Under periglacial conditions it is possible for valley-side asymmetry to develop, though it must not be thought that every valley with an asymmetric cross-profile is necessarily periglacial, because structural factors and moisture factors may also produce similar valleys. In north–south trending valleys any snow driven by westerly winds will tend to accumulate most thickly on the lee slopes, i.e. those facing east. The west-facing slope will have a thinner snow cover which will thaw more quickly and allow intense periglacial action, while the east-facing slope is still protected. This could lead to the west-facing slope becoming the steeper of the two. In east–west valleys it is possible for either north-facing or south-facing slopes to be steepened, depending on the exact nature of the climate and the vegetation cover. If the south-facing slope has its vegetation cover broken by mass movements initiated by thawing so that rill action can take place on bare ground, it is liable to become the steeper of the two. But, because of insolation variations, valley depth differences, snow cover variations and the ensuing critical effects on vegetation cover, the periglacial environment is liable to produce variable valley-side asymmetry; the whole subject has been recently investigated in detail by Kennedy (1969).

The prevalence of solifluxion was undoubtedly responsible for the removal of much of the weathered material formed in the warmer climates before the Ice Age. Thus it could have been instrumental in exhuming the granite tors of such places as Dartmoor, which on one hypothesis may represent the less weathered parts of the waste mantle.

Certain major slope processes have probably been accelerated in the cold periods of the Ice Age. In areas of approximately horizontal rocks where competent beds, such as sandstones and limestones, rest on incompetent clays, the hard beds are often cambered down towards the valleys and cracked open by joints, known as gulls, parallel to the contours. Hollingworth *et al.* (1944) described these features in detail from the Northamptonshire ironfield, where the rock disposition is very favourable (fig. 4.9A).

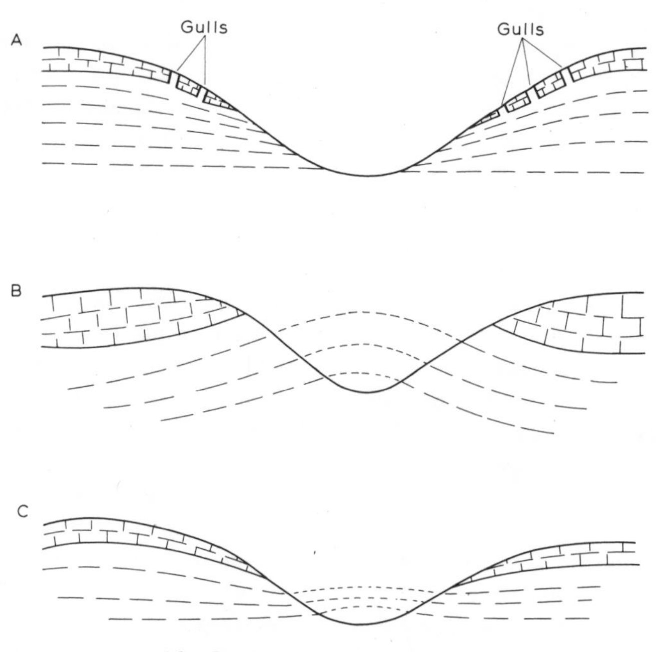

FIG. 4.9 Cambering, gulls and valley bulges.

Possibly allied is the feature of valley bulging (fig. 4.9B). Such features may be due to the early formation of ice in the saturated sediments of the valley bottoms. The expansion resulting from this could lead to valley bulging, followed by collapse and cambering when thaw later intervened. A snag with this hypothesis is how to explain why valley bulges are preserved in some valleys. Alternatively, thorough wetting in periglacial conditions may have greatly reduced the shear strength of the clays and allowed them to be extruded in plastic fashion into the valleys, thus giving rise to valley bulges.

If material is extruded in this fashion it is possible to conceive downwarping or cambering of the upper beds associated with bulging in the valley bottoms, as shown in fig. 4.9C.

Another feature caused by the failure of the underlying clays is the rotational type of landslip mentioned in Chapter 3 (fig. 3.9A). These features are common on various escarpments, for example both the Lower and the Upper Greensand escarpments of the Weald, and they occasionally occur at the present time. They reach considerable sizes, for example the slipped mass at Tilburstow Hill in Surrey is about 800 m long and 150 m wide. Wooldridge (1950) has suggested that many features he interpreted in the same way in the western Weald around Hindhead and the Vale of Fernhurst are Pleistocene features facilitated by periglacial action. Some of these slips are very large indeed, that on the eastern side of Blackdown Hill being over 3 km long and 400 m wide with a downward displacement of 30–60 m. The lithological conditions are, of course, ideal with the permeable, competent Hythe Beds of the Lower Greensand resting on impermeable Atherfield and Weald Clays.

Finally, there are altiplanation terraces, which are irregular terrace features on highlands ranging up to almost 1000 m in length parallel to the contours, to several hundred metres transverse to the contours and separated by scarps 5–10 m high. These have been variously explained by different authors, but all have expressed their belief that periglacial conditions can produce terraced patterns, which bear no relation to spasmodic river downcutting or falls in sea level. It was originally thought that they were mostly constructional features built of rock waste levelled out by frost-heaving movements, but they are now considered to be essentially denudation features. It seems most likely that they are related to pre-existing hollows and steps, where snow accumulation was encouraged and erosion of jointed rocks by freeze–thaw set in. The debris formed as the snowbank ate into the steeper slope above it may well have been removed by solifluxion. Their origin on this hypothesis is very similar to that of nivation hollows. Such terraces have irregular developments and slopes but seem to reach widths of almost 100 m with separating scarps 10 m high in Exmoor and Dartmoor, where they were first reported and where they have been most commonly described.

In a fairly recent paper Williams (1968) has attempted to estimate the amount of erosion likely to have taken place in various parts of southern Britain under periglacial conditions. Such estimates are very difficult to make, because they can only be reasonably accurate when the physical conditions are such that the total of moved material can be calculated and also the area from which it was derived. A raised beach backed by a degraded cliff, from which solifluxion has removed material, can allow a rough calculation to be made. Such situations occur in various places in south-western

England and it seems that between 7·5 and 9 m of erosion may have taken place on shales, slates and schists. A similar situation occurs near Portsmouth where a broad spread of solifluxion must have been derived from the straight southern slopes of Portsdown Hill: this had been previously estimated to be equal to about 10·5 m of erosion on those slopes.

Other calculations can be made where a stream of boulders has been moved from the source of the rock by periglacial action. Two examples are quoted by Williams. The first consists of the granite block streams, derived from the Dartmoor granite and spread downvalley on to the adjacent Culm Measures. A study of these suggests an average removal of 3–4·5 m from the slopes, a figure which is compatible with that for schists and shales quoted above in view of the greater resistance of the granite. Again in the Marlborough Downs the sarsen-filled valleys can be used to make a similar calculation. Originally the sarsens (silcrete boulders), which on the average weigh between one third and one half of a ton, developed on the watersheds and have since been moved into the valleys in solifluxion sheets. It seems likely that periglacial stripping of the Chalk slopes here of up to 9 m may have been involved, a figure close to that estimated for the south side of Portsdown Hill.

One has to be careful with the interpretation of these figures. They probably represent the maximum values of periglacial wasting on suitable slopes. They generally tend to indicate more solifluxion, presumably due to more severe periglacial climates, in central southern England than in the south-west. It could be argued that this is because Chalk is a much more susceptible rock than the igneous and slightly metamorphosed rocks of the south-west. Against this it could be held that there are many more steep slopes in the latter area and that these would have aided periglacial processes and so offset the greater rock resistance.

Another interesting and illuminating set of figures concerns the distances moved by solifluxion debris and the slope over which it has moved. In Dartmoor the greatest distance moved is slightly less than a kilometre and the slopes vary from 6° to 11½°. (The way in which the solifluxion deposits are confined near to their sources probably accounts for the apparent widespread nature of such deposits in the south-west, whereas in the south-east they seem to have flowed over lower slopes and ultimately to have become lost and reworked in the general mass of valley deposits.) In the Marlborough Downs the maximum distance travelled seems to have been about 4 km over slopes usually between 1½° and 3°. Solifluxion deposits derived from the Hythe Beds escarpment in west Kent and from the Chiltern escarpment seem to have flowed over slopes averaging about 1° for distances of 6·4 and 8·8 km respectively. Finally, deposits which Williams interprets as periglacial solifluxion, have flowed up to 10·3 km over slopes of about half a degree in the Cambridge area. All this seems to add up to a strong increase of mobility eastwards and is consistent with the much greater development of other periglacial features, such as ice mounds and polygonal patterns in the east.

Thus, the effects of the Ice Age were almost as profound outside the ice limits as within them. Nowhere in Britain can one safely neglect the possibility that landforms are relic features of the Ice Age.

5

Stratigraphy of Ice Age Deposits in Britain

The superficial deposits which cover much of Britain have been grouped together as drift, in contrast to the more ancient solid rocks beneath them. The drift deposits are usually unconsolidated – they can be best examined with trowel and spade, in contrast with the necessity of a geological hammer for dealing with older rocks. These superficial deposits are generally of Pleistocene age, and most of them were formed during the cold periods of the Ice Age directly by the agency of advancing or melting ice or by freeze–thaw sludging processes in the area beyond the glacier fronts. The extensive nature of glaciation in Europe is shown in fig. 5.1.

The existence and extent of drift deposits intrigued geologists from the eighteenth century onwards. Various hypotheses were put forward to explain their presence, more particularly to explain the presence in the drift of 'erratic' rocks found in areas miles away from their outcrop (fig. 5.2). The earliest explanation used the Biblical or some other flood as an agency for the distribution of the drift with its erratics. Supporters of the flood or diluvial explanation persisted to the end of the last century. A great submergence of land was postulated by some to explain the height at which drift was found up mountains. But with the observations of Agassiz and others on erratics in the Swiss Valleys and their relation to the former extent of glaciers, the glacial theory of the origin of drift gained gradual recognition and during the latter half of the last century won general acceptance. Nevertheless, the term diluvium, referring to glacial drift as a product of flood, has persisted on the Continent.

At first it was thought there was only one glaciation in the Ice Age. But it was soon shown, in the middle of the last century, that the drift could be

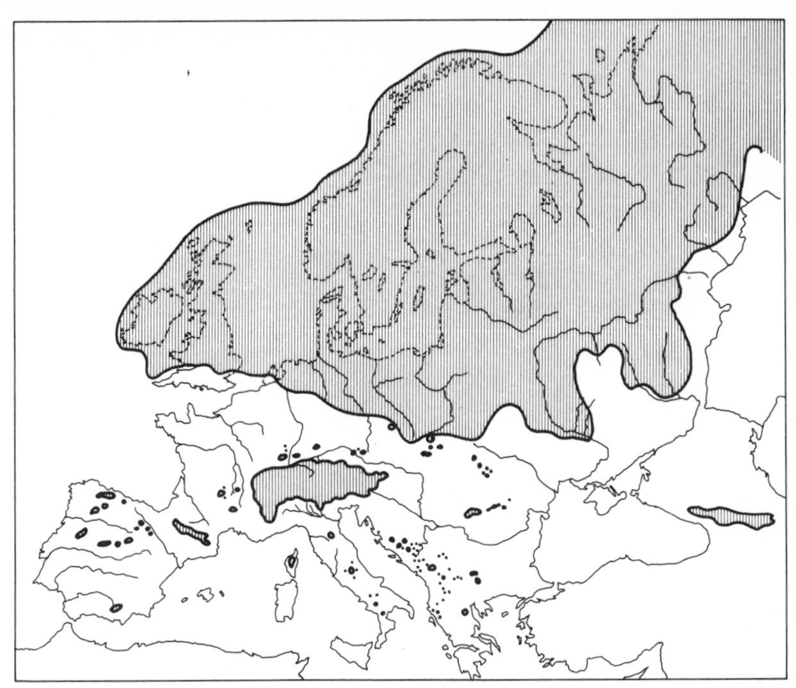

FIG. 5.1 The maximum extent of glaciers and ice sheets in Europe during the Ice Age (after R. F. Flint, *Glacial and Pleistocene Geology*, New York, Wiley, 1957, pl. 5).

subdivided. For example, on the coast at Corton near Lowestoft, two boulder clays or tills deposited from ice were described, and can still be seen, separated by fluviatile sands. The supporters of the monoglacial theory (see Chapter 1) were eclipsed by polyglacialists, though some survived till a few years ago. The limits of the successive glaciations of this polyglacial sequence, imperfectly known at present, are shown in fig. 5.19.

The evidence for polyglacialism lay not so much in evidence for different end-moraines of successive ice advances, for such are very difficult to discern in Britain, but in the finding of non-glacial sediments between glacial deposits. If such non-glacial sediments bear evidence of temperate climatic conditions, we then can demonstrate temperate interglacial conditions between glacial advances. We then deduce the alternation of cold and temperate climates which characterizes the Ice Age.

In these temperate stages between the cold stages, conditions very much like today prevailed, if we subtract man's activities in affecting the environment by forest clearance and in other ways. Forests clothed the land, sea levels were at least as high relative to the land as at present and rich vertebrate faunas developed.

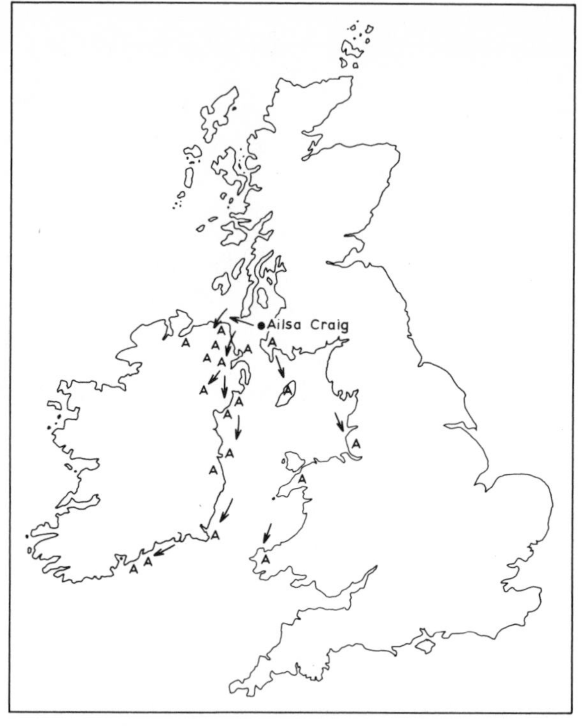

FIG. 5.2 Map showing the movement by ice of erratics (A) from the outcrop of riebeckite-eurite on Ailsa Craig (after Wright, 1937, fig. 37).

On the other hand, during the cold stages sea levels were much lower, at the most about 120 m lower than present, because the water abstracted from the sea during the build-up of the world's ice sheets caused a world-wide (eustatic) lowering of sea level (see Chapter 1). In these cold periods we see evidence for the predominance of herbaceous vegetation, for the instability of soils, especially on slopes, for the spring floods of meltwater, with the more active rivers grading to relatively much lower sea levels, and the formation of deep (now buried) channels, as at the mouth of the Thames.

The eustatic changes of sea level associated with expansion and melting of ice sheets leave their mark in the form of the raised beaches of times during earlier temperate stages when sea levels were higher relatively than at present. Another type of raised beach results from uplift of land after the removal of the ice load by melting, in areas central to the dispersion of ice (see Chapter 1). An extreme case is uplift in the northern part of the Gulf of Bothnia, an area under the former massive Scandinavian ice sheet, which is rising at the rate of a metre a century. Such effects, due to release of load and local upwarping, are termed isostatic. In Britain they are present in the north, presumably where ice thicknesses were greatest, e.g. in western Scotland (fig. 5.3).

Another kind of earth movement is important to the study of the Ice Age in Britain. The long-continued sinking of the southern part of the North Sea basin (fig. 5.3) has led to the accumulation of several thousand feet of Pleistocene deposits in this area. The low level of the Netherlands is a result of this sinking. East Anglia is on the western margin of this area, and as a result there are marine deposits in a basin formed of Tertiary and Cretaceous rocks. These marine deposits are the shelly Crag of East Anglia. It was formerly thought that these were pre-Ice Age and thus Tertiary in age. But,

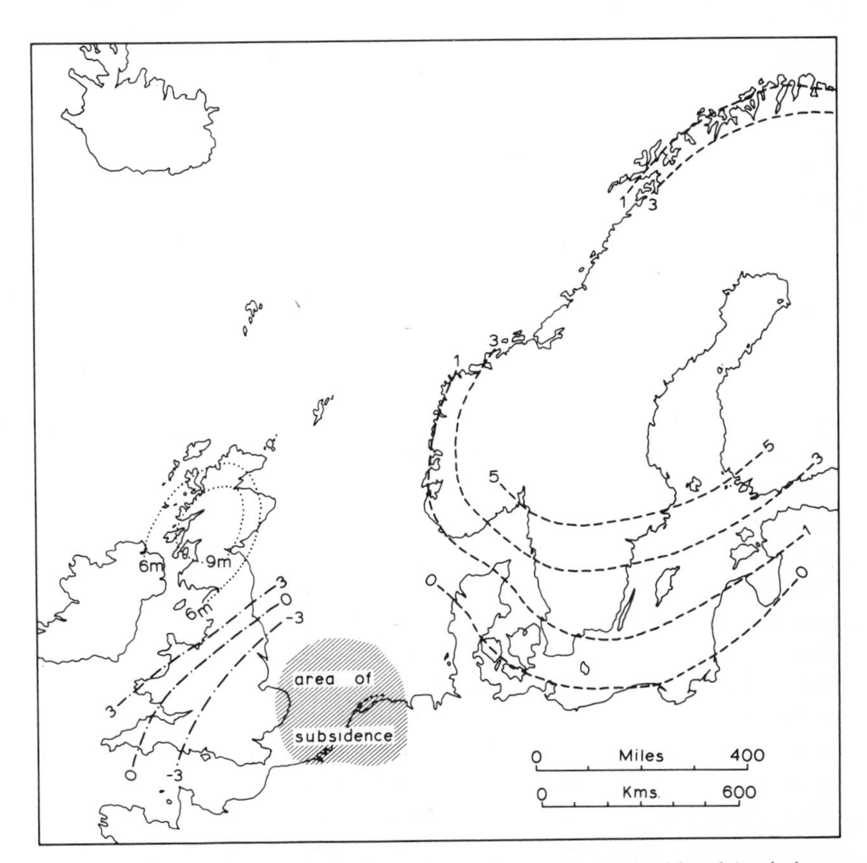

FIG. 5.3 Uplift and downwarping in north-west Europe. In Scotland 9 and 6 m isobases are shown for the main Flandrian shore-line (after Sissons, 1967). In England and Wales, recent subsidence in mm/year of the order shown has been suggested by Churchill. Isobases of recent uplift in mm/year are also shown in Scandinavia.

on the definition that the beginning of the Ice Age is marked by climatic deterioration, and because we find evidence of such deterioration in the marine sediments of the Crag basin, the lower boundary of the Ice Age is placed near the base of the Crag.

All these processes of the Ice Age demonstrate the instability of the environment characteristic of the period. We see great changes of sea level, with the consequent changes of river regimes, isostatic upwarping, climatic change, huge changes in vegetation, the advances and retreats of ice, and the development and melting of permafrost. In all this mêlée of universal change how can we expect to build up a sequential story of events during the Ice Age?

Building up the sequence

Knowledge of the detail of the geological sequence of the Ice Age is fundamental to the study of all aspects of the Ice Age, whether we are interested in the history of man, history of the fauna and flora, the climatic changes of the past, the changing geography of the Ice Age or the causes of ice ages. In pits, temporary sections such as gas pipe or sewer trenches and in boreholes we find sequences of sediments which provide the key to the geological history of the Ice Age. Sections giving information are very often created; anyone digging a hole is very likely, in lowland Britain anyway, to be digging in the drift deposits which blanket the surface of the solid rock. But frequent though they are, not one in a hundred is seen by a geologically-minded observer and much must be lost to science in this way.

A careful description of the section is the first prerequisite for study, in order to distinguish the sequence of different types of sediment. One of the difficulties of Ice Age geology is the rapid lateral variation of sediment type, as well as the vertical variation normally seen in a section. Careful description is thus all the more necessary. We have to distinguish glacial sediments, such as till or boulder clay (pl. 16), glaciofluvial deposits (pl. 18), from non-glacial sediments such as lake deposits, diatomite, organic muds and peats (pls. 32 and 30). A great deal of evidence regarding environmental history can be got from field examination of sediments. It is usually possible to distinguish wind-deposited sediments (pl. 23) from fluviatile ones, periglacial sludge deposits from tills, freshwater from marine deposits and the various types of biogenic deposit (e.g. terrestrial peats, limnic muds) from one another. Thick biogenic deposits signify a long period of time of high organic productivity, which will need a temperate climate. Three types of occurrence of biogenic deposit are of particular value. In fig. 5.4 each section shows the relation of temperate biogenic deposits to more inorganic deposits formed during cold periods (as in C) or by fluviatile or marine conditions (as in A and B) replacing quietwater environments in which organic remains can accumulate. Section A shows a marine transgression over freshwater deposits, section B shows organic sediments within a fluviatile terrace and C a period of temperate organic deposition in a lake between two cold periods during which till and solifluxion deposits were formed, the former during an ice advance, the latter during a period of periglacial climate. Section A gives information about changes of sea level in the past, section B about conditions in a river valley during terrace formation, and section C about climatic change.

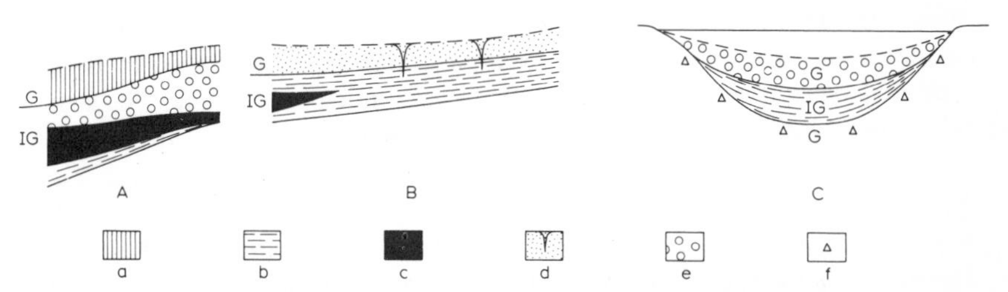

FIG. 5.4 Three sections giving evidence of climatic change in the Ice Age. A, coastal sequence; B, terrace sequence; C, lake sequence; G, glacial or periglacial sediment; IG, temperate (interglacial) sediment; a, loess; b, freshwater sediment; c, marine sediment; d, fluviatile sand with ice-wedge casts; e, gravel, beach in A, solifluxion in C; f, till.

All this results from a study of sediment. But in many sediments fossils occur. If the sediments are glacial, the fossils must be reworked by the ice from older horizons. If the sediments are non-glacial then the collections of plants and animals can reveal a large amount of information about past conditions of life and environment, as discussed later in Chapters 6 and 7. We can then characterize particular temperate periods by their vegetational history and associated fauna, and distinguish between interstadial and interglacial episodes, the former showing slight amelioration of climate from the glacial climate, the latter more temperate conditions with mixed oak forest as the natural vegetation.

Correlation of one section with another leads to the identification of sheets of till which have a wide spread over the country, and can be ascribed to one glacial advance. If we can then relate these till spreads to interglacial episodes of organic deposition, in lakes, for example (fig. 5.4C), we can build up a sequence of glacial and interglacial climatic stages.

It is these climatic stages, characterized by cold or temperate conditions, which form the basis for Ice Age stratigraphy. Even where we have no true glacial deposits, as in the marine Lower Pleistocene of East Anglia, they can be identified by fossil content, the fossil assemblages indicating cold or temperate conditions. In order to make certain that each climatic stage has a definitive climatic expression and stratigraphical position, there should be a type site for it, and the stage is then named after this type site or region in which it is found. Thus the Hoxnian temperate stage is typified by the organic interglacial deposits at Hoxne in Suffolk (pl. 32), Cromerian by organic deposits beneath the oldest till on the Norfolk coast near Cromer, and the Anglian glacial stage by glacial deposits on the coast of East Anglia at Corton near Lowestoft.

If there is then any argument over the stratigraphical position of any stage or its climatic and vegetational character, the type deposit can be used as the reference point. Much confusion has resulted in Europe through lack of type sites for climatic episodes. For example, there do not appear to be any type sites designated for major glacial stages in northern Europe. The lack of reference points means that the bases for particular arguments are not clear.

The stages of the British Pleistocene so far identified are shown in Table 5.1, with notes on the type sites and nature of deposits concerned. We shall refer to this table in the following discussion on the geological history of the Ice Age in Britain, starting first with the temperate stages, in each of which the climate, at the optimum, appears to have been much the same as today. Table 5.1 also shows the probable correlations of the glacial stages of north-west Europe with those of Britain.

Temperate stages

Temperate stages are periods of high organic productivity, giving sediments rich in fossils, both plant and animal. In a sequence of such sediments, the succession of fossil assemblages gives the clue to changing climates. In each temperate stage we then see changes of fauna and flora associated with the development of a period of optimum climatic conditions, followed by a period of climatic deterioration. The climatic optimum of the postglacial (Flandrian) period was first discovered by the Irish naturalist Lloyd Praeger (1896), who observed that estuarine clays and raised beaches formed in northern Ireland in mid-Flandrian times five to six thousand years ago contained a mollusc fauna with species not living so far north at the present time. Since then evidence has accumulated that in the optimum period the average July temperatures were 1° or 2°C warmer than at present. Similar climatic optimum periods have since been found in the preceding temperate stages. There is no evidence to suggest that these earlier temperate stages had optima much warmer than the Flandrian climatic optimum.

The details of change within a temperate stage are best seen in the pollen diagrams obtained from thick sequences of lake or river deposits. The vegetational sequences from each temperate stage, in south-east England at least, appear to be distinctly different. The different sequences are discussed

Table 5.1. *Stages of the Pleistocene*

	Stage	*Climate*	*Type site or area*	*Correlation of glacial stages with north-west Europe*
Upper Pleistocene	Flandrian (postglacial)	temperate	postglacial peats and lake deposits of Britain	
	Devensian (last glaciation)	cold, glacial	glacial deposits of Cheshire Plain	Weichselian
	Ipswichian (last interglacial)	temperate	interglacial lake deposits at Bobbitshole, Ipswich	
Middle Pleistocene	Wolstonian	cold, glacial	glacial deposits at Wolston, Warwickshire	Saalian
	Hoxnian	temperate	interglacial lake deposits at Hoxne, Suffolk	
	Anglian	cold, glacial	glacial deposits on the coast at Corton, Suffolk	Elsterian
	Cromerian	temperate	lake deposits at West Runton, Norfolk	
	Beestonian	cold	silts and fluviatile gravels at Beeston, Norfolk	
	Pastonian	temperate	tidal deposits at Paston, Norfolk	
Lower Pleistocene	Baventian	cold	marine sands and silts on the coast at Easton Bavents, Suffolk	
	Antian	temperate	marine deposits in borehole at Ludham, Norfolk	
	Thurnian	cold		
	Ludhamian	temperate		
	Waltonian		crag on coast at Walton-on-the-Naze, Essex	

in more detail in Chapter 6. The uniformity of regional vegetational history at different sites whose stratigraphical relationships are known, suggests that we can use this uniformity as a basis for correlation where the stratigraphical relations are obscure. Thus Hoxnian interglacial deposits in East Anglia are found in hollows on the surface of the Lowestoft Till and they contain similar vegetational sequences different from those of the demonstrably older Cromerian or younger Ipswichian deposits. If a vegetational sequence in the Midlands is then found which compares well with the Hoxnian of East Anglia, we have good grounds for a correlation between the two. We can then also make a correlation of the glacial deposits under and overlying the interglacial deposits in the Midlands and East Anglia.

The process of correlation via interglacial deposits has been a development of the last twenty years, and has allowed much more convincing long-range correlations to be put forward than was previously possible. Unfortunately, temperate deposits are not very frequent in northern and western Britain, and this results in difficulties in correlation of glacial deposits in these areas with those to the south and east. Thus the most detailed sequences known in Britain come from East Anglia, where interglacial deposits with clear stratigraphical relationships are not uncommon, and where there is a very considerable thickness of the preglacial marine and estuarine deposits revealing climatic change through long sequences of vegetational history.

We can divide the occurrence of temperate deposits into two categories, those formed in preglacial times, as in the East Anglian Crag basin, and those formed in interglacial times. The preglacial temperate stages are best preserved in East Anglia. The deposits take the form of marine or estuarine silts, together with freshwater muds, all containing the forest floras typical of temperate stages. The sediments of the Ludhamian and Antian temperate stages were revealed in deep boreholes (e.g. fig. 6.8) in the Crag basin and rarely outcrop on the coast of East Anglia and inland. They are characterized vegetationally by the presence of forest with Hemlock spruce (*Tsuga*). Fig. 5.5 shows localities for these stages.

The Pastonian and Cromerian stages are identified from deposits exposed in the Cromer Forest Bed Series of the coast of East Anglia. Each contains both freshwater and marine or estuarine deposits, and both show the development of mixed oak forest. The geological section at West Runton Gap (Woman Hythe), near Cromer, demonstrates the stratigraphical evidence for these temperate stages. Between the Chalk on the foreshore and the glacial deposits of the cliff-face, there is a sequence of fossiliferous sediments. In fig. 5.6, a sketch of the section, the tidal silt *d* contains a temperate forest flora, recording a time of tidal conditions with temperate climatic conditions, the Pastonian temperate stage. The sand and gravel *e* contain evidence of cold

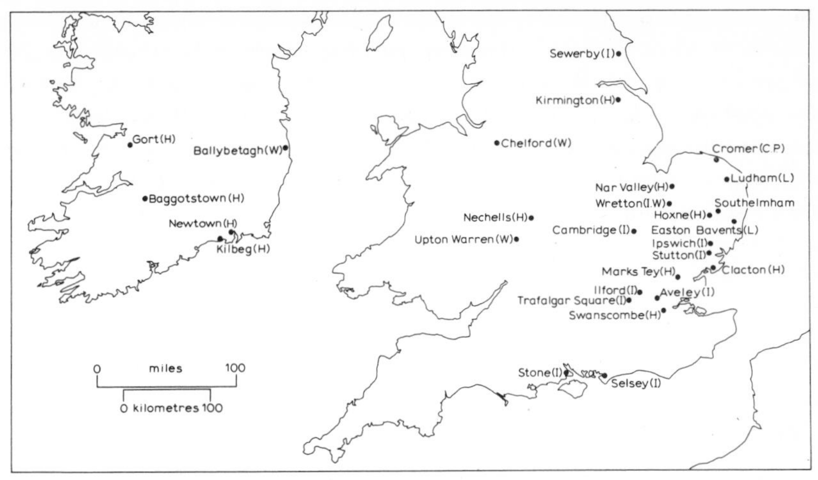

FIG. 5.5 Sites with evidence for temperate stages and interstadials in southern Britain and Ireland. L, Lower Pleistocene; P, Pastonian; C, Cromerian; H, Hoxnian; I, Ipswichian; W, interstadials in the last (Devensian) glaciation.

climate climatic conditions, in their fossil flora and periglacial phenomena. The freshwater mud *g* occurs in a channel cut into *e*, and contains a temperate forest flora; it is overlain to the west by gravel *h* with the marine mollusc *Mya truncata* in the position of life, then the tidal silt *i* with a forest flora; both *h* and *i* were formed as a result of marine transgression over the earlier freshwater beds. Beds *g*, *h* and *i* are all in one temperate stage, the Cromerian. This stage was succeeded by the glacial stage which deposited the tills of the cliff-face. This sequence at West Runton gives the clearest demonstration in the British Isles of climatic changes of immediately pre-glacial times.

The other group of temperate deposits are those which occur within the glacial sequence. The Hoxnian temperate interglacial deposits, with their characteristic vegetational succession, lie on tills of the earliest of the well-attested glacial stages, usually in hollows or troughs in till. The deposits may be very thick, reaching some 40 m. At Marks Tey, near Colchester, a complete record of vegetational history has been obtained from lake sediments of such thickness (fig. 6.7). The sites where Hoxnian interglacial deposits are found in Britain, and the correlative Gortian interglacial in Ireland, are shown in fig. 5.5.

The deposits of the Ipswichian (last) Interglacial occur in present river valleys often associated with terraces. It is curious that none have yet been discovered in kettle holes of the previous glaciation. Perhaps the mode of melting of the ice of this previous glaciation was not favourable to kettle-hole formation, or the extent of this glaciation is very much smaller than we think it is. A typical occurrence of Ipswichian interglacial deposits is shown in fig. 5.14. Localities for evidence of vegetational history for this interglacial are shown in fig. 5.5. We do not include sites of terraces containing bones of the ***Hippopotamus***, considered a characteristic animal of the Ipswichian interglacial in south-east England. It is remarkable that no site of this age with vegetational history has yet been discovered in Ireland or certainly in Wales or Scotland.

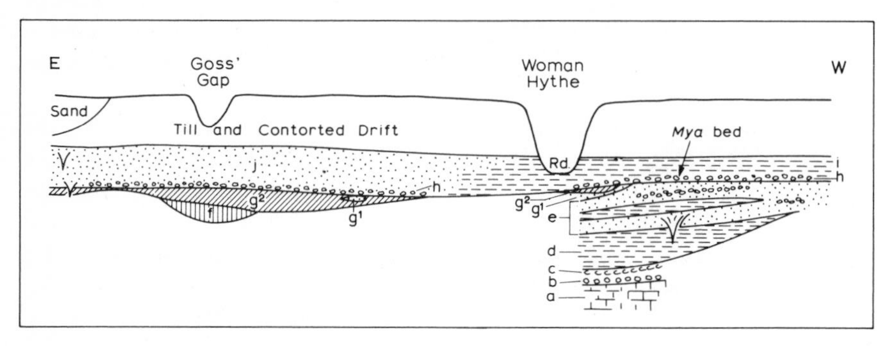

FIG. 5.6 Sketch section at West Runton, Norfolk, about 450 m length and 7 m depth. a, chalk; b, stone bed; c, Weybourne Crag (Baventian, Pastonian); d, tidal silts (Pastonian); e, silt, sand and gravel with permafrost features and arctic flora (Beestonian); f, 'late-glacial' marl (late Beestonian); g^1, sands with temperate non-marine molluscs (Cromerian); g^2, freshwater mud (Cromerian); h, marine gravel with *Mya truncata* (Cromerian); i, tidal silt (Cromerian); j, fluviatile sands (early glacial, i.e. early Anglian).

The deposits of the present temperate stage, the Flandrian, are widely distributed over the British Isles, in alluvial and coastal basins, in lakes and fens and as the terrestrial peats of raised bogs (pl. 33) and blanket bog (pl. 34). The wealth of such organic deposits allows very detailed reconstructions of vegetational, faunal and climatic history in the Flandrian.

Cold stages

Evidence for the cold stages comprises the geological evidence of glacial deposits and periglacial phenomena and the biological evidence of fossils, especially beetle assemblages and plant assemblages indicating treeless herbaceous vegetation of tundra aspect.

The glacial deposits are the most obvious evidence of former cold climates. They are widespread in central and northern Britain and in Ireland. But though they are so obvious, it is by no means certain that they represent deposition over long periods within a cold stage. For example, in the last cold stage, the Devensian, midland Britain appears to have been extensively glaciated for only a short time, as we shall see in the later discussion on the last glaciation. Outside the glaciated area we rely on evidence from periglacial phenomena and fossils for our interpretation of climatic events within a cold stage.

The geological and biological evidence shows that the cold stages do not merely comprise a single ice advance. The situation is far more complex than that. Slight ameliorations of climate are observed shown by changing plant and animal assemblages, but none so great as to suggest the degree of climatic amelioration seen in an interglacial. These minor ameliorations of climate within a cold stage have been termed interstadials, the term stadial referring to a period of ice advance.

Whereas the subdivision of interglacials is a relatively simple matter, the subdivision of cold stages is far more difficult. Are we to use a curve of climatic change synthesized from biological evidence as a basis? The difficulty here is that different aspects of the biological evidence, e.g. that from beetles and plants, may offer a different view of the changing climate. Or are we to take periods of ice advance and retreat as the basis for division. Presumably climatic change (temperature and rainfall) is related to biological change and ice sheet behaviour, but the connection is not as clear as might be thought.

In fact we take at present the occurrence of interstadial-type fluctuations, whether demonstrated by plants or beetles, as the basis for subdivision, interpolating into the sequence so obtained the periods of glaciation and permafrost. We will discuss later the results of this method of subdivision in regard to the last glaciation. Before doing this we should consider the geological evidence from glaciated and periglacial areas which is important in stratigraphical studies.

THE GLACIATED AREA

The extent and direction of movement of a former ice sheet may be studied in terms of its deposits or in terms of the landforms left on its retreat. Both approaches are complementary.

The deposits are formed during the course of ice advance and retreat and their constitution is directly related to the course the ice advance has taken in its advance. Where it is eroding it will pick up debris and carry it forward to be deposited as conditions come to favour deposition rather than erosion. Such conditions include the slowing of the ice advance and the release of the basal debris load, the melting of debris-rich ice as general downmelting starts, and the deposition of glaciofluvial sediments from streams issuing forth from the ice front. Deposits formed from ice and from meltstreams differ in the degree of sorting. The former are hardly sorted, containing particles of very different sizes, from boulders to clay (pl. 16). Thus we have the term boulder clay to describe the principal deposit formed from ice. Boulder clay is a particular type of till, a term which includes all relatively unsorted deposits originating close to the ice. A particular till can be characterized by the set of erratic rocks it contains, which is related to the outcrops eroded by the ice of the ice sheet which deposited it. Thus the grey-blue chalky boulder clay of East Anglia derives its colour from the Mesozoic clays picked up by the ice as it traversed the Midlands, and its chalk by passage over the Chalk escarpment in the west of East Anglia.

A till is also characterized by the arrangement of the boulders within it. Although till may appear in section to be heterogeneous, it has in fact a fabric, with the particles preferentially arranged in certain ways. The forces associated with ice movement as till is deposited result in preferred orientation of stones in till. A common result is that the long axes come to lie parallel with the direction of motion of ice which deposited the till. It is therefore possible to reconstruct movements of past ice sheets by such studies as these (fig. 5.7). However, other processes, such as solifluxion down slopes, also give preferred orientations of long axes of contained stones, and some types of till, such as those formed by melting-out of debris-rich ice which is accompanied by much flowage, may not show well-developed fabric.

In contrast, the deposits formed by meltwater streams are usually well-sorted (pl. 18), though an intermediate category, that of ice-contact sediments, occurs between these glaciofluvial deposits and till. Meltwater streams may contribute their load to build outwash plains and valley trains or into lakes to form deltas. The coarser grades are deposited first nearest the ice, and the finer grades, such as silt and clay, are carried away to sediment in lakes or the sea. Here they may form varved clays (pl. 35), sediments laminated because of seasonal deposition of silts, and perhaps some sand during the summer melting period and the clays during the winter period when melting is not active and the clay settles.

The extent of former ice sheets and their direction of movement can thus be studied by the areal distribution of tills and associated outwash, till fabric and erratic content. The distribution of landforms associated with glaciation, already discussed in Chapter 3, greatly assists in these studies. The marginal regions of former ice sheets may be marked by a concentration of glacial landforms such as end-moraines, eskers and kames, in contrast to the flatter till plains of the hinterland. Such landforms are well seen in northern Britain associated with the melting-out of the last glaciation. They are much rarer in the areas glaciated in the older glacial stages, partly because the landforms have been degraded by periglacial action in later cold stages. But the presence of constructional glacial landforms is not a clear indication of last glaciation age. Constructional landforms are known from the older glaciations as well, for example the kames of north Norfolk (see Chapter 3).

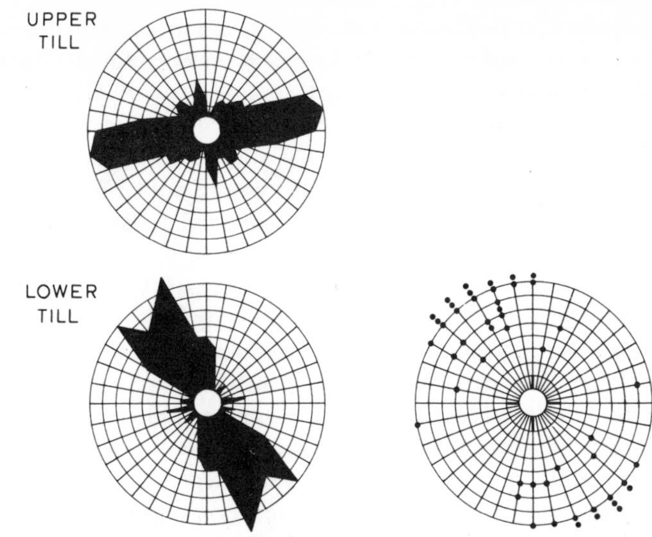

FIG. 5.7 Stone orientation in tills. On the left are rose diagrams showing the directions in 10° classes of the long axes of stones in a count of stones of an upper and a lower till at Scratby, Norfolk. The upper till belongs to the Lowestoft ice advance, the lower to the advance which deposited the till named Norwich Brickearth. The number of stones in each class is shown by the distance of each peak from the innermost circle, each circle representing one stone in the upper till and half a stone in the lower till, with half the number of stones in each class being placed on opposite sides of the centre of the diagram. Magnetic north at the top.

On the right is a dip diagram from the lower till at Scratby. Each dot represents one stone. The direction of the long axis is shown by the direction of the radius on which the dot lies, north at the top, and the dip by the distance of the dot from the outermost circle of the diagram, the concentric circles indicating 10° dip, 0° on the outermost circle, 80° on the innermost. The absence of a marked dip direction preference in the diagram suggests the deposit is not formed by solifluxion. (After R. G. West and J. J. Donner, in *Quarterly Journal of the Geological Society, London,* 1956, Vol. 112, figs. 1 and 2.)

Conversely, the absence of such landforms does not necessarily mean an age older than the last glaciation. The process of ice-melting may not favour their production, and a till plain may be the only resulting landform.

THE PERIGLACIAL AREA

The processes, sediments and structures which characterize the periglacial area, have already been discussed in Chapter 4. They concern the terrestrial periglacial land surface. We must also consider the lower Pleistocene marine deposits, which at present provide the main evidence for the earlier cold stages of the Pleistocene in East Anglia.

It is in the periglacial area that we frequently find deposits, with fossils, and allied with periglacial phenomena, that allow a synthesis of the climatic reconstruction of a glacial stage. Although we are not so fortunate to have found in this country a long sequence throughout the last glaciation, as has been found in the Mediterranean, the sequence of deposits is sufficient to determine the outline course of last glaciation climatic events (fig. 6.15). The fossiliferous deposits most frequently occur in river terraces of last glaciation age, as still-water deposits in abandoned channels of braided rivers. The thin lenses of organic deposits formed in such places contain an abundant record of the so-called 'full-glacial' flora and fauna, 'full-glacial' as a contrast to more temperate interstadial conditions. Organic deposits of full-glacial type are also found as erratics in last glaciation river gravels, presumably cut by erosion from an old fill exposed by a meandering river.

The relationship of these fossiliferous beds to periglacial structures such as ice-wedge casts (pl. 26) and involutions (pl. 25) makes it possible to relate times of permafrost and soil-freezing action to the fossiliferous beds. The difficulty of periglacial stratigraphy is the correlation of isolated occurrences of fossiliferrous beds in different parts of the same terrace and between terraces. In the last glaciation relative age determination can be based on radiocarbon dating, and it is solely by this means that a chronology of geological and biological change has been built up in the last glaciation in this country.

When we get back to the previous cold stages, beyond the range of radiocarbon dating, it is not possible to produce detailed sequences of climatic change. Such few fossiliferous deposits that have been found occur most often in terraces, and they show similar full-glacial characteristics of flora and fauna. Floras of this type have been found in the Wolstonian, Anglian and Beestonian cold stages. In the earlier cold stages, the evidence for climate comes from pollen analysis of marine clays and silts found in the East Anglian Crag basin. In such deposits as these it is not possible to obtain the detail of climatic change seen in the terrestrial record. The pollen content is derived largely from rivers flowing into the sea, and consequently the picture of local and regional vegetation obtained from terrestrial pollen analysis is very largely blurred.

River terrace sequences

Investigating the history of rivers plays an important part in unravelling the environmental changes of the Pleistocene. An aggrading river carrying a load

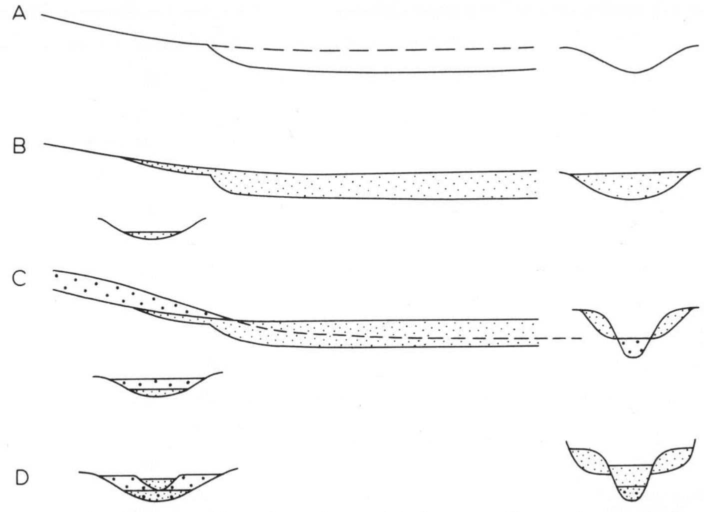

FIG. 5.8 Long profiles of a river with down-cutting and aggradation phases associated with changing sea levels. Fine dots, alluvium of temperate stages; large dots, sands and gravels of cold stages. A, preglacial valley, river rejuvenated by low sea level of glacial times. Nick-point marks the head of rejuvenation. B, aggradation as a result of a rising interglacial sea level; C, a further glacial low sea level results in down-cutting in the lower parts of the valley and aggradation of outwash and weathering debris in the upper part; D, further aggradation during a second interglacial stage of higher sea level. (After West, 1968, fig. 8.5.)

of sediment will deposit fluviatile sediments and floodplain alluvium in its valley to the height of its floodplain. At the seaward end of the river the level to which aggradation takes place will be determined by sea level. If sea level falls, as during a cold stage, incision of the alluvial plain will result and we are left with a terrace marking the former level of aggradation. If we consider a series of cold and temperate stages, accompanied by falling and rising eustatic sea levels respectively, we may find a flight of terraces in a river valley, each graded to a particular level, that level marking the height reached by sea level during a temperate interglacial stage (fig. 5.8). There is evidence that interglacial sea levels have fallen from one interglacial to the next (see Chapter 1). If this is so, then in a flight of terraces the highest will belong to the oldest and lowest to the youngest.

This is an oversimplification concerning terrace sequences in the lower parts of rivers for a number of reasons. One is that uplift or downwarping of the land may be another complicating level of aggradation. A second is that sea level change will respond to ice sheet formation, which itself may not be related only to temperature but also to precipitation changes. We may therefore find aggradation of terrace deposits under cold conditions, especially because climatic deterioration will be the cause of a greatly increased load on the river, resulting from solifluxion and loss of soil-stabilizing forest. In the headward parts of rivers, where eustatic changes of sea level will not affect the river gradient, aggradation will depend on sufficient load and low gradient. Terraces will again record the height of former aggradations, but with different factors controlling aggradation than are seen in seaward parts of the valleys.

Within the area covered by the last glaciation ice, the river valleys may carry simple terrace sequences related to river downcutting and aggradation during the melting phase and in the Flandrian. In the great river valleys beyond the extent of the ice, such as the Thames and the Severn and the rivers of southern England, the terrace sequences record a much greater span of time, covering several climatic stages. In the intermediate area between the limits of the earlier glaciations and the last glaciation, the terrace sequence records events in river history since the retreat of the last major glaciation, the Wolstonian.

The interpretation of terrace sequences is fraught with difficulty. There are two aspects of the study. One is the interpretation of the terrace landforms. Are the surfaces constructional or erosional? Can discontinuous terrace features down a river valley be clearly correlated? Does the gradient so deduced conform to a particular eustatic level, or have tectonic effects obscured this relationship? How can each terrace be related to a particular stage of the Pleistocene?

The last question can best be resolved by the second aspect of terrace study, the investigation of the terrace sediments themselves and their contained fauna and flora. Fossiliferous sections in terraces are all too rare, and when they are seen, they often demonstrate that the apparent simple terrace surface is underlain by deposits of more than one climatic stage. This situation is seen in many terraces of the Upper Thames (fig. 5.16). Another example is seen in the low terrace of the river Wissey at Wretton, Norfolk (fig. 5.14) where last interglacial deposits underlie the main aggradation fluviatile sands of the last glaciation. If fossils were procured only from the interglacial deposit, it might be thought the terrace was interglacial in age. Such is not the case at Wretton.

The most satisfactory way of substantiating a terrace sequence and correlating it to climatic stages is to have enough fossils to make possible correlations with interglacials and glacials. Unfortunately fossils are rarely found, so that no terrace sequence in the south of England has yet been correlated in detail and to everyone's satisfaction to the climatic stages of the Pleistocene. The best-known sequences are those of the Thames and the Severn; that of the Thames will be discussed in more detail later in the chapter.

Raised beaches and wave-cut platforms

The high eustatic sea levels of the temperate interglacials, already discussed in Chapter 1, are deduced from the occurrence of ancient beaches at various heights above the beaches of the present day. In certain areas sequences of raised beaches are found at different levels. How can we use these indications of former high sea levels to interpret sea level change in the past, and how can they be fitted to the stratigraphical sequence?

The first question is the relation of an old beach to its contemporary sea level. Storm beaches may form at considerable heights above mean sea levels. Then there is the possibility that during fall of eustatic sea level during the latter part of a temperate interglacial beaches will be formed, so that a sequence of, say, three beaches may not necessarily mark the heights reached during three successive interglacial high sea levels. Thirdly, tectonic movements over a long period may have affected the height of a raised beach, as in northern Britain, so that it does not now mark the former high sea level.

At one time it was thought that a eustatic chronology of the Pleistocene, based on former world-wide sea level changes, might be the basis for world-wide correlation of Pleistocene events, effected through correlation of raised beaches at particular heights. But for this to be possible, crustal stability is required, and it might well be asked if there are any areas of crustal stability sufficient to give reliability to such a method of correlation over a long period of the Pleistocene. The answer to this question at present appears to be that such eustatic correlation may be meaningful for beaches formed in the last 120,000 years, that is, during the last interglacial and since, but not so for older beaches.

During a stillstand of sea level, a wave-cut platform may be formed at about sea level if the rock is sufficiently soft for wave action to work effectively in the time available. A notch may be present on the back of the platform marking high spring tide levels. Such wave-cut platforms are known around the coasts of western Britain and Ireland. Some are near present sea level; others are raised. How can these platforms be related to the Pleistocene sequence? A possibility exists if they are covered by fossiliferous deposits which can be correlated with a climatic stage. Thus if the platform is overlain by beach gravels, these gravels might be correlated with a particular temperate stage by means of a contained fauna and flora. But there is no guarantee that the platform below the beach will belong to the same stage. It might well be much older and 'reoccupied' by beach gravel of a later transgression or regression, just as certain old wave-cut platforms today are overlain by the present beach in south-west England.

This difficulty of dating platforms and beaches is one of the prime difficulties of interpreting the changes of Pleistocene sea level and the Pleistocene sequence itself in southern and western Britain and in Ireland, as we shall see. In many areas, the old beaches are overlain by glacial deposits such as tills. If the beaches could be dated, there would be an effective basis for correlation of the tills.

The most certain method of correlation of sea levels with the Pleistocene climatic sequence is through the discovery of fossiliferous sections which can be correlated with interglacial deposits elsewhere. For example, at Selsey Bill in Sussex, Ipswichian freshwater organic deposits rich in pollen are overlain by estuarine clays. By levelling in the transgression contact and relating it to a particular point in a pollen diagram, it is possible to date transgression (or vice versa, regression) levels to particular times in an interglacial. This method has been particularly useful in the Flandrian, where transgression and regression contacts can be dated by the radiocarbon method (see Chapter 9). As a result it has been possible to work out the eustatic curve of sea level change in the last 10,000 years in great detail (fig. 9.9). In the Ipswichian interglacial it has been possible to work out eustatic changes of sea level to some extent, but matters are complicated by downwarping since that time. There is a further discussion of Ipswichian sea levels in Chapter 10.

In the Hoxnian temperate stage there is good evidence for higher sea levels than in the Ipswichian. In the Nar valley, near King's Lynn, marine deposits of this interglacial are found to a height of 20 m O.D. (Ordnance Datum, i.e. the official mean sea level for Britain). But as we get farther back in time, although transgression and regression contacts of Cromerian and Pastonian age are known in Norfolk, their significance for revealing contemporary eustatic sea levels are obscured by tectonic changes since that time. Thus the Cromerian and Pastonian transgression contacts and tidal sediments are very near present sea level, but in the Mediterranean, the supposedly correlative raised beaches indicate sea levels that are tens of metres above present sea level. Which of these gives a true indication of contemporary sea level? Probably neither, in view of the crustal instability of both areas.

The raised beaches of the British Isles are of two types: those in the south that are related to high interglacial sea levels, and those in the north that are more recent in age, younger than the last glaciation, and have been uplifted since the time of their formation by the isostatic recovery of the earth's crust after the melting of the load of ice present during the last glaciation. Such are the so-called 'Neolithic' or '25 ft' raised beaches of northern Britain, and also the so-called '100 ft' and '50 ft' raised beaches of Scotland, formed in Late Devensian times. The warping of beaches of these ages is shown in fig. 5.20.

Stratigraphy of the Ice Age in Britain

We have now considered the characteristics of cold and temperate stages in the Pleistocene and of river terraces and evidence of former high sea levels. We are now able to look in rather more detail at the sequences which have been discovered and consider how they lead to the framework of stages already mentioned (Table 5.1). It is to such a framework that we have to relate glacial events, sea level changes, river terraces, and all the other expressions of environmental change which can be observed in Britain.

Up to some fifteen years ago, there was no generally accepted framework of climatic stages. Correlation from one part of the country to another and to the European sequence was difficult, if not impossible. There was considerable argument over correlations of the major glacial stages. One particular difficulty was the desire to force the British sequence into the framework of glacials and interglacials described by Penck and Bruckner with reference to the Alps, which resulted in unsatisfactory correlations. With the characterization of interglacial stages by their vegetational history in East Anglia (discussed in Chapter 6), it then became possible to correlate associated glacial deposits. The detailed study of interglacial deposits was first developed in East Anglia, a region rich in such deposits. The reason for this richness may lie in the peripheral position of East Anglia in relation to the extent of the former ice sheets. It is not too near the centre of glaciation for glacial erosion to have removed interglacial deposits, and yet is in the glaciated region, so that the glacial stratigraphy can be related to interglacial deposits. Moreover, many of the tills are boulder clays able to hold up lakes in which interglacial deposits have accumulated. In addition, the East Anglian Crag basin holds the most complete record in the British Isles of Lower Pleistocene events. Thus East Anglia can be treated as the type area for the British Pleistocene.

A detailed treatment of stratigraphy in all regions of the British Isles is neither possible nor desirable here. It will be best to treat East Anglia in more detail, not only because it is a type area but also because Pleistocene studies largely developed in this part of England and because the short distance to the Continent makes it the connecting link between the British and Continental Pleistocene sequences. It is thus a key area for the British Pleistocene. It will also be useful to discuss the Thames valley in more detail. This again is a classical area for Pleistocene studies, is rich in archaeological remains, and provides a view of Pleistocene stratigraphy in an area immediately outside the glaciated area. Finally, we can briefly discuss the Pleistocene sequences of other parts of Britain in relation to the East Anglian and Thames regions.

EAST ANGLIA

Many of the interglacial deposits in East Anglia were described by nineteenth-century geologists. A study of the old literature, including the *Geological Survey Memoirs* written in the last century, reveals many very clear descriptions. Subsequent pollen-analytical studies in the last fifteen years have yielded the details which have made it possible to erect an outline stratigraphical framework. Much of this sort of investigation remains to be done, however, and it is certain that the framework will be improved and made more precise and more complicated by future discoveries and investigations.

The Pleistocene deposits of East Anglia are divisible into three series. The

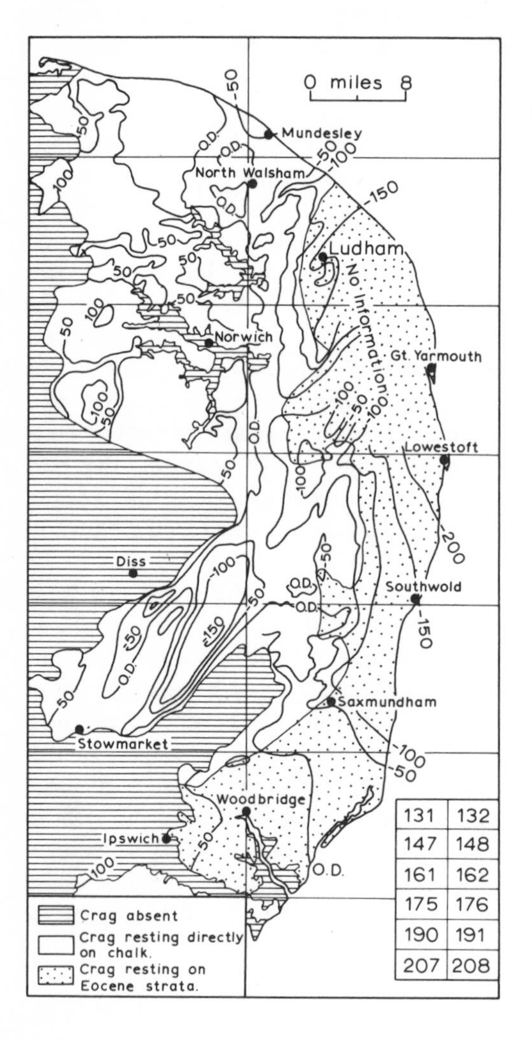

FIG. 5.9 Contours of the base of the Crag in East Anglia (from West, 1968, fig. 12.5, and by permission of the Institute of Geological Sciences; after internal report by J. H. Price and J. Tuson, *The Hydrology of the Pleistocene Deposits of East Anglia*, and A. H. Woodland, *Water Supply from Underground Sources of Cambridge–Ipswich District*, Wartime Pamphlet 20, Pt X, fig. 7). The key, bottom right, shows the numbers of 1″ Geological Survey maps, the areas of which are shown by the rectangular grid on the map. (Crown Copyright Geological Survey diagram. Reproduced by permission of the Controller of Her Majesty's Stationery Office.)

lowest is of marine and freshwater deposits formed in the Crag basin (fig. 5.9); the middle series is of glacial and interglacial deposits, and the uppermost of terrace and ailuvial deposits. The members of these three series are shown in fig. 5.10.

Let us first consider the deposits filling the Crag basin. These are principally marine sands and clays, often very shelly, which constitute the Crag. Near the top of the succession freshwater sediments occur intercalated in the marine sequence. These comprise the Cromer Forest Bed Series. The Crag and Cromer Forest Bed Series are to be seen in outcrop at the base of the cliffs and below the glacial deposits around the coast of East Anglia, outcropping in the valleys of east East Anglia and in deep boreholes into the Crag basin.

The shelly Crag was a favourite subject of study in the last century with various conchologists, notably S. V. Wood, father and son. But no convincing history of environmental events associated with the filling of the Crag basin could be made out. As a result of borehole sequences, it has been possible to start such reconstructions, but so far only the outlines are known and much remains to be done. In the borehole sequence (e.g. at Ludham, fig. 6.8), it has been possible to distinguish a sequence of climatic stages, two of temperate type, the Ludhamian and Antian, and two of cold type, Thurnian and Baventian. These stages are characterized by their palaeobotany. The foraminifers also give similar indications of climatic change. But the marine molluscs do not appear to be such good indicators of climatic change. Their distribution is subject more to local ecological conditions, and most of the changes in the mollusc assemblages seem to be more determined by the nature of bottom sediments and offshore habitats, than directly by climatic change.

These Crag stages are only known from a few boreholes and outcrops. In order to see more clearly the processes of formation and filling of the Crag basin, it will be necessary to relate the classic Crag sequence, described by Prestwich, Harmer and others, on the basis of surface outcrop studies, to the stages so far identified. Harmer's divisions of the Crag were based on molluscs, and were mapped by the Geological Survey (fig. 5.11). The Pliocene Coralline Crag of the Orford area in Suffolk was seen to be overlain by Red Crag, with a rich mollusc fauna, found at heights to 45 m (150 ft) O.D. in southern East Anglia and even higher in isolated localities in Hertfordshire and Surrey. Farther north in the Crag basin the Red Crag was replaced by the Norwich Crag with a smaller fauna of fresher water aspect. To the north of the Crag basin, in the Cromer area, the Norwich Crag was replaced by the Weybourne Crag, characterized by a marine mollusc, *Macoma balthica*, with an even more restricted fauna considered to represent much colder conditions.

By further palaeobotanical studies of the Norwich and Weybourne Crags,

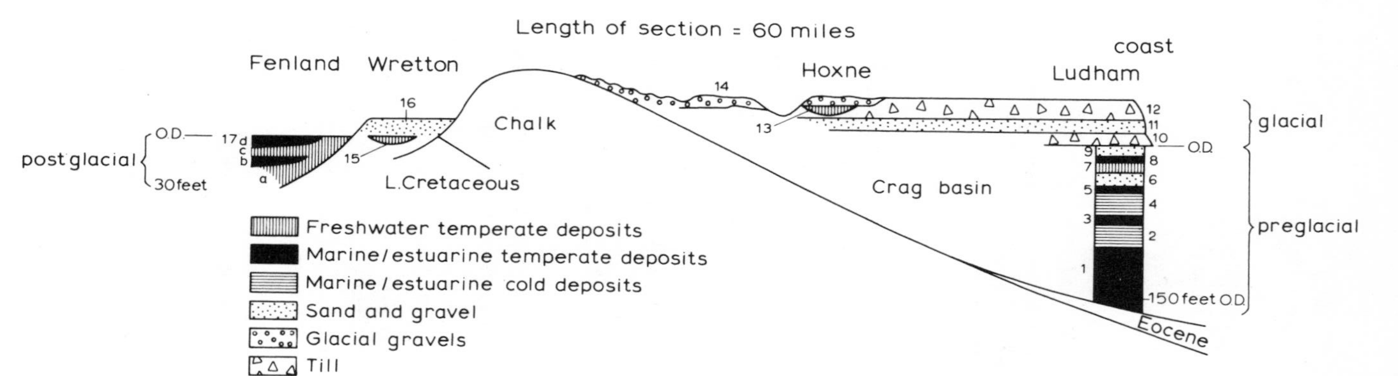

FIG. 5.10 Sketch section E–W through the Pleistocene of East Anglia. 1, Ludhamian Crag; 2, Thurnian Crag; 3, Antian Crag; 4, Baventian sand and silt; 5, Pastonian tidal silts; 6, Beestonian sands and gravels; 7, Cromerian freshwater muds; 8, Cromerian tidal silts; 9, early Anglian sands and silts; 10, Cromer till; 11, Corton sands; 12, Lowestoft till; 13, Hoxnian clay muds; 14, Wolstonian gravels; 15, Ipswichian muds; 16, Devensian terrace sands and gravels; 17, Flandrian alluvium of Fenland: a, lower peat, b, Fen clay (estuarine), c, upper peat, d, Romano-British silt.

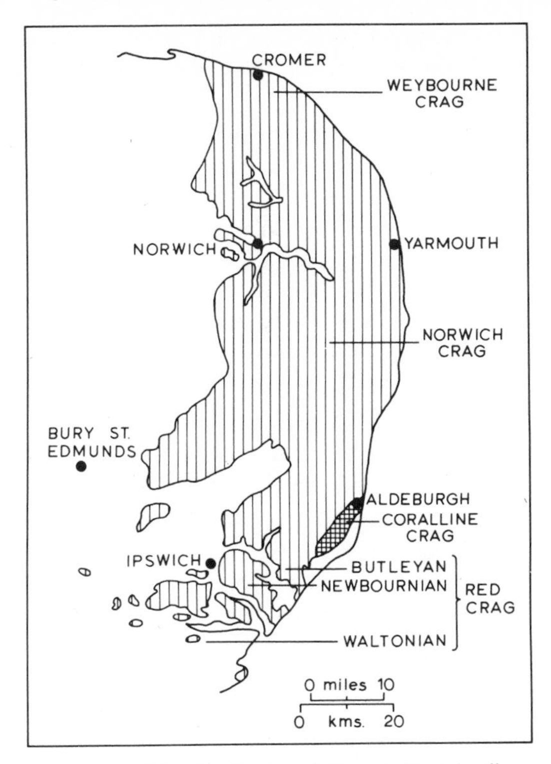

FIG. 5.11 The distribution of Crags in East Anglia.

it is now known that the former type of Crag is of late Ludhamian and later in age and the latter type is Baventian and later in age. The Red Crag faunas are believed to be lower Ludhamian and earlier in age, belonging to the oldest part of the Pleistocene Crag. If these correlations are correct, then we may envisage the following sequence of events.

The Pliocene Coralline Crag was deposited in warm temperate seas. The fauna became impoverished at the beginning of the Pleistocene and the Red Crag, a marine offshore sediment, was formed. Subsequently relative uplift to the south isolated the North Sea from the Atlantic. The Red Crag was raised to considerable heights in the south and the further impoverished Norwich Crag was deposited, filling the Crag basin. Finally fluctuations of relative land–sea level resulted in the intercalation of freshwater sediments such as we see in the Cromer Forest Bed Series. The section at West Runton revealing these changes has already been described.

Much further research needs to be done on the Crag basin before its detailed history can be worked out. It will involve many more borehole sequences and more definite correlations with Harmer's Crag horizons. We may expect much more detail of Lower Pleistocene tectonic movements as a result of these studies.

The stratigraphical basis for the Baventian to Lowestoftian stages in East Anglia has already been described, based on sections at West Runton (fig. 5.6, p. 132).

Above the Cromerian temperate sediments at West Runton, there is a sequence, some 30 m thick, of glacial deposits. This brings us to the middle series of Pleistocene deposits in East Anglia, that of the glacial and interglacial deposits. The number of glacial advances in East Anglia is not easy to decipher. At West Runton the Cromerian is overlain by a thickness of some 30 m of glacial deposits, heavily contorted by ice-pushing and by movement during ice-melting. Farther south-east, at Corton Cliff near Lowestoft, we see uncontorted tills overlying the Cromer Forest Bed Series. There are two till sheets separated by stratified sands. The lower till sheet is the North Sea Drift, so-named because the presence of Scandinavian erratics suggests an ice movement over the North Sea from the north. The North Sea Drift at Corton is thought to be part of a widespread till sheet, which includes two till types, a bluish Cromer till and a more weathered till, the Norwich Brickearth, well-exposed in the Norwich area. The exact relation between the two tills is unclear, but since their stone orientation in north Suffolk and north-east Norfolk and their stratigraphical positions are similar, it is thought they belong to the same period of glacial advance.

The upper of the two tills at Corton is a blue-grey chalky boulder clay, the Lowestoft till, widely present in East Suffolk and South Norfolk. It is the thickest and most prominent of the East Anglian tills. It is rich in Cretaceous and Jurassic erratics picked up to the west of East Anglia, and both the erratic content and the stone orientation of the till indicate a west to east movement of the ice over East Anglia (fig. 5.12). To the north of East Anglia, the tills become more chalky, and the ice advance is thought to be represented by the Marly Drift of north Norfolk. At Corton, and elsewhere in eastern East Anglia, the Lowestoft till is separated from the North Sea Drift by the Corton Beds, sands which contain both thin silt lenses containing a cold flora and also abundant marine shells. It is probable that Corton Beds represent a period of high water level during a period of readjustment of ice lobes, in which the earlier ice of the North Sea Drift melted and was replaced by Lowestoft ice. There is no clear evidence that an amelioration of interstadial magnitude occurred in the interval.

A number of buried channels are known beneath the Lowestoft till in East Anglia. They appear to be gouged subglacially during this ice advance. At Marks Tey, Essex, interglacial deposits of the subsequent temperate stage, the Hoxnian, fill such a deep channel. Several other sites with Hoxnian

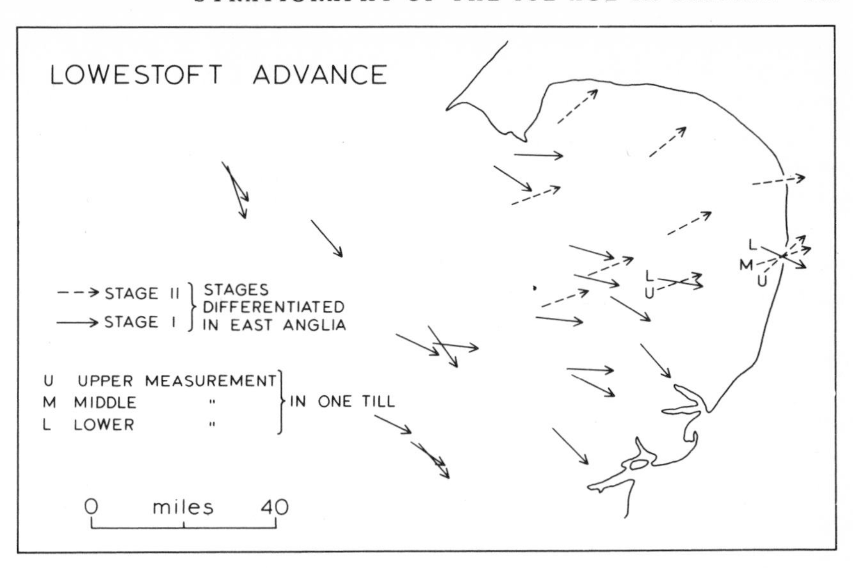

FIG. 5.12 Stone orientation of till of the Lowestoft ice advance (after R. G. West, in *Quarterly Journal of the Geological Society, London*, 1956, Vol. 112, fig. 6).

sediments are known in East Anglia. At the type site at Hoxne Brickworks, near Diss, a lake basin of irregular form in the Lowestoft till is filled with interglacial deposits. The stratigraphy, determined by boreholes and open sections, is shown in fig. 5.13. As well as lake deposits, marine and estuarine deposits of this interglacial occur at Clacton and in the Nar valley, rising to heights of 27 ft. O.D. at Clacton and 65 ft. O.D. in the Nar valley. The interglacial marine transgression thus rose to heights well above present sea level at a time, determined by pollen analysis of the transgression contacts, late in the interglacial (zone Ho III).

The glaciation of East Anglia during the glacial stage (Wolstonian) after the Hoxnian interglacial is difficult to interpret. A till younger than the Lowestoft till, the Gipping till, has been described from the western part of East Anglia. It is a paler and more chalky till with igneous erratics from the north of England. Both the stone orientation results and the erratics indicate an ice movement from the north. The distribution of the till is irregular. It has not been certainly found on top of a Hoxnian interglacial deposit. At Hoxne, sands and gravels probably deposited near the ice margin and containing Gipping erratics rest on the interglacial deposits. At Mildenhall in west Suffolk, interstadial silts are associated with a pale chalky boulder clay of Gipping type. In north Norfolk there are glacial gravels in the valleys and on the divides. These gravels, of 'cannon-shot' type, lie over Hoxnian deposits in the Nar valley and under Ipswichian interglacial deposits in the

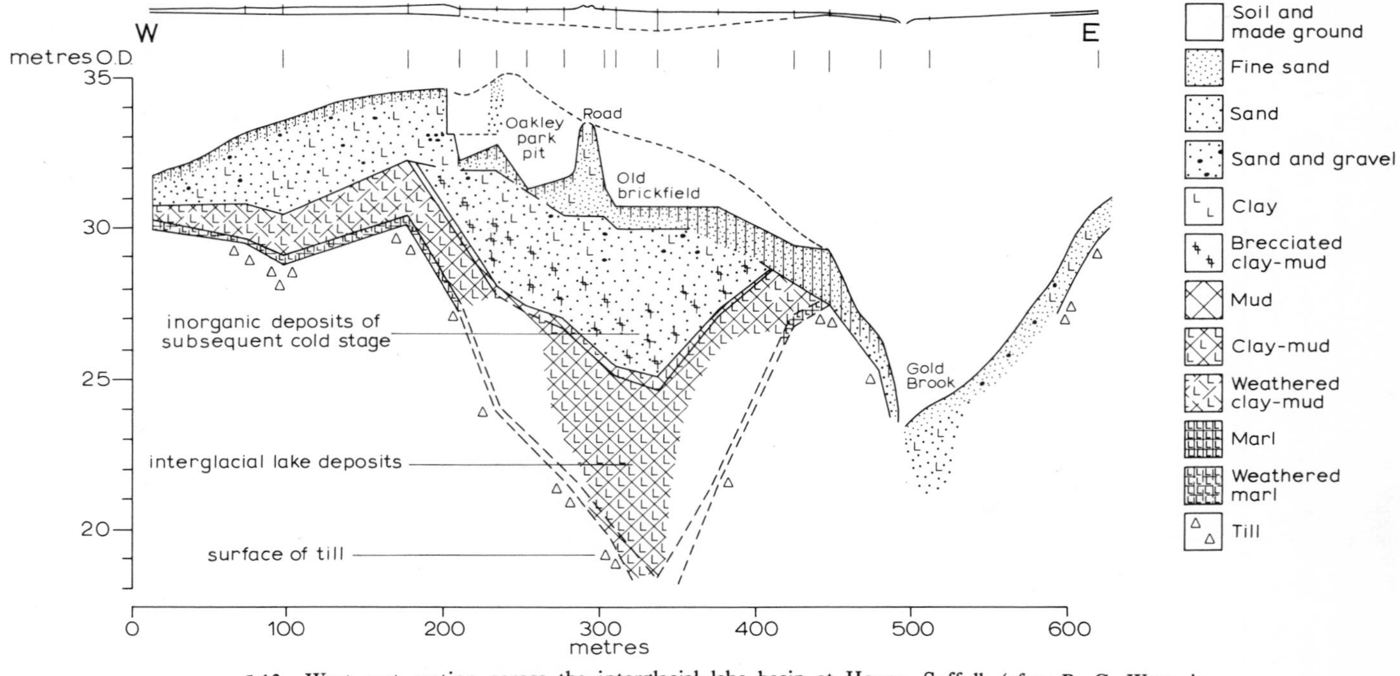

FIG. 5.13 West–east section across the interglacial lake basin at Hoxne, Suffolk (after R. G. West, in *Philosophical Transactions of the Royal Society*, 1956, **B**, **239**, fig. 8).

upper Wensum valley, and are probably to be referred to the Wolstonian glacial stage. The morainic features found on the north coast of Norfolk between Holt and the coast, including outwash plains, kames and crevasse-fillings, appear to belong to a retreat phase of this glacial, as do kame-like gravels in valleys farther south, for example in the Waveney and Cam valleys. The deposits and sequence of events in the Wolstonian glacial stage in East Anglia indeed require much clarification.

The deposits of the Ipswichian (last) interglacial are found in river valleys associated with terraces. By the beginning of the interglacial, the present valleys were clearly in existence, and during the interglacial alluvial deposition occurred. Thus interglacial deposits are known from the valleys of the Cam (High Terrace), Wissey, Waveney, Wensum, Orwell and Stour. Most of the deposits are freshwater, but there are indications of brackish incursions in the Wissey deposit at Wretton (fig. 5.14) and at Stutton on the northern side of the Stour estuary. Farther west, in the centre of the Fenland, the marine March gravels are probably deposits of the same marine transgression. This transgression is considered in more detail in Chapter 10.

The last glaciation is represented in East Anglia by a till, the Hunstanton Brown Boulder Clay, along the north and north-west coast of Norfolk. This till, seen for example at Holkham Brickworks, is very similar to the uppermost till of the Lincolnshire and Holderness coast, the Hessle till. It is associated with a few fresh morainic features, such as the esker in Hunstanton Park, and contains a suite of erratics from northern England. It appears that a lobe of ice moved down the east coast of England, to end in the Wash and against the rising ground now forming the north Norfolk coast.

Farther south, there are very extensive spreads of fluviatile gravels of the rivers draining into the Fenland. These are important from the point of view of last glaciation chronology and life, for they contain lenses with rich beetle faunas and plant remains, with possibilities for radiocarbon dates. There are also abundant periglacial phenomena of last glacial age, such as the stripes and polygons of the Breckland, ground ice mounds, and ice-wedge casts, discussed in Chapter 4.

The deposits of our present temperate stage, the Flandrian, form the alluvium of the Fenland, Broadland and the river valleys, built up as the sea level rose relative to the land after the last glaciation. The record of sea level change since the last glaciation is well-seen in both Broadland and Fenland. Boreholes show depths of up to 10 m, and occasionally more, of alluvium, with a lower Flandrian freshwater peat, a middle Flandrian estuarine silt or clay, an upper Flandrian freshwater peat, overlain by estuarine silts. The Fenland succession, with radiocarbon dates, is shown in fig. 5.15. An extensive marine transgression took place in mid-Flandrian times; its age can be

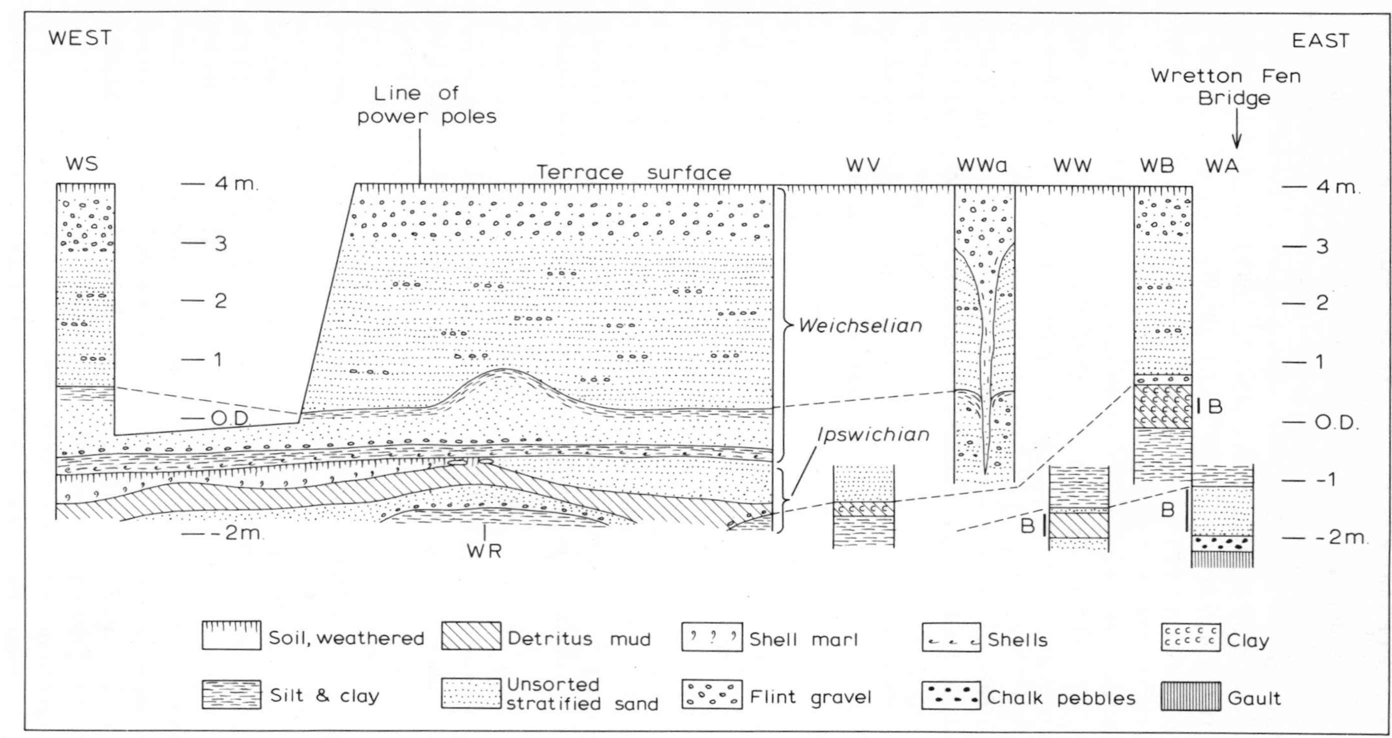

FIG. 5.14 West–east section through the low terrace at Wretton, Norfolk. B, brackish horizon identified from molluscs. WA, WB, WR, WS, WV, WW, WWa are the notations of different parts of the Wretton sections. (After B. W. Sparks and R. G. West, in *Philosophical Transactions of the Royal Society*, 1970, **B**, **258**, fig. 3.)

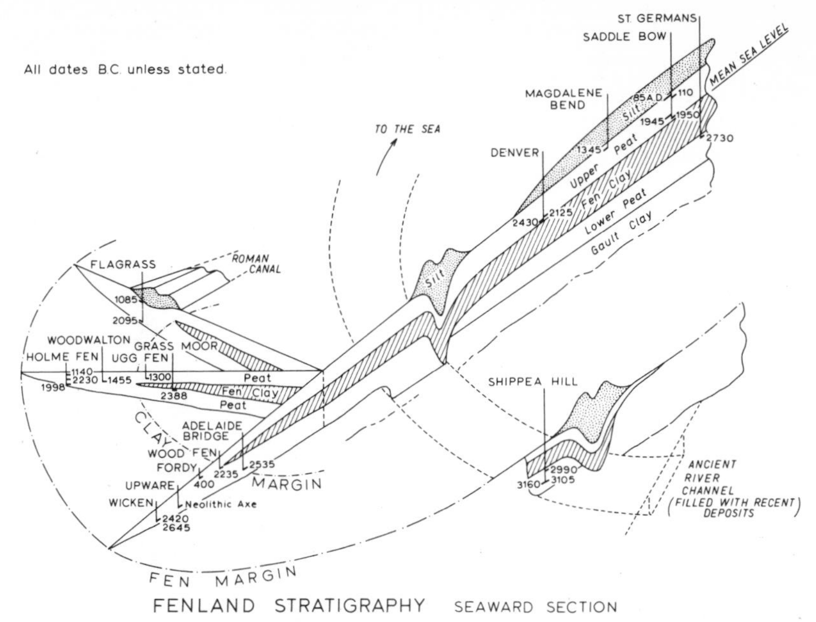

FIG. 5.15 Sketch section through the Fenland Flandrian deposits, indicating the main marine transgressions and their radiocarbon ages (after E. H. Willis, in the *Annals of the New York Academy of Sciences*, 1961, Vol. 95, figs. 5 and 6).

determined relatively to other deposits by means of pollen analysis and absolutely by radiocarbon measurements of the transgression and regression contacts.

THE THAMES VALLEY

The Thames valley lies at the southern margin of the glaciated area of the Midlands and East Anglia. It has a long history of terrace formation (fig. 5.16). The terraces are well-developed and the sediments are remarkable for their content of faunal remains, principally bones, and archaeological remains. The history of the river can be reconstructed from a study of the terraces, their contained sediments, and their faunas. The correlation with the climatic stages of East Anglia is not easy, because of the general absence of plant remains from the terrace deposits. But the presence of cold and temperate indicating faunas makes a degree of correlation possible.

The terrace sequences of the Upper Thames are separated from those of the Middle and Lower Thames by the Goring Gap, where the Thames cuts through the Chalk escarpment. In the Upper Thames a series of terraces have been distinguished, rising to 40–50 m above the present river level. The highest terraces are found to contain striated erratics worked from the

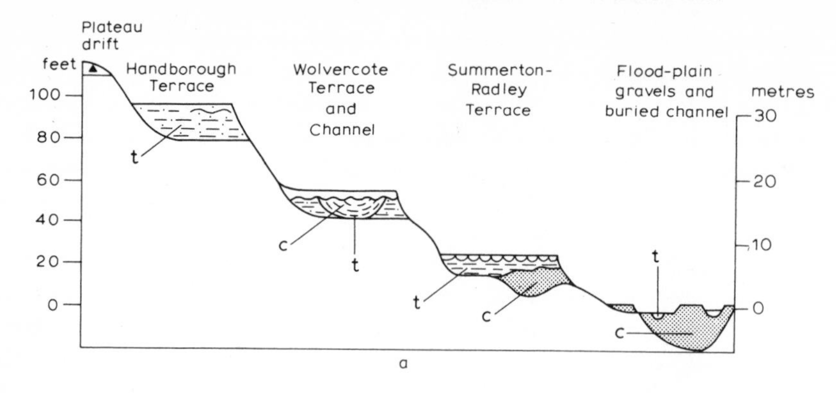

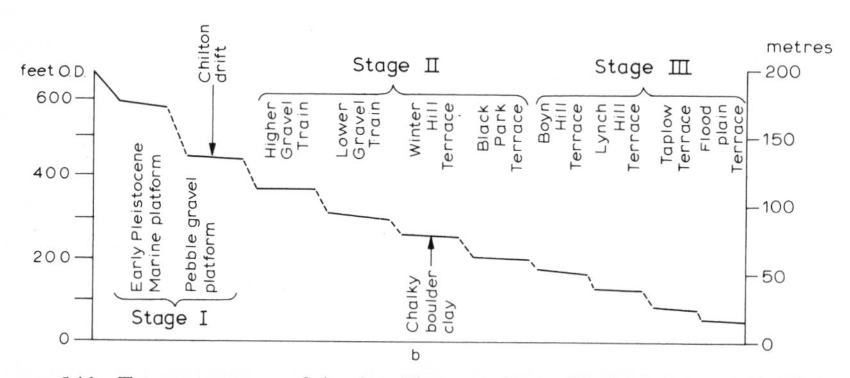

FIG. 5.16 Terrace sequences of the river Thames. a, Upper Thames (after K. S. Sandford, *The Oxford Region*, London, Oxford University Press, 1954, fig. 7). b, Middle and Lower Thames (after S. W. Wooldridge, in *Proceedings of the Geological Association*, 1960, Vol. 71, fig. 2).

Plateau Drift of the Oxford region. Such terraces occur at Coombe and Freeland, at heights between 40 and 53 m, and they are thought to be related to fluvial action during an early glacial stage, probably the Anglian. The course of the river in earlier times has not been deciphered, but evidently by Anglian times the present valley had been initiated. The next highest terrace is the Handborough Terrace, reaching a height of 30 m above river level, containing in contrast to the higher terraces, predominantly local Jurassic pebbles and a warm fauna with *Elephas antiquus.* The Wolvercote Terrace lies at a height of 12–15 m above river level. Little is known of its fauna. A channel occurs near Wolvercote which has a fossiliferous filling, with a temperate fauna at the base and a cold fauna and flora higher up. The Wolvercote Terrace contains erratics readily explained as derived from outwash of the Midlands Chalky Boulder Clay ice sheets 30 kilometres to the north, and is included in the Wolstonian glacial stage. The Wolvercote channel filling has been correlated with both the Ipswichian and Hoxnian interglacials by different geologists. The problem encountered is one characteristic of Pleistocene terrace interpretation. Due to great lateral variations in sediment, single exposures even a short distance apart are difficult to correlate. It is not certain whether the channel is cut in the terrace with a younger (Ipswichian) filling, or is separated from it and is older (i.e. Hoxnian) than the terrace.

The Summertown–Radley Terrace reaches a height of 8 m above river level. It is a composite terrace, the basal gravels containing a cold fauna and flora of apparently Wolstonian age, and the upper part a temperate fauna with *Hippopotamus*, correlated with the Ipswichian (last) interglacial. The lowest terrace is near to the present floodplain and conceals a buried channel of Devensian (last) glacial age. The buried channel gravels contain a cold flora and a fauna with mammoth. Evidently downcutting occurred between the last interglacial and the deposition of the gravels with mammoth.

In the Middle and Lower Thames the terrace sequence is rather more complicated (see fig. 5.16). The course of the Thames has been diverted south by glacial advances. This affects the terrace distribution in the Middle Thames, with the older terraces being distributed to the north of the present valley. In the Lower Thames, the effects of aggradation in high interglacial sea levels are seen in thick accumulations of alluvial 'brickearths'.

The terraces of the Middle and Lower Thames have been grouped into three stages, shown in fig. 5.16. The earliest (Lower Pleistocene) stage comprises marine high level deposits and pebble gravels, antedating the formation of the present Thames valley, and indicating an early course north-eastwards (A in fig. 5.17), developed during a relative lowering of sea level. The Chiltern drift, with many erratic pebbles, is associated with a slightly lower platform. This drift may be correlated with a Chiltern glaciation earlier than the Anglian.

The terraces of stage II contain gravels with numerous Bunter quartzites and other erratics. The distribution of the terraces indicates an eastwards course of the Thames (B in fig. 5.17), the diversion from the earlier course being related to the Chiltern ice advance. The stage II gravels have been differentiated into a Higher (107–114 m O.D.) and Lower (93–96 m) Gravel Trains, the Winter Hill Terrace (79 m O.D. at Slough) and Black Park Terrace (64 m O.D. at Slough). The gravel trains occur in a belt from Beaconsfield to Watford and Ware. Their erratics suggest they are associated with ice advances from the north-west. The Winter Hill Terrace follows the present course of the Middle Thames but eastwards appears to be associated with the same course as the Lower Gravel Train. It has been correlated with the time when ice of the Anglian stage deposited boulder clay in the Vale of St Albans and in the Finchley area. The Black Park Terrace has been recog-

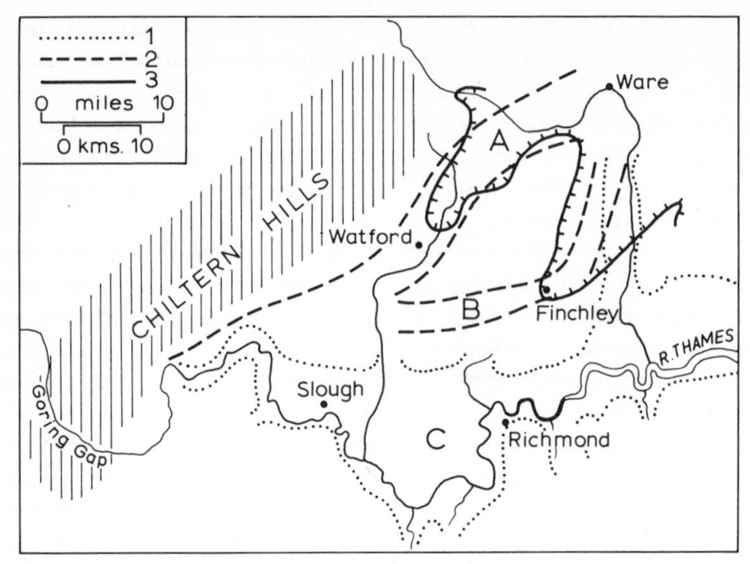

FIG. 5.17 Diversions of the river Thames. A, Pliocene or Lower Pleistocene course; B, course after diversion south by Chilterns ice; C, present-day valley, following diversion from B by chalky boulder clay ice. 1, limits of present-day valley; 2, suggested limits of earlier courses; 3, southern limit of chalky boulder clay ice. (After S. W. Wooldridge, in *Proceedings of the Geological Association*, 1960, Vol. 71, fig. 3.)

nized eastwards to the Richmond–Wimbledon area in the present Thames valley, so that it appears that the present course (C in fig. 5.17) of the Thames was initiated just before this time. The movement south from course B to course C is related to the Anglian ice advance which blocked the northerly course. It has also been suggested that the diversion was caused by capture by a pre-Thames river near Staines.

The final group of terraces are aligned along the present course of the river. They are rich in archaeological remains and are better known than those of the earlier stages. The Boyn Hill Terrace at 30 m above present river level contains an abundant vertebrate and mollusc fauna and lower Palaeolithic artefacts at Swanscombe, Kent, at the site where the Swanscombe skull was found. The Palaeolithic hand axe industry resembles that at Hoxne. The terrace gravel is thought to lie on till correlated with the Anglian stage at Hornchurch, Essex. All the evidence points to the correlation of the temperate terrace faunas and archaeological remains of the Boyn Hill Terrace with the Hoxnian temperate stage.

In the Lower Thames, the chalky solifluxion deposit known as the Main Coombe Rock forms a solifluxion sheet over the Boyn Hill gravels. It falls below the present-day sea level and is thus associated with a glacial time of low sea level (Wolstonian), prior to the aggradation of the Taplow Terrace. An important Palaeolithic industry is associated with the Main Coombe Rock. The Taplow Terrace, the so-called '50-ft' terrace of the Thames, is well developed in the Middle Thames. It contains a cold fauna and is correlated with the Wolstonian glacial stage, with aggradation of coarse gravels under cold climatic conditions.

During the Ipswichian interglacial, extensive aggradation of alluvium took place in the Lower Thames. A flora of this age is known from the Upper Floodplain Terrace in Trafalgar Square in London, and from Ilford and Aveley on the north bank in Essex. Alluviation took place to a height of some 15 m O.D. during this period. On the south bank similar alluvium occurs at the same height at Northfleet. Here again there is an important archaeological site associated with the interglacial sediments. The presence of these fossiliferous horizons of the Lower Thames does allow the valley sequence to be deciphered in detail. But if we look at the terrace surfaces, we find it difficult to distinguish discrete terraces over long distances. Upper and Lower Floodplain Terraces have been described from the Lower Thames, at heights of about 9 and 3–5 m O.D. respectively. Their deposits are associated with further spreads of Coombe rock and buried channel filling, all of them forming a complex difficult to decipher. Presumably the buried channel filling, which contains a cold fauna and flora, is of Devensian age, as are the associated Coombe rocks. Much more detailed stratigraphical work on the low terraces is required before a detailed reconstruction of the recent history of the Thames can be made.

The seaward part of the valley is filled now by Flandrian alluvium. In the excavations for Tilbury Docks, 14 m of peat and alluvium were proved. Alternations of freshwater peats and estuarine silt gave the possibility of tracing the rise of sea level in the Flandrian. Radiocarbon dates of 7000 ± 120 B.P. and 5790 ± 120 B.P. have been obtained from peaty sediments at -35 ft O.D. and -27 ft O.D. giving some idea of the rate of accumulation of these Flandrian deposits. Unfortunately the environment of deposition of the sediments in terms of marine or freshwater is not clear, so that it is difficult to use the dates to interpret rates of rise of sea level.

OTHER REGIONS OF THE BRITISH ISLES

It is impossible here to present a detailed picture of the stratigraphy of other regions of the British Isles. Two examples have been given. We have described a stratigraphical framework from East Anglia, and shown the kind of evidence available to interpret the sequence of temperate and cold stages. Then in the Thames valley a terrace sequence has been described and possibilities of correlation with East Anglia considered. These are examples of the stratigraphy of a glaciated and periglacial area. In other areas of the British

Isles, the stratigraphical record is not so well-known, and in most areas it is not so extensive.

In the Midlands the presence of a few Hoxnian interglacial deposits makes it possible to relate the glacial deposits to glacial stages before and after the Hoxnian. In northern England, Wales and Scotland, there are no interglacial deposits which can be referred to particular temperate stages known to be intercalated into the glacial sequence, and the great problem is the separation of last glaciation tills from those of earlier glaciations. Let us first consider very briefly the situation in Scotland and Wales. Both are mountainous areas in which ice sheets developed. Both show coastal raised beaches and platforms, but those of Scotland show more evidence of tilting related to the isostatic recovery after ice-melting. In Scotland, great ice sheets must have developed during each glacial stage, but only the glacial deposits of the most recent glacial stage are widely recognized. The glacial deposits are very extensive, stretching to the Outer Hebrides and the Shetlands. The general directions of ice movement are shown in fig. 3.11 (see p. 58). The ice divide was in the high mountains of the west. Apart from the mountainous areas, with all the characters of mountain glaciation described in Chapter 3, the scenery of the glaciated areas is of two main types: the smooth and often driftless areas of north-west Scotland, and the drift-covered areas of the east and south, where the hummocky landforms characteristic of recent glaciation are well developed. There are extensive drumlin fields in the Central Valley and in Galloway and a number of end-moraines of retreat stages of the last glaciation. These are shown on fig. 5.18.

Although a number of interglacial deposits are described from Scotland, none show convincing evidence for dividing the glacial sequence into stages correlatable with those farther south in East Anglia. There are some areas where till sequences are well developed, such as in the Aberdeen region. Fig. 5.19 shows a reconstruction of the glacial episodes of this area.

In addition to the glacial drifts, Scotland has abundant evidence of changes of relative land–sea levels during the Pleistocene. Late Devensian and Flandrian raised beaches, and marine clays of the west and east coasts have been radiocarbon-dated and correlated with pollen zones, so that rates of uplift since the melting of the ice of the last glaciation are known to some extent. Fig. 5.20 shows the warping of various beaches in south-east Scotland. The older, late-glacial, beaches are strongly tilted. The younger, Flandrian beaches are moderately tilted. The 6 and 9 m isobases for the uplift of the main Flandrian shoreline formed in 8000–6000 B.P. are shown in fig. 5.3. The age of the so-called preglacial wave-cut platforms of the Western Isles is not known.

In Wales the drift sequence is very complex. Ice sheets developed in the various mountain centres. In the west Wales was overrun by ice from the

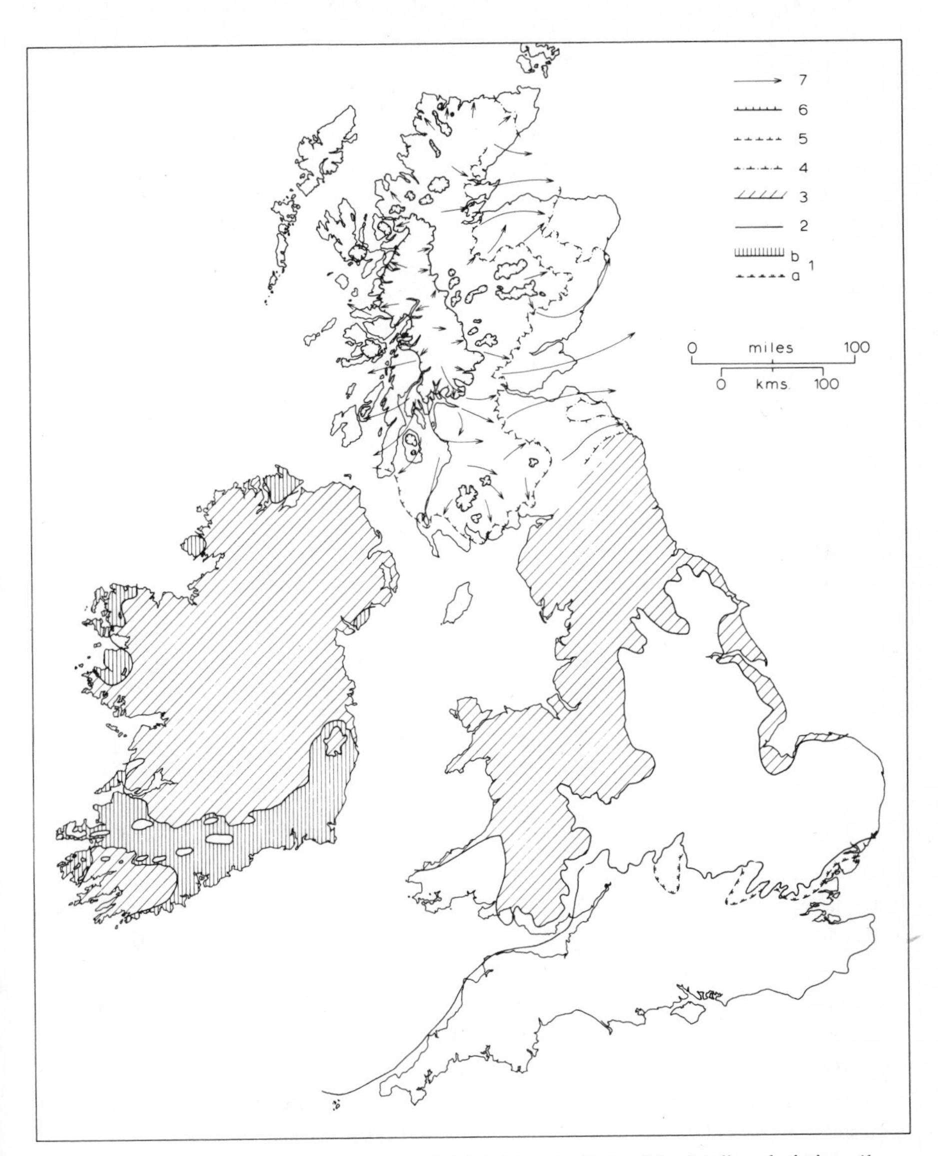

FIG. 5.18 Limits of ice advances in the British Isles. 1a, limit of the Anglian glaciation; 1b, limit of pre-last glaciation ice in Ireland; 2, limit of Wolstonian glaciation; 3, limit of the Devensian (last) glaciation; 4–6, limits of readvances during the retreat of the last glaciation; 4, Aberdeen–Lammermuir Readvance limit; 5, Perth Readvance limit; 6, Loch Lomond Readvance limit (correlated with pollen zone III); 7, Directions of the movement of the last glaciation readvances.

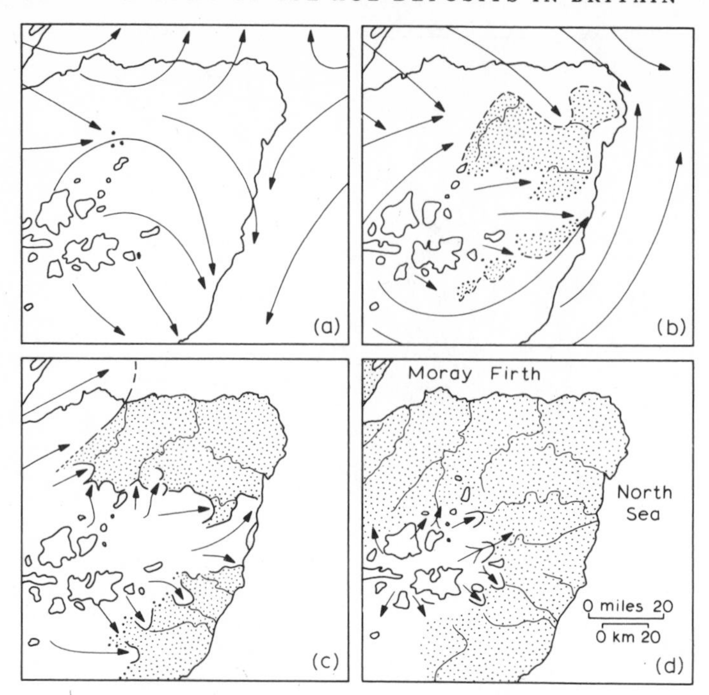

FIG. 5.19 The glacial sequence in north-east Scotland. (a) Greater Highland glaciation; (b) Moray Firth–Strathmore glaciation; (c) Aberdeen Readvance; (d) Dinnet Readvance. (b), (c) and (d) are phases of the last glaciation; (a) is believed to belong to an older glaciation. (After F. M. Synge, in the *Scottish Geographical Magazine*, 1956, Vol. 72, fig. 5.)

Irish Sea ice sheet, with an ice lobe which at its maximum stretched south to the Scilly Isles. In the glacial sequence there are no organic interglacial deposits present which can be correlated to those farther east. It is generally thought that two, or possibly three, major glacial stages are represented, at least the Wolstonian and Devensian. The older glaciation appears to have covered Wales almost completely except for a few summits projecting above the ice (nunataks) in the south. The boundary of the east glaciation is in great dispute. It may be in south or north Pembrokeshire or farther north.

The oldest Pleistocene deposits are seen on the coast associated with a raised beach. The beach may be covered by solifluxion deposits or till. The correlation of the beach is vital to the correlation of the glacial sequence, but unfortunately it has not yet proved possible to relate it certainly to a particular interglacial. Another source of the difficulty is the determination as to whether particular deposits seen in thick coastal sections are solifluxion deposits or till. All these difficulties make the interpretation of the Welsh

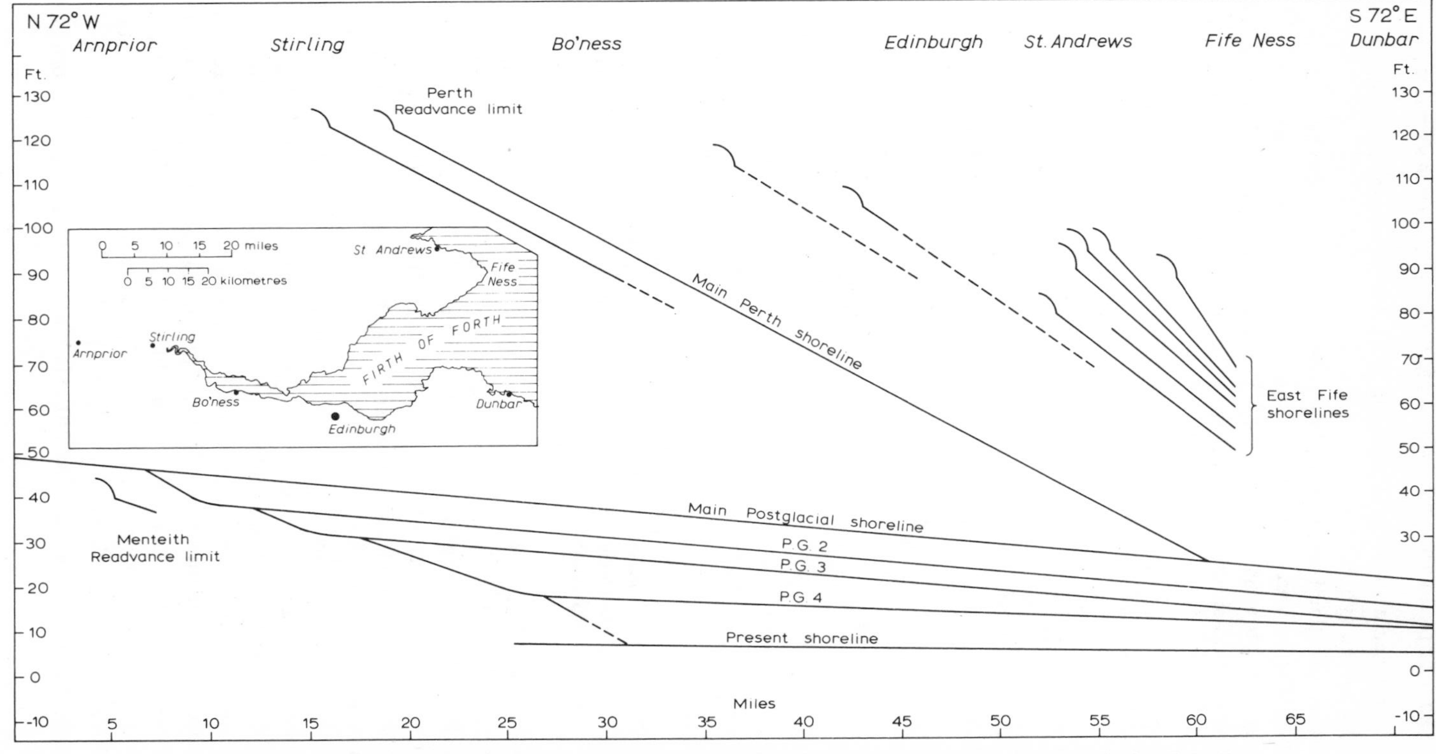

FIG. 5.20 The shoreline sequence in south-east Scotland. (After J. B. Sissons, R. A. Cullingford and D. E. Smith, *Transactions of the Institute of British Geographers*, 1966, Vol. 39, fig. 1.)

glacial sequence a great problem, and it is not yet possible to arrive at a generally agreed synthesis.

Northern England was subject to glaciation by a local ice sheet forming in the Lake District and by smaller ice sheets of the Pennines, and to glaciation by ice streams flowing from centres in the southern Uplands of Scotland (fig. 5.21). Several till sheets have been identified, but the only certain interglacial

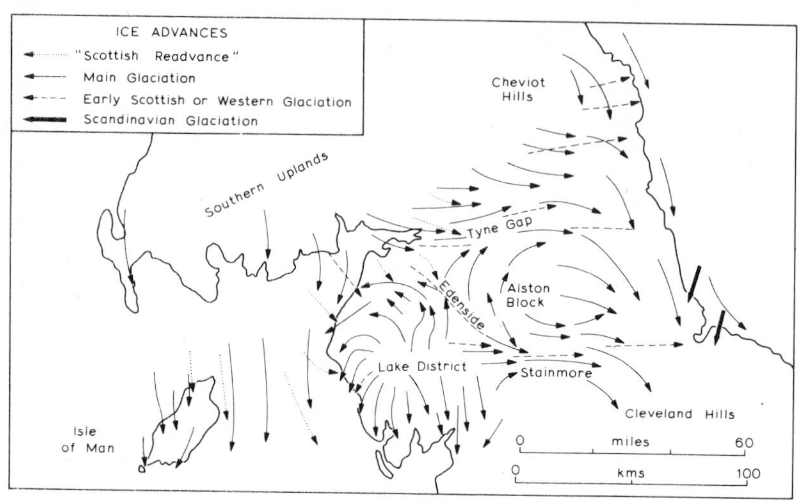

FIG. 5.21 Ice streams in northern England (after T. Eastwood, *British Regional Geology – Northern England*, London, H.M.S.O., 1953, fig. 26). (Crown Copyright Geological Survey map. Reproduced by permission of the Controller of Her Majesty's Stationery Office.)

deposit known is Ipswichian and was found as a raft incorporated in the uppermost of the tills. Two glacial stages are thought to be represented. To the earlier one belongs a till with erratics from Scandinavia, believed to have been formed by an ice sheet extended from Scandinavia to Durham. The tills of the upper glacial stage are complex, due to the complicated interplay of ice streams in the region.

In the Midlands, there are tills from all the glacial stages of the Pleistocene. The area is a complex one with ice streams, each producing its own till type, entering from Wales, from the Irish Sea via the Cheshire Plain, and from the north-east, to produce chalky boulder clay. These advances each produced their own tills. Hoxnian interglacial deposits are present and divide a fragmentary sequence of an older glacial stage (Anglian) from a widely-known and extensive later stage (Wolstonian). During the advance of the Wolstonian ice a large lake, Lake Harrison, was ponded up in the Midlands (fig. 5.22). The mapping of the laminated clays of this lake provided a solution to the interpretation of the drift sequence in the Midlands, which had for long been a great problem.

The West Midlands area is a critical one for Devensian stratigraphy. The limit of the last glaciation is argued, but is now thought to be near Wolverhampton. Radiocarbon assays of organic deposits associated with tills of this latest glacial stage have helped in the interpretation of the sequence.

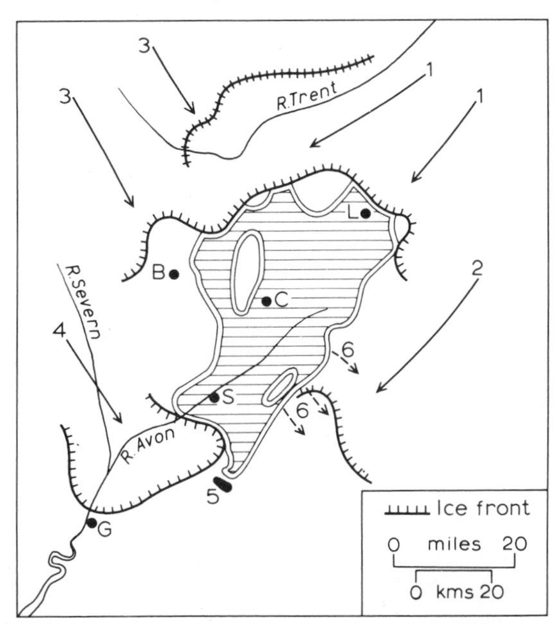

FIG. 5.22 The Wolstonian glaciation in the Midlands. Lake Harrison at its maximum size is shaded. B, Birmingham; C, Coventry; G, Gloucester; L, Leicester; S, Stratford. 1, eastern ice movement and margin at Lake Harrison maximum; 2, later stage of eastern ice; 3, northern ice; 4, Severn valley ice; 5, Moreton moraine; 6, Lake Harrison overflows. (After F. W. Shotton, in the *Philosophical Transactions of the Royal Society*, 1953, **B**, **237**, fig. 7.)

It appears that a major glacial advance occurred late in the last glaciation after 26,000 years B.P. There is evidence for a similar late advance from a radiocarbon date of 18,000 B.P. from below tills in Holderness, east Yorkshire.

The West Midlands area drained south into the Severn, which contains an extensive suite of terraces which have been correlated with the Middle and Upper Pleistocene glacial and interglacial stages. A prominent terrace is the

Main Terrace of the Severn, correlated with a period of major outwash formation from the Cheshire Plain lake of the Irish Sea ice of the last glaciation. The terrace deposits on tributary rivers, correlated to the last glaciation terrace of the Severn, are rich in organic remains and their faunas and floras have been the basis for the reconstruction of the climatic cycle of the last glaciation.

In southern England, we are mainly outside the glaciated area, though tills are found at Fremington in Devon, and in the Scilly Isles. But nevertheless there is considerable evidence of Pleistocene climatic and sea level changes ranging from Lower Pleistocene to Flandrian. Many of the larger rivers show terrace systems which provide some evidence of sequences of climatic change. Periglacial climates are recorded by solifluxion deposits and patterned ground and by terrace faunas and floras, indicating cold conditions. Temperate deposits of Ipswichian age are known from the Solent area, formed near the sea in response to eustatic rise of sea level in the interglacial. Changes of relative land–sea level are recorded by raised beaches and other marine deposits above present sea level. Fragmentary high level marine deposits in south-east England are of Lower Pleistocene age, and occur at 131–183 m O.D. A series of raised beaches occurs at lower levels on the south coast. The best known are those at about 30 m O.D. in Sussex, probably Hoxnian in age, and a younger series at 5–8 m O.D. associated with Ipswichian organic deposits, found in Sussex and traced to south-west England.

Finally, we must briefly consider Ireland. At the time of maximum glaciation and low sea level Ireland must have been joined to Britain by ice sheets. But even in the absence of ice sheets, and with a sea level of – 100 m, Ireland would still be joined to Britain only by a narrow isthmus to the north (fig. 5.23). Thus Ireland occupied an isolated position both at times of low and high sea level.

Except for small nunatak areas in the south, Ireland was blanketed by ice during maximum glaciation. The ice was mostly of Irish origin, but near the east and north-east coasts Irish Sea ice moved inland at various times. The ice limits and movements of the two major glacial stages are shown in fig. 5.18. These are correlated to the Wolstonian and Devensian glacial stages of Britain. Only one fragmentary deposit of last interglacial age is known, but several substantial deposits of Hoxnian age have been discovered. These are remarkable in containing a flora rich in evergreen oceanic species such as *Rhododendron*, *Ilex* and *Taxus*, distinct from the floras of this age known from East Anglia. Certain of the so-called Lusitanian species are also present, e.g. *Erica mackaiana* and *Daboecia cantabrica*. In view of the isolation of Ireland from Britain it is quite possible these 'Lusitanian' species are of ancient origin in Ireland and that the Irish flora has occupied an isolated

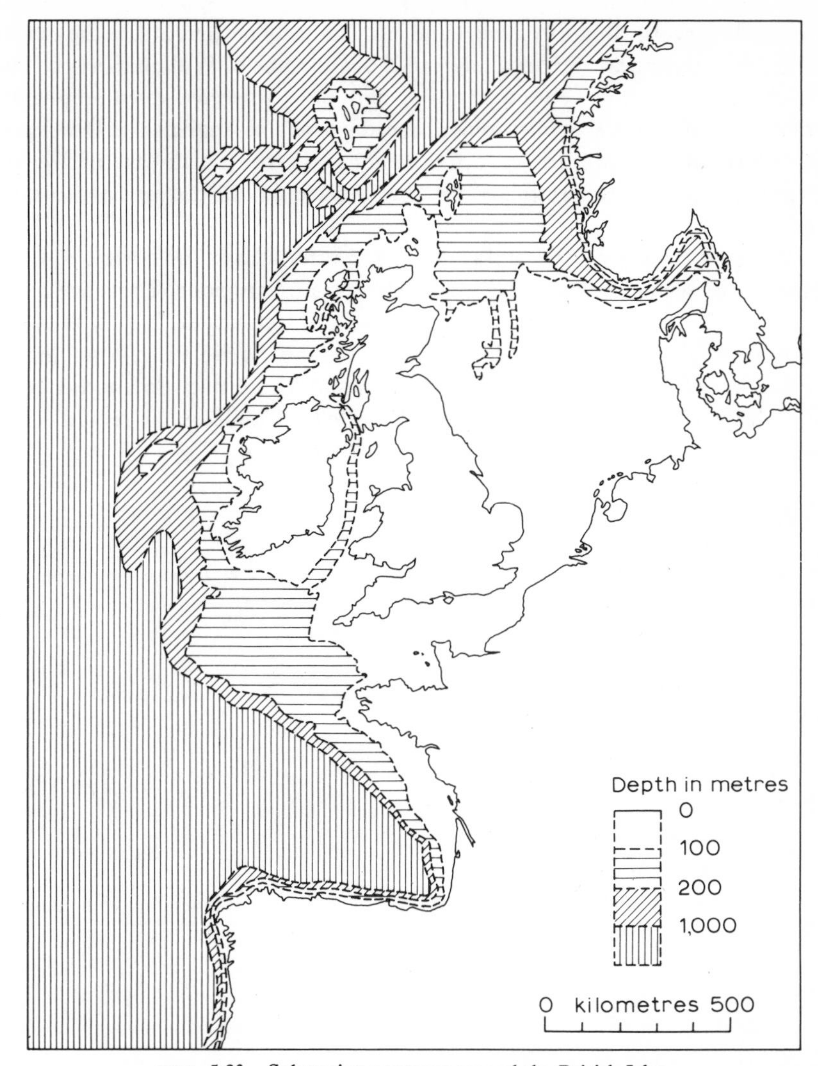

FIG. 5.23 Submarine contours around the British Isles.

position throughout the Pleistocene and perhaps late Tertiary. Survival of oceanic species may have been possible through a glacial stage because of the continuing proximity of the Atlantic. In the Flandrian, many British species are absent from Ireland, e.g. ***Tilia cordata, Fagus sylvatica, Carpinus betulus*** and ***Paris quadrifolia.*** This absence has been ascribed to the barrier to migration formed by the Irish Sea, which reached its present height sufficiently early in the Flandrian to act as a barrier for these species.

Around the Irish coast an important wave-cut platform occurs, overlain by beach gravels, then often head and till. The tills occur outside the extent of the last glaciation and are therefore thought to be of Wolstonian age. The beach gravels below contain erratics, and their presence must imply the reworking of a glacial deposit older than the beach of Anglian age. Otherwise pre-Hoxnian deposits are very fragmentary. In the north-east of Ireland Flandrian raised beaches occur, for here we are within the area of isostatic uplift.

Whereas the interglacial floras of Ireland offer a basis for correlation, the faunas which might also assist in correlation are very poor. There are a few records of a cold fauna from the Weichselian, but earlier faunas, showing ancient species of *Elephas*, *Rhinoceros* are absent. So are Palaeolithic industries. The earliest industries are Flandrian and Mesolithic. So another basis for correlation is lost. The poor archaeological and faunal record again emphasizes the isolation of Ireland in the Pleistocene.

6

The Ice Age Botany of Britain

Submerged forests around our coasts and also peat bogs (pl. 37) have long been known to yield tree stumps and bog oaks, giving evidence of former extensive forest where it no longer exists. Such fossil plant remains are obvious to the naked eye, and the existence of forests in the past, now drowned or cleared, is a natural and ready conclusion. These fossils are, however, merely the tip of an iceberg of information about vegetational history which lies concealed in peat deposits and lake muds.

If we take a small sample of peat, say a few hundred grams, we may extract from it large numbers of smaller plant fossils not easily visible to the naked eye. These smaller fossils fall into two classes; on the one hand the macroscopic fossils from a half to a few mm in size, such as seeds, fruits, leaves and twigs (fig. 6.1), and on the other hand microscopic fossils such as the pollen grains and spores of plants, of size 10–150 microns (μm) (fig. 6.2). It is the presence of these kinds of plant remains in quantity that allows detailed reconstructions of the past history of plants and vegetation.

Plant remains are preserved when they escape oxidation by deposition in a waterlogged environment. In such an environment they remain unchanged and may be recognizable as belonging to particular species of plants, even though they may be millions of years old. Thus the principal places where plant remains are preserved are in lake sediments, shallow marine sediments and peat deposits formed in fens and bogs. Lake sediments receive macroscopic plant remains from water plants growing in the lake and from plants growing around the lake. They also receive pollen and spores from the atmosphere and from streams draining into the lake. Marine sediments may similarly contain macroscopic and microscopic plant fossils. Peat deposits contain the remains of plants growing on the surface of bog or fen, the

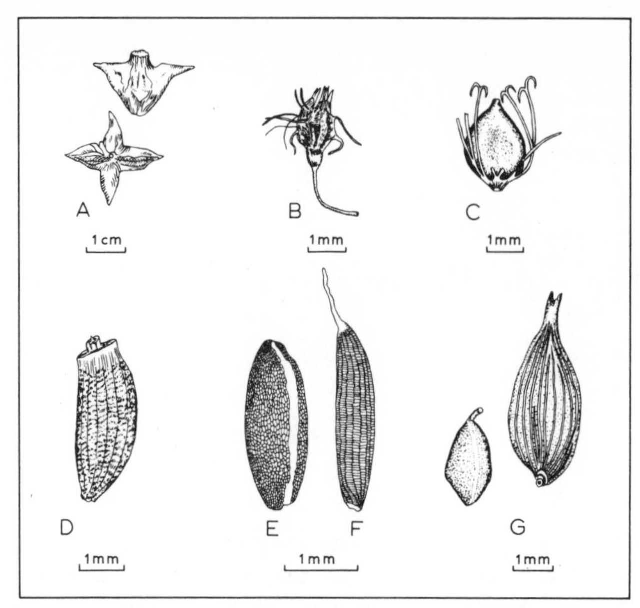

FIG. 6.1 Macroscopic plant remains. A, nut of *Trapa natans*, a thermophilous water plant, found in interglacial deposits; B, fruit of *Rumex maritimus*, a dock of damp places; C, fruit of *Polygonum lapathifolium*, a weed of waste places and damp ground; D, achene of *Cirsium arvense*, a thistle of pasture and waste places; E, fruit of *Najas flexilis*, and F, of *Najas minor*; these are common plant remains in lake deposits. The former is found under a wide range of climatic conditions, temperate and cold; the latter is confined to temperate interglacial deposits. G, the nut and fruit (nut within utricle) of *Carex riparia*, a fen sedge.

pollen and spores of these plants as well as pollen and spores received from the atmosphere.

The extraction of the plant fossils, their identification and the calculation of the frequency of species represented are preliminary processes which provide the data we use as a basis for reconstruction of the history of vegetation. But the interpretation of the data does not stop there. The geographical distribution of plants in the past was effected by environmental change, so that past environmental change can be reconstructed from changes in the distribution of plants. Thus, the presence of fossils representative of the species of deciduous forest, such as oak and elm, indicates a temperate climate at the time the fossils were deposited. But an assemblage of tundra species indicates a time of severe and cold conditions.

There are many pitfalls in the interpretation of the botanical data; most of

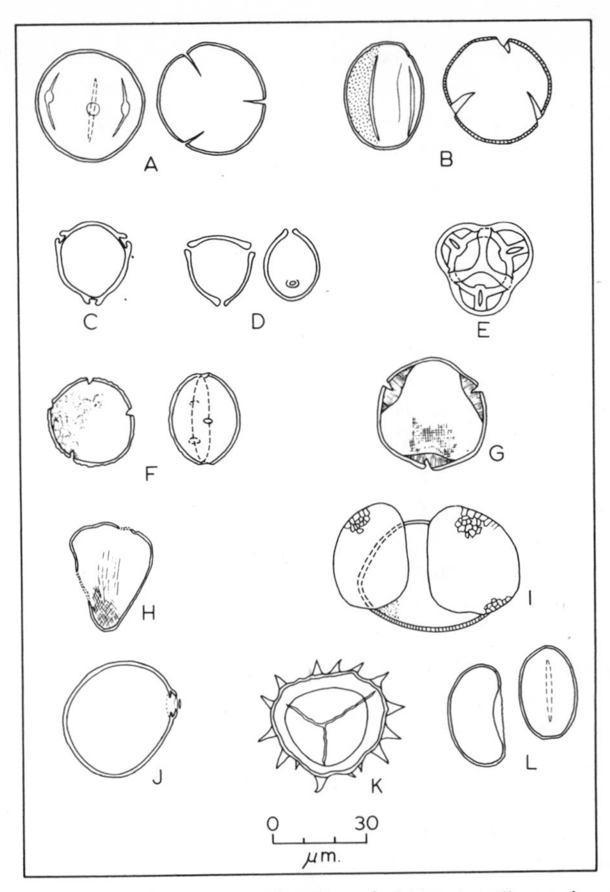

FIG. 6.2 Pollen grains and spores of plants. A, pollen grain, with three furrows each with a pore, of beech, ***Fagus sylvatica***, in equatorial and polar view; B, three-furrowed pollen grain of oak, ***Quercus***, in equatorial and polar view; C, three-pored pollen grain of birch, ***Betula***; D, three-pored pollen grain of hazel, ***Corylus avellana***, in polar and equatorial view; E, pollen tetrad, each grain with three furrows each with a pore, of Crowberry, ***Empetrum***; F, five-pored pollen grain of elm, ***Ulmus***, in polar and equatorial view; G, pollen grain of lime, ***Tilia***, in polar view, with three short furrows each with a pore; H, pollen grain of sedge, ***Carex***, with poorly defined pores; I, winged pollen grain of Scots pine, ***Pinus sylvestris***; J, single-pored pollen grain of a grass; K, spore of ***Selaginella***, with a triradiate scar; L, spore of ***Athyrium***, a fern, with a single scar, in two views.

them are similar in nature to those dealt with in the following chapter on Ice Age zoology. But before considering further these matters, we should now discuss the two principal approaches to past vegetational history, one through the macroscopic remains, and the other through the microscopic remains. The two approaches are complementary, each giving a different view of vegetational and environmental history.

Macroscopic plant remains

Sediments differ greatly in the number of plant remains they contain. It is often not possible to see how abundant they are in a hand specimen, except that in laminated deposits rich in plant remains breaking open along the laminae will reveal leaves, seeds and other remains. To obtain a sufficient quantity of identifiable remains it is necessary to take a sample of a few hundred grams, soak it in a pudding basin with dilute nitric acid or potassium hydroxide, then leave till the sediment can be easily broken down during the course of stirring. Some seeds and fruits will rise to the surface of the liquid during this process and can be removed with a paint brush. The contents of the bowl are then sieved (meshes between 0·5 and 0·25 mm) and washed on the sieves by a spray of water. The sieved fractions are then placed in a soup plate or other flat container with a white background and identifiable remains are removed with a paintbrush. Examples of plant macrofossils isolated in this way are shown in fig. 6.1.

The problem here is recognizing the limits of what may be identifiable and what may not be identifiable. To begin with, this is a matter of experience. The basic requirement is for a reference collection of fruits, seeds and leaves of species of the present flora, so that comparison may be made between fossil and living species. Atlases of drawings will assist in running down the fossils, but they are not a substitute for a reference collection. Shape, size and surface (epidermal) detail are the properties compared. The success of the identification depends on powers of observation and on the accuracy of the taxonomy of the reference collection. Here there is another difficulty, that of being certain that the identification of the reference material is correct. This is especially apparent in the identification of fossils of difficult taxonomy or of species with a number of subspecies. The identification will only be as good as the taxonomy of the reference material available will permit.

Considerable judgement is required to determine what limit of identification should be made. On the one hand, we have seeds and fruits which may be easily identified. On the other hand we have pieces of epidermis which may take a very considerable amount of effort to identify. The lengths to which an investigator will go to obtain identification depends on the time available and the significance of any result which he hopes to obtain. Obviously it will be worth spending time on the identification of a fossil seed which it is thought might be a rare subspecies of very limited distribution in Britain today. But it will not be so important to have an identification of a piece of epidermis in a sedge peat containing fruits of sedge more easily identifiable.

In the old days during the development of the subject it was customary to produce a list of species found at a particular site, and such a list was used to infer environmental conditions at the time the deposit was formed. A far better picture can be obtained, however, by a quantitative approach. Thus we may take a vertical series of 10 cm segments from section or borehole (pl. 36), identify the plant remains from each sample, calculate the frequency of each species as a percentage of the total number of remains identified, and plot the frequency changes seen in the vertical time sequence. Such a plot is shown in fig. 6.3, and we can use it, assuming that the ecology of the species concerned is known, to infer the detail of changing environment, both in terms of local conditions such as water level changes, or of more regional conditions such as temperature changes. Conclusions reached from such studies of the last interglacial are considered in the final chapter.

An assemblage of plant fossils in a particular sample may show ecological homogeneity or heterogeneity. The former is found usually in lake sediments of the temperate periods, such as a collection of remains of temperate aquatic species, or in terrestrial peats, formed in fens or bogs during temperate periods. In these examples the plant remains are derived from a restricted group of temperate plant communities. Heterogeneity is obtained by derivation of plant remains from a more diverse area of plant communities. In an extreme case, it is found in the seams of drift mud seen in the fluviatile terrace gravels of the glacial stages. These drift muds, formed in pools on the aggrading terrace surface, contain aquatic and terrestrial species, the latter perhaps coming some distance with the aid of solifluxion processes and seasonal flood waters of the periglacial environment.

These processes leading to heterogeneity may also favour the incorporation of fossils of earlier stages by, say, the erosion of interglacial deposits. This will introduce temperate species into a periglacial flora. Thus at the base of the last glacial terrace gravels in the Fens, nuts of the hornbeam, ***Carpinus betulus***, may be found. These nuts are commonly found in sediments of zone III of the Ipswichian temperate stage, but not later in the interglacial when climatic deterioration sets in. It is very likely that the nuts in the terrace gravels are derived from those beds. This problem of derivation is more difficult to solve in other examples where a fossil may have a more extensive distribution in an interglacial. Similar problems exist with faunas and are discussed in greater detail in the following chapter.

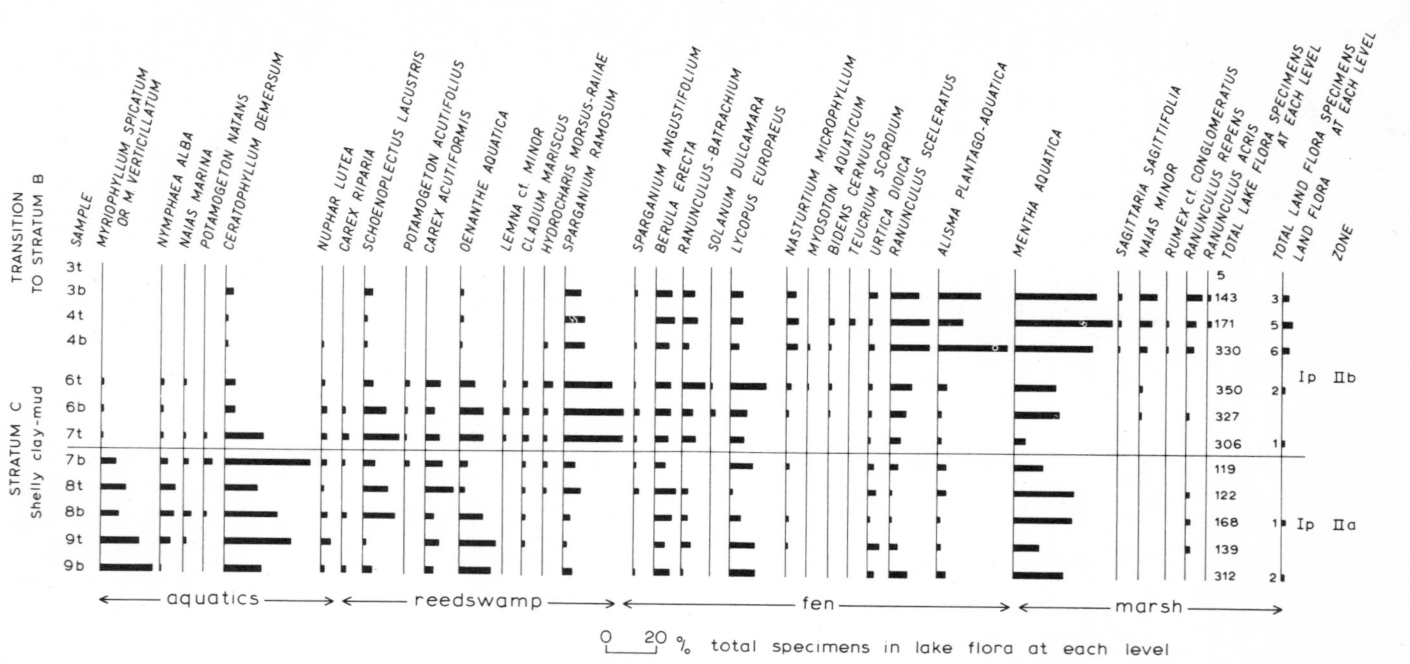

FIG. 6.3 Diagram of the changing frequency of various macroscopic remains in the Ipswichian interglacial lake deposits at Bobbitshole, near Ipswich. Aquatic species to the left, reedswamp species in the centre and marsh plants to the right. The sequence of changes of the plant remains indicates changes from open water to reedswamp to marsh, corresponding to the change from lacustrine clay-mud to mottled alluvial clay (stratum B). (After R. G. West, in the *Philosophical Transactions of the Royal Society*, 1957, **B**, **241**, fig. 7.)

Pollen analysis

In macrofossil investigations we are dealing with plant remains which have usually not come far and so give a picture of the local vegetation. In pollen analysis we are dealing with much smaller fossils which may have come a considerable distance through the air, as well as from local vegetation. Pollen analysis will thus contribute a different, more regional, picture of vegetational history than with macrofossil analysis.

Pollen analysis is the process of identifying and counting the frequency of types of pollen grains and spores in a sediment. Each species of higher plant produces pollen grains (pl. 38) of a particular morphology, and likewise, each species of fern, moss or liverwort produces a particular type of spore. Thus the parent plants may be identified through the morphology of their pollen grains or spores. The degree of identification varies. In some genera it is possible to identify all the species in the genus from their pollen grains, as for example in the British species of *Plantago* (plantain). Or it may only be possible to identify to the generic or family level because of the similarities of pollen morphology between species in a genus or genera in a family.

Two properties of pollen grains and spores make pollen analysis possible. One is their great morphological diversity, making their identification possible. The other is the very resistant nature of the pollen grain and spore wall, which promotes their survival in sediments throughout geological time. The wall is composed of a complex carbohydrate, chemically resistant except to oxidation. So that in unweathered sediments of lakes or seas pollen grains and spores up to hundreds of millions of years old may be preserved.

The morphological diversity of pollen grains and spores is a result of the variation of the type and number of apertures (pores, furrows, pores with furrows), the structure of the outer layer of the wall (exine), whether it is single or many-layered, and the sculpturing of the exine into spikes, knobs, reticulation, striation and other patterns. Fig. 6.2 shows some common British pollen grains and spores, with variations in aperture type and number and surface pattern.

The identification of pollen grains and spores must be done by comparison with a reliable reference collection from living species. The same principles apply that were discussed in connection with macrofossil identification. Before identification can take place, a slide preparation is made by taking the sample of sediment and treating it chemically to remove cellulose, lignins and inorganic content. The resulting preparation is stained and mounted on a microscope slide. The slide is traversed under a binocular microscope, with a magnification of about × 400, a tally being made of all pollen grains and spores identified. The total number of grains counted in each analysis will vary between 150 and 2000, depending on the number available to count and the total desirable on statistical grounds to reduce to a minimum the sampling error.

The resultant analysis can be expressed in several ways. The frequency of each pollen taxon (the word refers to a discrete group without implication of its status, e.g. genus, species, etc.) can be expressed as a percentage of the total pollen counted or as a particular part of the total, such as a percentage of total tree pollen. The total used depends on the problem in hand; one method may give a more useful portrayal of vegetational history than another. Alternatively, the frequency of a pollen taxon may be expressed in terms of the number of grains deposited over a certain surface area (cm^2) in unit time (year). To determine this needs complicated preparation techniques which relate the volume of the original sample used to the sample on the slide, together with a dating control. This latter method of expression may give more information during a period of rapidly changing vegetation, when percentage frequency changes might conceal numerical changes in the frequency of each taxon.

The results of pollen analyses are made up into a pollen diagram which shows the frequency changes of each taxon at each level. An example from the Flandrian is shown in fig. 6.4. Once the diagram has been constructed, the difficult work of interpretation starts. First the diagram is zoned into a series of pollen assemblage zones, each one characterized by a particular assemblage of pollen taxa. The series of zones gives a good idea of a broad vegetational sequence in the area from which the sediment received its pollen content. In the Flandrian a birch zone is succeeded by a pine–hazel zone, then a mixed oak forest zone. In an interglacial (fig. 6.7) the same sequence is seen, but the mixed oak forest is succeeded by a boreal forest zone with pine.

The more detailed interpretation of pollen diagrams is only possible after a consideration of the meaning of the frequency changes of each taxon in terms of the ecological preference of that taxon, and after a consideration of the possible origins of the pollen rain in the area and of the pollen-producing properties of the plants in question, which vary very much from species to species. Each pine tree may produce fifteen times as much pollen as a beech tree, and thus a pollen diagram with equal amounts of pine and beech pollen will mean very different representation of the trees in the surrounding forest. We have to assume that species have not changed their ecological tolerance and that the relative pollen production figures of different plants as measured now are applicable to the past. A knowledge of the relation between present vegetation in an area and the pollen rain it produces is also a great help in the interpretation of pollen diagrams. With these considerations in mind it is possible to carry the interpretation of pollen diagrams a considerable distance.

Errors in interpretation may arise in several ways. Weathering of sedi-

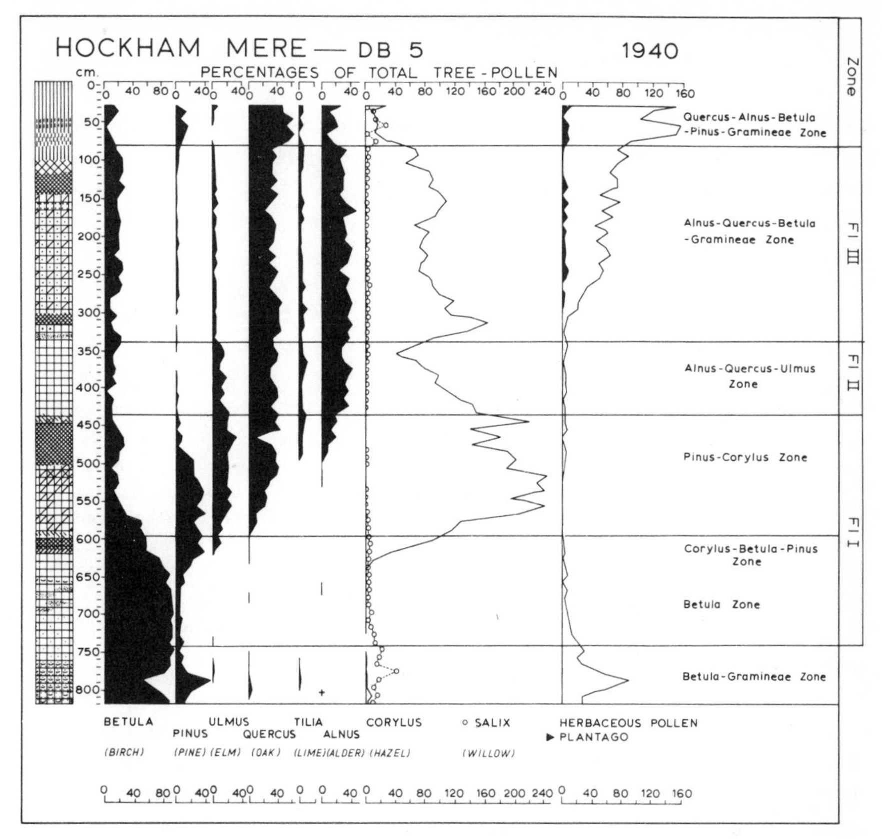

FIG. 6.4 Pollen diagram through a lake deposit at Hockham Mere, Norfolk. The symbols in the column on the left refer to various types of limnic mud up to 100 cm depth and to fen peat above this level. (After H. Godwin and P. A. Tallantire, in the *Journal of Ecology*, 1951, Vol. 39, fig. 3.)

ments causes differential destruction of pollen grains. The pollen grain of *Quercus* is far more sensitive to weathering than that of *Pinus*. The presence of pollen derived by erosion of older sediments and re-introduced into a sediment may interfere with the interpretation. Here a consideration of the origin of pollen in a sediment will be of help. Fig. 6.5 shows the various ways pollen may become incorporated into a sediment. Derived pollen is likely to be present in lacustrine and marine deposits receiving drainage water containing pollen grains removed from older sediments. Some estimate of what is derived may be obtained by pollen analysis of these older sediments through which the water is draining. On the other hand, in raised bogs (pl. 33) and blanket bogs (pl. 34), where the growth of the peat surface is dependent on high rainfall and the surface of growth is above the regional water table, all pollen present comes from the local plants or by more distant air transport from regional vegetation so that derived pollen is hardly likely to be present.

Each site of pollen analytical study has its own particular conditions of pollen receipt. The determination of local and regional components through

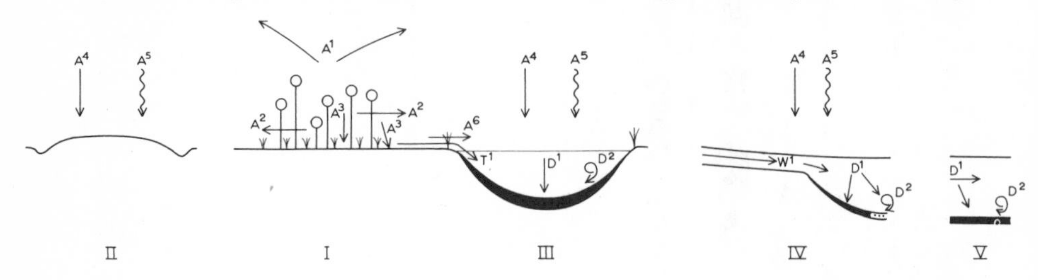

FIG. 6.5 Paths of pollen dispersal and deposition in different environments. I, pollen dispersal from vegetation; II, pollen deposition on a growing raised bog surface; III, pollen deposition in a lake; IV, pollen dispersal in an estuary; V, pollen deposition in a sea distant from land, low atmospheric contribution. A^1, atmospheric dispersal; A^2, dispersal in the trunk space; A^3, dispersal to the ground; A^4, atmospheric fall-out; A^5, atmospheric rain-out; A^6, filtering effect in lateral dispersal; T^1, terrestrial dispersal of component A^3 (and components A^4 and A^5); W^1, transport by water in river; D^1, deposition in water, accompanied by sorting by current action and differential flotation; D^2, recirculation of pollen, already once deposited, by reworking of sediments.

knowledge of the more locally derived macroscopic plant remains will aid in the interpretation of vegetational history. Thus we see the complementary nature of macrofossil and microfossil studies.

Interpretation of palaeobotanical results

From our analyses of the macrofossils and microfossils we produce lists of species in particular deposits and pollen diagrams from these deposits. If the deposits are securely dated, then this information can be used to reconstruct the history of particular species. The reconstruction of vegetation is more difficult, as we have already discussed. The delineation of particular plant communities in the past is hindered by uncertain relations between plant communities and the identifiable fossils they produce. Thus the woodland communities of the mixed oak forests in the past cannot be distinguished in the regional pollen rain produced. Ecological groupings of species can be made which give a broad indication of the detail of vegetation in an area, as we shall see in the discussion on periglacial floras. In some cases, it appears that groupings occurred in the past which do not appear today. Illustration of this can be demonstrated in the periglacial floras. It is not surprising that with the great and repeated climatic changes of the past, redistribution of species occurred to produce combinations which we no longer know today, as, for example, in the periglacial floras of low latitudes where a temperate climate reigns today.

The changes of species distribution and frequency which the palaeobotanical record shows are a result of competition between species during times of environmental change. Competition and migration are two factors involved in vegetational change. How far can we identify the role of changing climate from the palaeobotanical record? The gross change between the mixed oak forest of the temperate stages and the tundra vegetation of the cold stages must be climatically controlled. We are able to ascertain climatic change more closely if we observe the changes in distribution of particular species whose distribution is known to be controlled by particular climatic parameters. This is more difficult than might appear, because only in a few cases do we know the relation between distribution and climate in more than a general way. The present distribution of a plant may not be as extensive as climate would allow (as in plants introduced from North America thriving in Europe), or historical factors such as deforestation may have assisted in the spread of a species beyond its natural limits. Climatic deductions from plant distribution are therefore fraught with difficulty. Nevertheless some ideas of temperature change may be gleaned for example from the fossil occurrence of warmth-loving water plants which require certain minimum summer temperature to fruit, and some idea of precipitation may be gleaned from the behaviour of species or groups of species with an oceanic or continental distribution at the present time. Many of the points concerning the interpretation of fossil plant finds are equally applicable to fossil animals and they are discussed at greater length in the following chapter.

Thus the palaeobotanical record provides a basis for interpreting climatic change as well as vegetational history. The application of the methods of analysis described above has led to great advances in our knowledge of Ice Age environments.

History of the flora and vegetation

During the Ice Age the two principal environments for plant growth were temperate, with the growth of forest, and cold, with the growth of tundra vegetation. The alternation of these temperate and cold conditions happened, as we have seen, several times. The forest sequences of the temperate stages differ from one stage to another. The cold floras have, as might be expected, a greater uniformity. The two types of flora are linked by the common possession of certain tolerant species, occurring in both temperate and cold stages.

The length of the palaeobotanical record back through the Ice Age

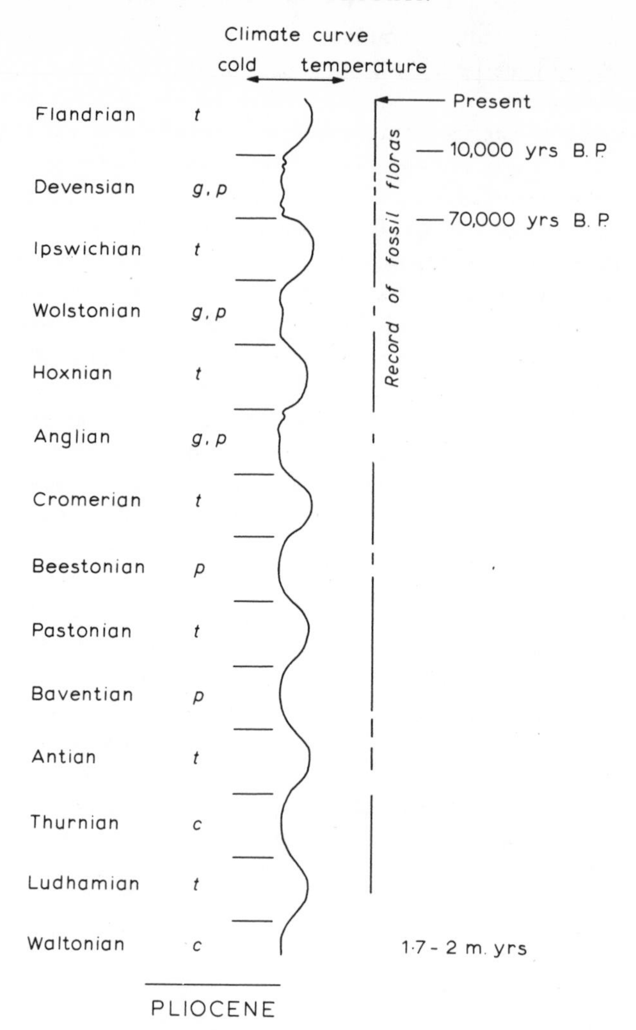

FIG. 6.6 The sequence of stages in the Pleistocene in relation to climatic change and the fossil record. t, temperate; g, glacial; p, permafrost; c, other evidence of cold conditions.

(fig. 6.6) permits us to see the effects of the climatic fluctuations on the flora and vegetation in considerable detail, in terms of extinction, migration and competition. It allows us to explain distribution patterns we see at the present day, putting them in a historical perspective, and demonstrates the remarkable sensitivity of plants to environmental change, both natural and man-made.

THE TEMPERATE STAGES

Pollen diagrams from the temperate stages of the Middle and Upper Pleistocene show the appearance and development of deciduous forest, and its subsequent retrogression to boreal forest, which finally gives way to herbaceous vegetation as cold climatic conditions come to prevail. (A detailed description of the course of such events in the last interglacial is given in Chapter 10.) A comparison of pollen assemblage zones in the last three temperate (interglacial) stages shows that a broad division may be applied to all of them. There is an early zone (I) characterized by the presence of boreal genera *Betula* and *Pinus*, with significant pollen frequencies of light-demanding herbs and shrubs. The second zone (II) is characterized by the pollen dominance of trees of the mixed oak forest, *Quercus*, *Ulmus*, *Fraxinus* and *Corylus*. The third zone (III) is characterized by the expansion of forest trees, *Abies*, *Picea* and *Carpinus*, at the expense of mixed oak forest genera already present. The final zone (IV) shows the predominance, among the trees, of boreal genera, *Pinus*, *Betula* and *Picea*, together with an expansion of herbaceous pollen types, indicative of a thinning of the forest and, often, the expansion of heathland.

Such a sequence as this is seen in the Ipswichian temperate stage (fig. 10.2), the Hoxnian temperate stage (fig. 6.7) and the Cromerian. In the Flandrian (postglacial) we appear to have reached the beginning of an equivalent of zone III. Although this broad similarity in major pollen assemblages is seen among the temperate stages, there are considerable differences in the detail of forest history between them, so that it is possible to characterize each period of forest history in a particular way. Thus, the following characteristics of each stage have been noted in East Anglia:

Flandrian: High frequencies of *Colylus* pollen in zone Fl 1.

Ipswichian: High frequencies of *Corylus* pollen zone Ip II, and a well-marked *Carpinus* zone (Ip III).

Hoxnian: High frequencies of *Hippophaë* pollen in the late glacial (late Anglian). Considerable *Tilia* frequencies in late zone Ho II. *Corylus* maximum later in zone Ho II. *Abies*, *Picea*, *Carpinus* and *Pterocarya* in zone Ho III.

Cromerian: Low frequencies of *Corylus* pollen. High frequencies of *Ulmus* in zone Cr II, but *Tilia* frequencies low. *Abies*, *Picea* and *Carpinus* occur in zone Cr III.

Pastonian: *Carpinus* in zones Pa II and III. *Tsuga* in zone Pa III. No or very low *Abies* in zone Pa III.

This degree of detail has not yet been found in the older temperate stages, Antian and Ludhamian. The deposits of these stages are restricted to the

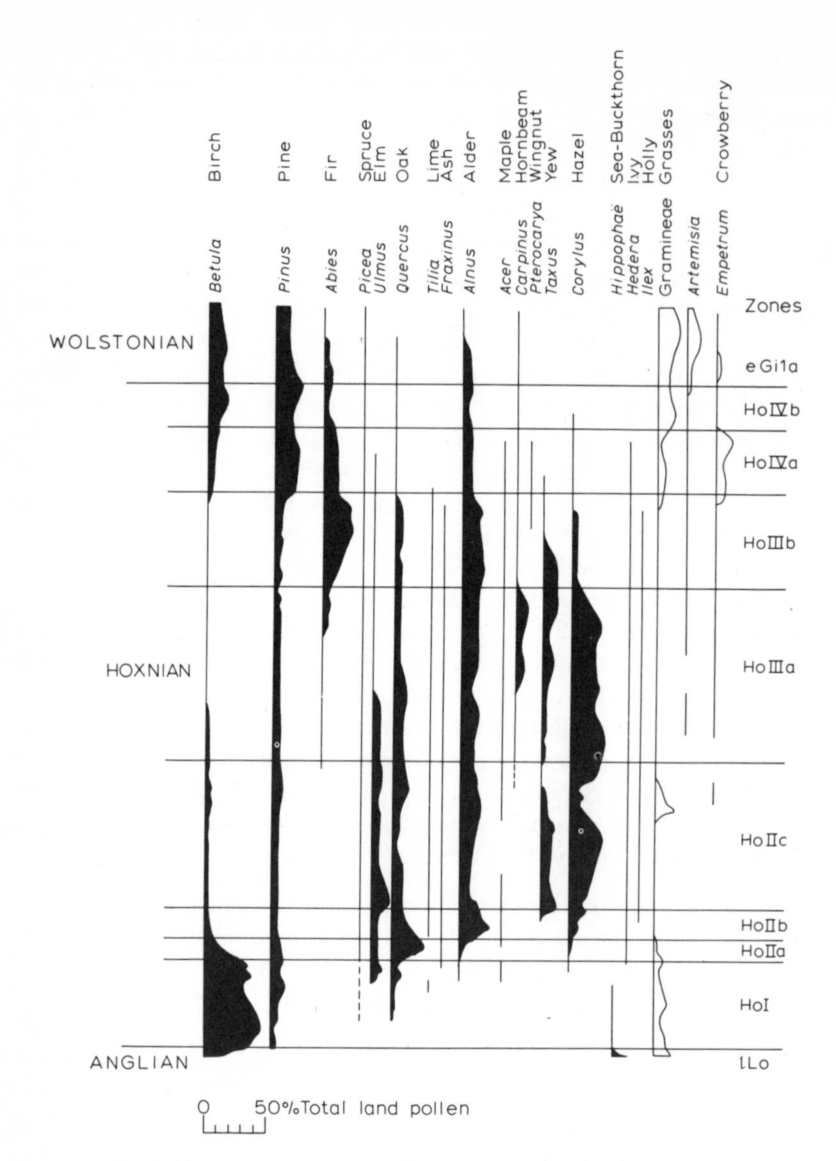

FIG. 6.7 Pollen diagram from the Hoxnian lake deposit at Marks Tey, Essex (after C. Turner, in the *Philosophical Transactions of the Royal Society*, 1970, **B**, **257**, fig. 15).

basin of Lower Pleistocene marine deposition in East Anglia (fig. 5.9). These older temperate stages are characterized by deciduous forest with *Tsuga* (fig. 6.8) and notable differences between the Antian and Ludhamian have not yet been distinguished.

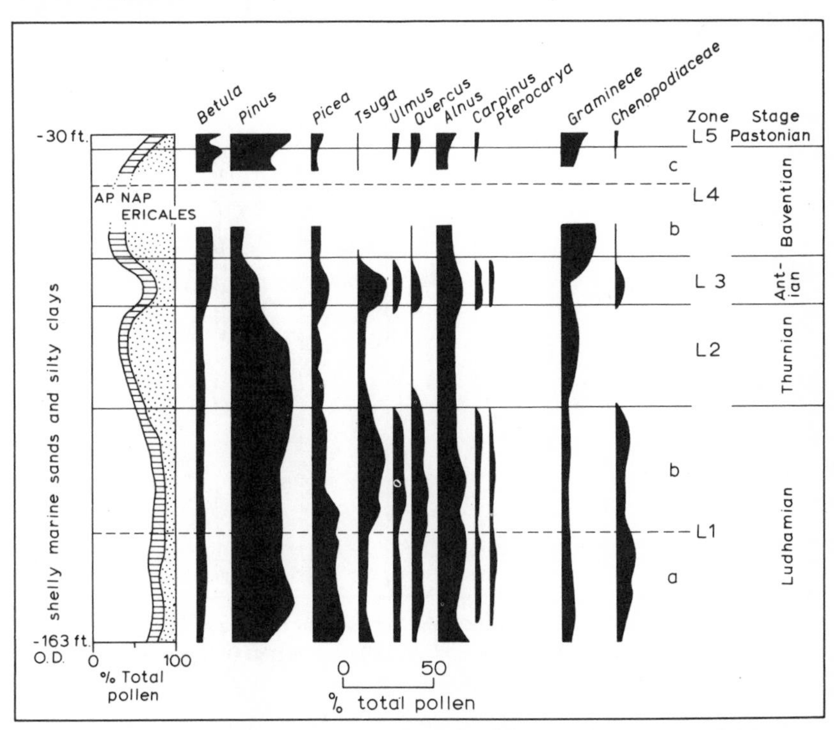

FIG. 6.8 Pollen diagram through the Lower Pleistocene Crag at Ludham, Norfolk (after R. G. West, in the *Proceedings of the Royal Society*, 1961, **B**, **155**, fig. 2).

The sequence of four zones already outlined can be related to environmental changes forming a cycle of change in the temperate stage (fig. 6.9). Zones I and II are an expression of the early expansion of forest genera in the order boreal to temperate. Zone I is characterized by light-demanding genera, zone II by the spreading of the temperate shade-giving genera *Ulmus*, *Quercus*, *Alnus* and *Corylus*. These changes must partly be a result of succession on soils improved from the raw, immediately postglacial state, to richer mull soils, of the times of immigration and expansion of the genera concerned, of competition between light-demanding and shade-tolerant genera, and of climatic amelioration. Some of the plants characteristic of zones II and III are shown in fig. 6.10.

The change from zone II to zone III is given by the expansion of genera, in particular *Carpinus*, *Abies* and *Picea* at the expense of mixed oak forest

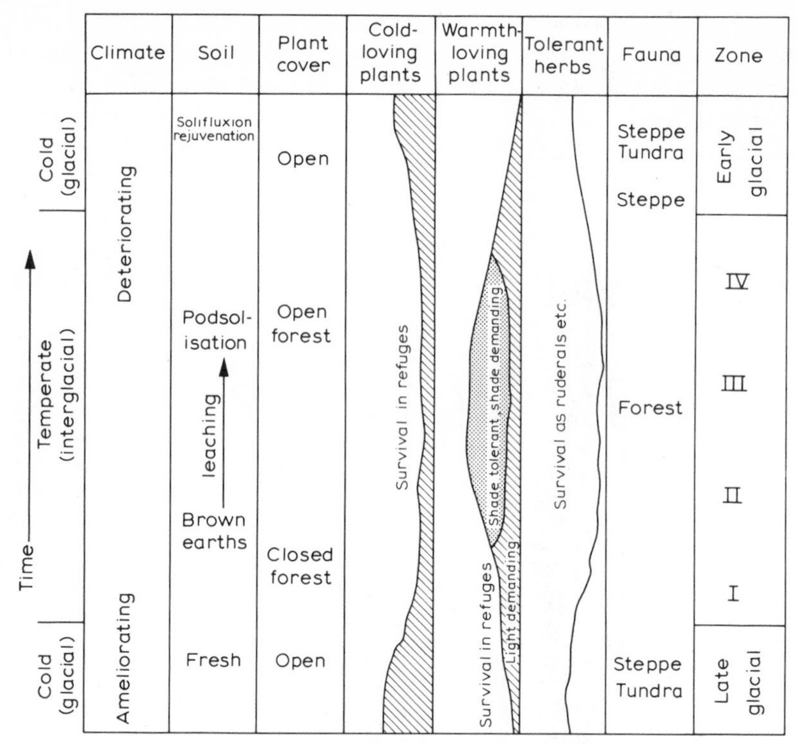

FIG. 6.9 A scheme of environmental and vegetational changes in Britain during a climatic oscillation from cold (glacial) to temperate (interglacial) and back to cold (glacial).

genera. The trees concerned may already be represented by low pollen frequencies in zone II. The change may be related to their successful competition, e.g. of *Carpinus* and *Abies* with *Quercus* and *Ulmus*, to climatic change, or to soil development from mild (mull) to acid (mor) humus. The growth of *Picea* encourages such soil changes, and the increased acidification may be associated with the spread of heathland, indicated by rises in pollen frequency of Ericales (Ericaceae and *Empetrum*). But climatic change towards oceanicity may also be responsible. It is probable that both climatic deterioration and soil changes played a part in the changes expressed by the vegetational history of zones III and IV. Thus, in summary in each temperate stage we see:

1. Climatic amelioration.
2. Dominance of light-demanding genera.
3. Soil improvement to mull condition and expansion of more shade-tolerant forest genera.

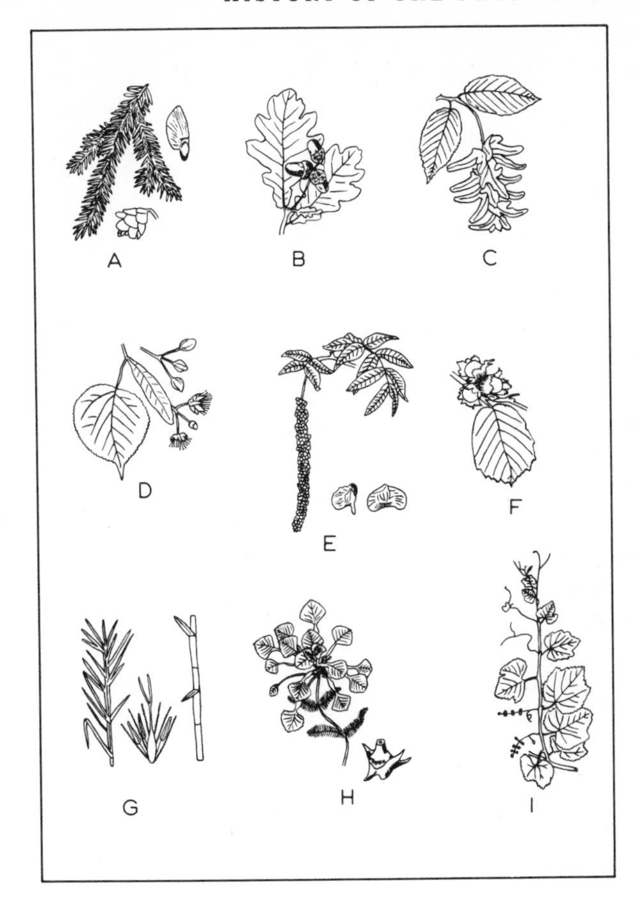

FIG. 6.10
Plants of the temperate stages in Britain. A, hemlock spruce, *Tsuga*, now extinct in Europe; B, oak, *Quercus*; C, hornbeam, *Carpinus betulus*; D, lime, *Tilia*; E, wing-nut, *Pterocarya fraxinifolia*; F, hazel, *Corylus avellana*; G, *Dulichium spathaceum*, an American and east Asian reedswamp plant, now extinct in Europe; H, water chestnut, *Trapa natans*, a European water plant not native to Britain; I, the grapevine, *Vitis*. (After W. Szafer, *Epoka Lodowa*, Warsaw, Panstowe Zaklady Wydawnictw Szkolnych, 1950, figs. 14 and 15.)

4. Soil deterioration to mor condition and/or expansion of late-arriving genera.
5. Climatic deterioration, restriction of thermophilous genera, expansion of heathland.

When we consider the detail of the differences already mentioned between the sequences in the successive temperate stages, we find many factors may be responsible. They include climatic differences between the stages, different barriers to migration during the stages – especially that formed by the present English Channel, differing distances of glacial refuges from which genera expanded, changes in ecological tolerance and variability within genera, and other changes consequent upon evolution or extinction. It is difficult to disentangle these factors. One difficulty is of inferring climate from vegetation or flora records. Perhaps a general trend of oceanicity or continentality may be discerned, for example, in the differences between the

Hoxnian and Ipswichian interglacials. The former has an abundance of *Taxus*, *Ilex* and *Alnus* pollen less well represented in the latter, while the latter contains records of many continental thermophilous genera indicating higher summer temperatures than at present. Again this particular element in the Ipswichian flora is lacking in the Flandrian, which suggests that the Ipswichian climatic optimum may have been warmer than that of the Flandrian. But apart from these indications, it is difficult to draw further or close conclusions from analyses of fossil floras.

Some conclusions may perhaps be drawn regarding the persistence of the Channel barrier during the temperate stages. The resemblance of the Hoxnian of East Anglia and Holsteinian interglacial of continental Europe suggests no barrier between them in the early part of the interglacial. On the other hand, the considerable differences between the Ipswichian and Eemian (notably the lack of development in the Ipswichian of zones with *Tilia* and *Picea*) suggest a barrier early in the interglacial, perhaps formed by the connection through the Channel of the last interglacial Eem Sea in the Netherlands and the Atlantic. Such contrasts in the last two interglacials are in accord with marine mollusc faunas. The Eemian fauna has connections with the Lusitanian fauna, while the Hoxnian fauna appears to have no such connection.

Differing distances of glacial refuges may account for the different behaviour of *Corylus* in the Hoxnian, Ipswichian and Flandrian stages (fig. 6.11).

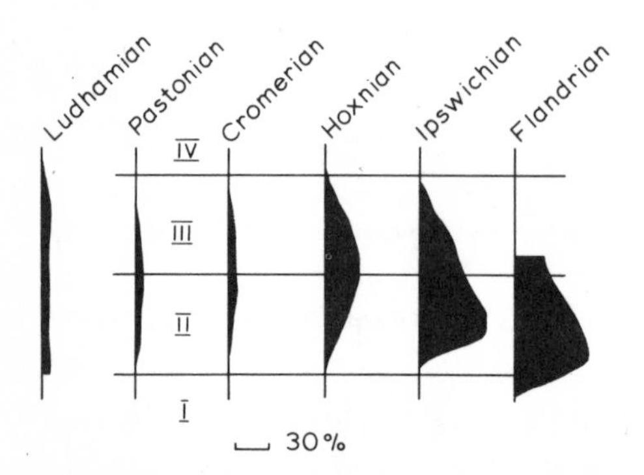

FIG. 6.11 Schematic *Corylus* pollen curves for temperate stages in East Anglia, presented as percentages of total tree pollen and related to the zone system described in the text (after Walker and West, 1970, fig. 3).

The notable difference is the progressively earlier time of expansion in England, late in the Hoxnian, at the time of the mixed oak forest in the Ipswichian, and before this in the Flandrian. This may be related to the increasing nearness of the glacial refuges to the ice-freed areas. If so, it may suggest a change in the ecological tolerances of *Corylus*, allowing it to survive in progressively nearer refuges, assuming that the intervening glacial climates were of a similar type. A difficulty here is the lack of knowledge of where the glacial refuges of thermophilous genera were. They may have been on the oceanic fringes of the continent or in small areas in southern Central Europe, where survival in particularly favourable habitats was possible. The possible paths of flora immigration at the end of a glacial stage are shown in fig. 6.12.

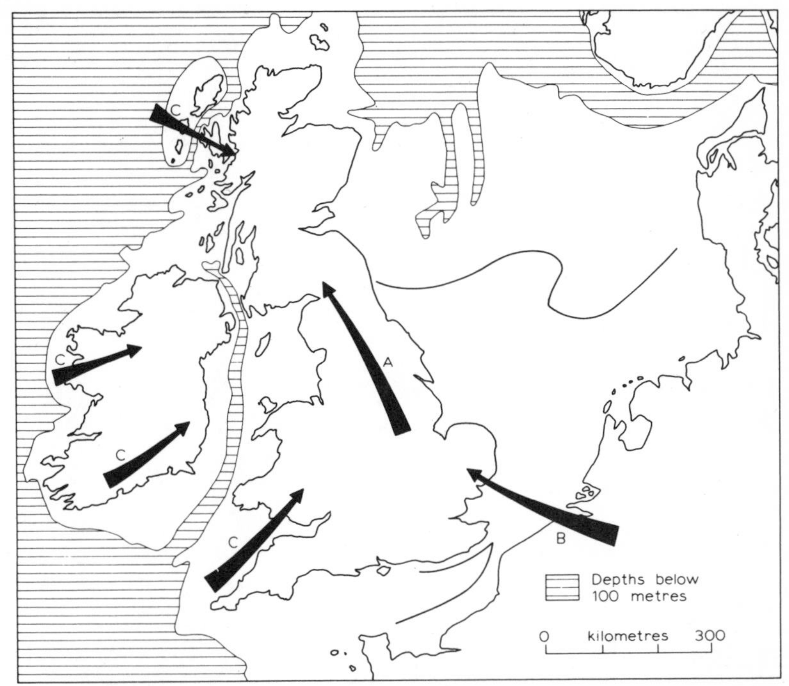

FIG. 6.12 Paths of flora migration accompanying the retreat of the ice sheets of a glacial and the accompanying climatic amelioration. Refugia in the west, north-west and to the south are shown as sources for the flora. The 100 m submarine contour is shown, and reveals that the near-isolation of Ireland must have taken place early in the Flandrian eustatic rise of sea level; the continuous lines in the Channel and in the North Sea indicate the approximate position of the shoreline about 9000 years ago.

A, movement north from refugia in the south of Britain; B, movement from the Continent; C, movement from western refugia.

Here we also come up against the question of changing biotypes or ecological tolerances within species and genera, discussed later in the chapter.

The vegetational history of the Flandrian (postglacial) temperate stage is naturally known in far greater regional detail than that of earlier temperate stages; many more deposits are known and they can be studied relatively easily. The sequence to the establishment of mixed oak forest in zone Fl II is shown in fig. 6.4, and regional distinctions are summarized in fig. 6.13. In the south *Pinus* plays a more important part than in the north and west. The

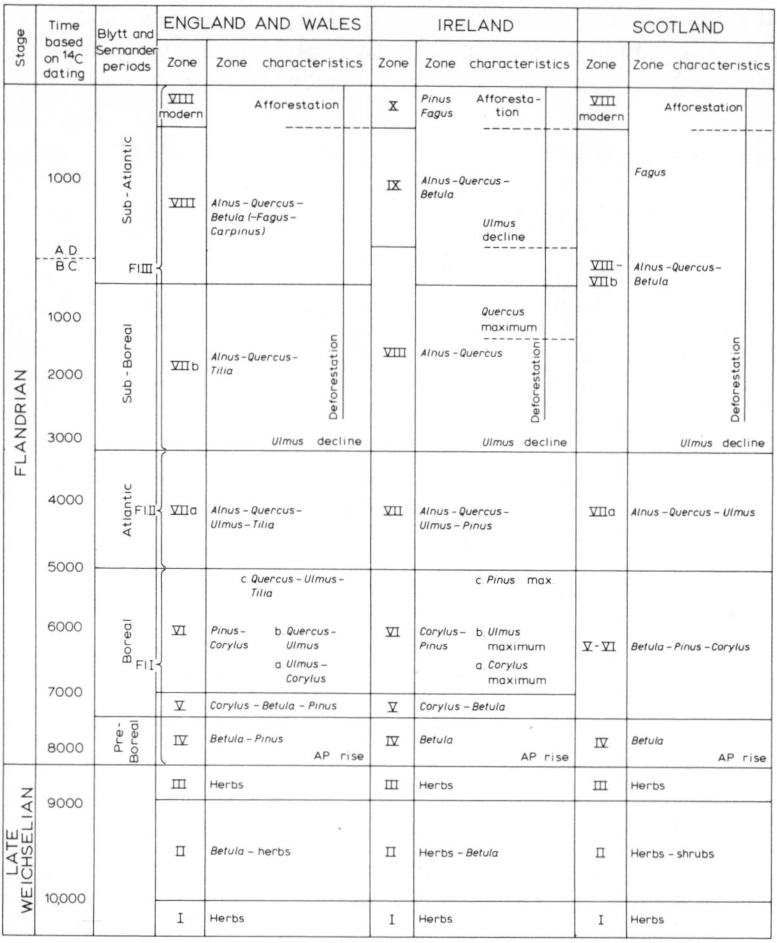

Stage	Time based on ^{14}C dating	Blytt and Sernander periods	ENGLAND AND WALES		IRELAND		SCOTLAND	
			Zone	Zone characteristics	Zone	Zone characteristics	Zone	Zone characteristics
FLANDRIAN			VIII modern	Afforestation	X	*Pinus* *Fagus*; Afforestation	VIII modern	Afforestation
	1000	Sub-Atlantic	VIII	*Alnus – Quercus – Betula* (–*Fagus – Carpinus*)	IX	*Alnus – Quercus – Betula*; *Ulmus* decline		*Fagus*
	A.D. / B.C.	Fl III					VIII – VIIb	*Alnus – Quercus – Betula*
	1000	Sub-Boreal				*Quercus* maximum		
	2000		VIIb	*Alnus – Quercus – Tilia*; Deforestation	VIII	*Alnus – Quercus*; Deforestation		Deforestation
	3000			*Ulmus* decline		*Ulmus* decline		*Ulmus* decline
	4000	Atlantic Fl II	VIIa	*Alnus – Quercus – Ulmus – Tilia*	VII	*Alnus – Quercus – Ulmus – Pinus*	VIIa	*Alnus – Quercus – Ulmus*
	5000	Boreal		c *Quercus – Ulmus – Tilia*		c *Pinus* max.		
	6000	Fl I	VI	*Pinus – Corylus*; b *Quercus – Ulmus*; a *Ulmus – Corylus*	VI	*Corylus – Pinus*; b *Ulmus* maximum; a *Corylus* maximum	V – VI	*Betula – Pinus – Corylus*
	7000		V	*Corylus – Betula – Pinus*	V	*Corylus – Betula*		
	8000	Pre-Boreal	IV	*Betula – Pinus*; AP rise	IV	*Betula*; AP rise	IV	*Betula*; AP rise
LATE WEICHSELIAN	9000		III	Herbs	III	Herbs	III	Herbs
			II	*Betula* – herbs	II	Herbs – *Betula*	II	Herbs – shrubs
	10,000		I	Herbs	I	Herbs	I	Herbs

FIG. 6.13 Sequence of main vegetational changes in the British Isles in the last 12,000 years.

mixed oak forest contains *Tilia* in the south, but this tree is absent from northernmost Britain and from Ireland. *Corylus* is more important in the west and north-west. As with the earlier temperate stages, these differences relate to the different climatic tolerances of the mixed oak forest trees, the diversity of regional climates and the positions of refuges for the genera in the Devensian (last) glacial stage. An important part of the differences between the Irish and British floras may be ascribed to early isolation of Ireland by rising sea level in the early Flandrian, earlier than the isolation of Britain from the Continent (see fig. 6.12 for the relevant submarine contour).

The natural development of the mixed oak forest of zone Fl II was abruptly halted by the forest clearances initiated by Neolithic man in many parts of the British Isles, at about 5000 B.C., and is expressed in pollen diagrams as a rise in non-tree pollen (fig. 6.4). Since then the progressive destruction of natural vegetation has continued so that our present landscape, in most areas, bears little relation to the natural forest vegetation which might have been expected in the absence of interference. The history of the destruction is discussed in more detail in Chapter 8.

THE COLD STAGES

The obvious evidence for cold climates in the Ice Age comes from the extensive glacial deposits covering much of midland and northern Britain.

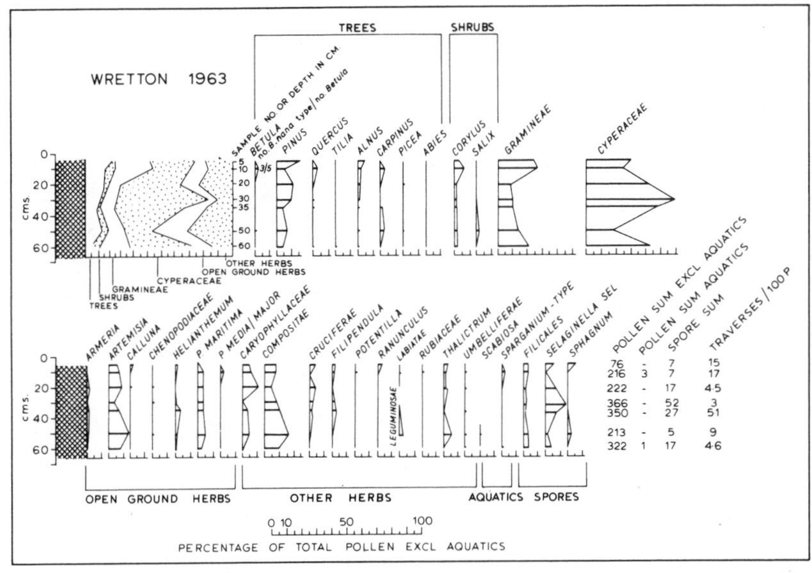

FIG. 6.14 Pollen diagram from organic mud of last glacial age filling a meander cut-off pool in a river plain at Wretton, Norfolk. The low frequency of tree pollen and the variety of herbaceous pollen types is characteristic of pollen spectra in periglacial areas in southern Britain.

To the south of the ice limits, in southern England, we find the periglacial accompaniment of solifluxion, frost-heaving (cryoturbation) and permafrost. In this periglacial area occur fossil floras which give rich evidence of tundra-like* flora and environment during the glacial stages. Similar cold-indicating floras are found in Lower Pleistocene cold stages. They contrast greatly with the floras of the temperate stages. No temperate forest trees occur. The plant

* tundra – vast, nearly level, treeless regions of northern Russia and similar areas in Siberia and Alaska (Lappish).

remains are mostly those of herbaceous vegetation and of a few shrubs, remains which give a tundra rather than a forest aspect to the flora. The pollen spectra show predominance of non-tree pollen, especially that of grasses and sedges, with a very wide variety of herbaceous pollen taxa (fig. 6.14).

The course of climatic change, as indicated by plant fossils, during a cold stage is not simple. Not only is there evidence for tundra, but also for vegetation of less severe climate, such as birch woodland or coniferous forest. These periods of an ameliorated climate are termed interstadial, in

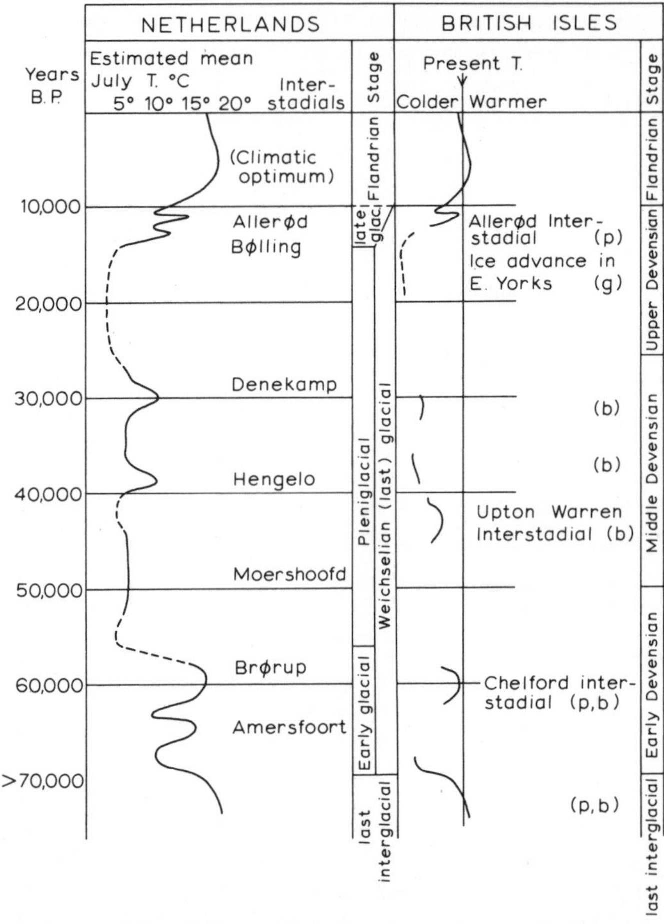

FIG. 6.15 The climatic sequence of the last glaciation in Britain compared with that of the Netherlands. The evidence for climate in Britain is indicated on the right: g, based on glacial deposits; b, based on beetle evidence; p, based on pollen evidence.

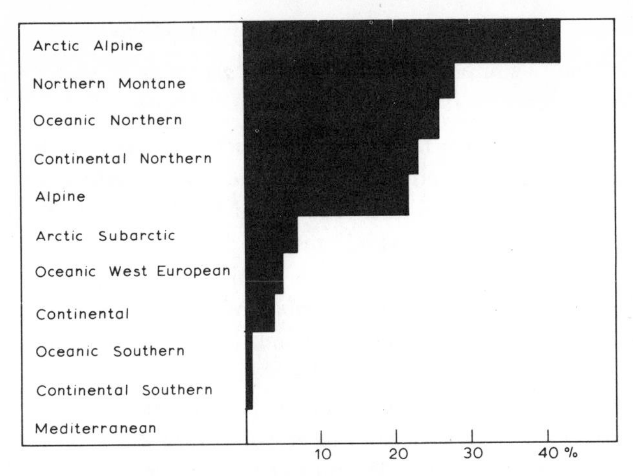

FIG. 6.16 Diagram showing the relative percentage of various geographical elements in the British Isles flora (Matthews, 1955), represented by records in the last glacial stage (Devensian, Weichselian). Species with wide extra-British range are not included in the analysis (after Godwin, 1956, fig. 108).

contrast to the stadials – periods of glacial advance or permafrost development thought to relate to the coldest parts of the cold stages.

From the botanical point of view there are two interstadial periods in the last glaciation in Britain (fig. 6.15). The one, characterized by coniferous forest with *Pinus* and *Picea*, is dated at about 60,000 years ago. Evidence for this has been found at two sites, one in East Anglia, the other in the Cheshire Plain. The other interstadial is much younger, dated about 11,000 years ago. This is the Allerød interstadial, named after a site in Denmark. During this interstadial, in southern and midland Britain, the landscape became more populated by birches, compared to the periods immediately before and after (pollen zones I and III of the Late-Devensian), when relatively treeless vegetation flourished.

The character of the stadial floras is an intriguing one. We do not see a flora of the type that is now characteristic of the northern tundra. Instead we see a flora of much more diverse phytogeographical affinities (fig. 6.16). A consideration of species in various categories as follows will make this clear. Some of the plants are illustrated in fig. 6.17.

Northern and montane plants: *Salix herbacea, Dryas octopetala, Draba incana, Thalictrum alpinum, Saxifraga oppositifolia.*

Maritime plants: *Armeria maritima, Atriplex hastata, Plantago maritima, Suaeda maritima.*

Weeds: *Polygonum aviculare, Potentilla anserina, Ranunculus repens.*

Plants of a more southern distribution: *Potamogeton crispus, Potamogeton densus, Ranunculus sceleratus, Damasonium alisma.*

This diversity of flora must be related to the habitat diversity and the lack of competition from shrubs and forest trees. We may imagine a treeless landscape, with shrub–tundra in the valleys and tundra on the uplands. The diversity of slope, aspect and soil will mean a richness of habitats. Instability of soil on slopes because of solifluxion would favour pioneer species and weeds. The maritime element is probably related to the development of saline soils under continental climatic conditions with frozen ground, as in Siberia at the present day. The southern element, which includes plants requiring warm summers for their survival, indicate summer (July) temperatures approaching 14–16°C. Cold winter conditions are indicated by periglacial freeze–thaw phenomena. We thus have indications that the climate in which these floras thrived was markedly continental – very different from what we have in southern Britain at the present day.

The origin of the heterogeneous flora of the stadials is indicated in

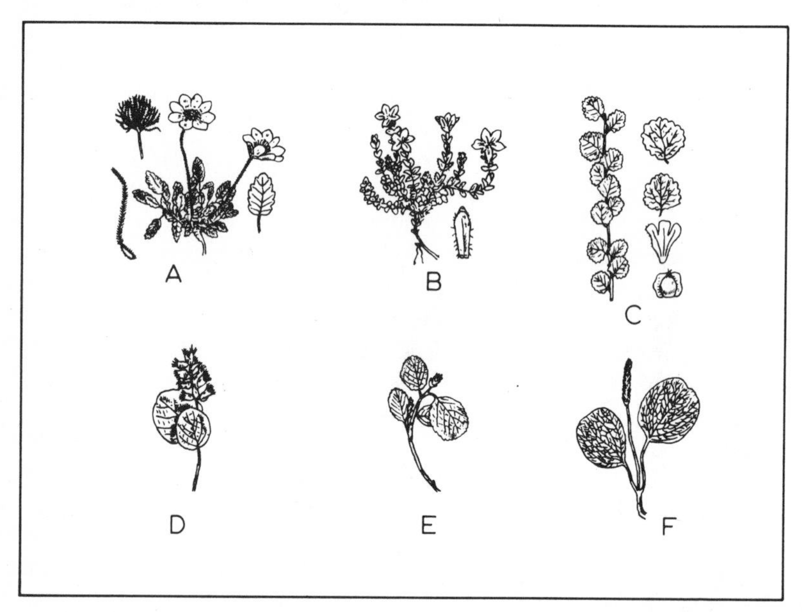

FIG. 6.17 Plants of the cold stages in Britain. A, mountain avens, *Dryas octopetala*; B, purple saxifrage, *Saxifraga oppositifolia*, with leaf; C, dwarf birch, *Betula nana*, with fruit and catkin scale, often found as fossils; D, polar willow, *Salix polaris*, extinct in Britain; E, least willow, *Salix herbacea*; F, reticulate willow, *Salix reticulata*. (After W. Szafer, *Epoka Lodowa*, Warsaw, Panstowe Zaklady Wydawnnictw Szkolnych, 1950, fig. 18.)

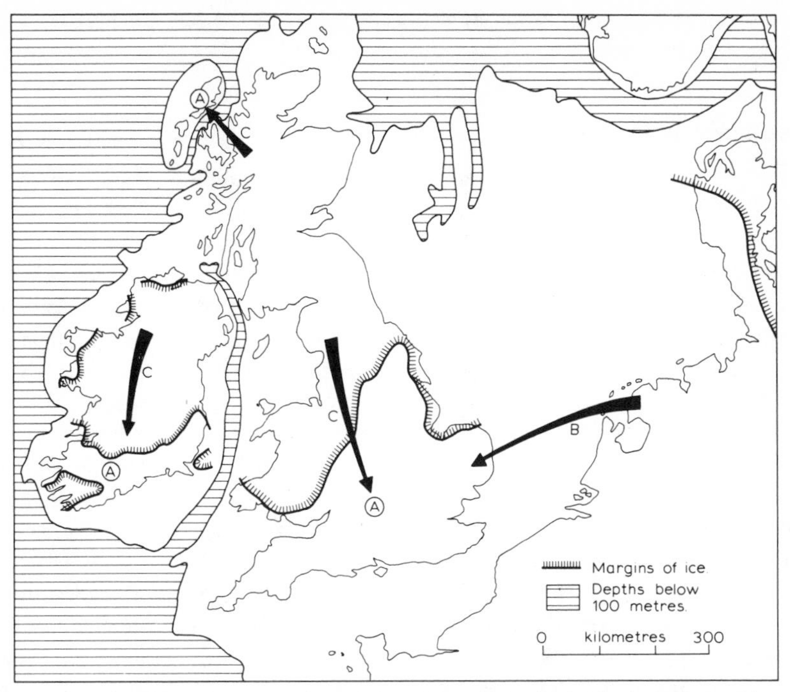

FIG. 6.18 Paths of flora migration accompanying the beginning of a cold (glacial) stage. The margin of the ice of the Devensian (last) glaciation is shown. C, movement south and west of plants from interglacial refugia and other habitats to refugia marked (A) outside the ice limits; B, movement across the dry North Sea of a continental element into the periglacial region.

fig. 6.18. On the extinction of forest, herbaceous species already present as weeds in temperate stages, would spread, and would be joined by more northern herbaceous species and shrubs migrating from northern habitats where they survived during the temperate stages. With a low sea level the periglacial flora would also be open to plant migration from the east, with the possibility of arrival of more continental genera, such as *Corispermum*, than might be present in the temperate stages.

A much discussed question has been the extent of survival of plants through the cold stages. At one extreme was the view that the flora disappeared. The stadial floras discussed above show that this was not the case. A peculiar tundra vegetation survived at times. But we still have no continuous record of vegetation through a cold stage. In the Netherlands a polar desert with extremely scarce vegetation is thought to have prevailed at various times during the last glaciation. The continuity of the tundra vegetation in southern Britain during a cold stage remains to be proved though the many occurrences of stadial floras at different times in the last glaciation might suggest such continuity.

Evolution and extinction

In a period of changing climate lasting about 2 million years, we might expect the flora to have undergone considerable evolution and extinction. Here is a chance of relating the passage of time and well-identified environmental changes to rates of evolution and causes of extinction. The raw data are the identified plant remains. We have only the seeds, fruits or leaves of particular species, not the whole plant. We have to assume that if a part can be identified, then that species was present. If we cannot match it with a living species, is it because our reference collections of living species is not complete or because it is an extinct species? The latter can be reasonably inferred if the part is identified to a genus with few species, species which it does not resemble. Thus there is an extinct species of *Pterocarya* which has been identified in the Lower Pleistocene, and extinct species of *Stratiotes* in the Tertiary. But there are few convincing examples.

Another approach is to examine the behaviour of forest tree genera during successive temperate stages in north-west Europe. Certain genera, such as *Carpinus*, *Tsuga* and *Picea*, show a restriction of distribution in successive temperate stages (fig. 6.19), perhaps caused by loss of biotypes or of genetic variability during enforced migration and survival during cold stages. Perhaps the degree of genetical variation within temperate genera is related to ease of survival in cold stages and ability to recolonize in the temperate stages. Conversely, the great expansion in the Pleistocene of the area suitable

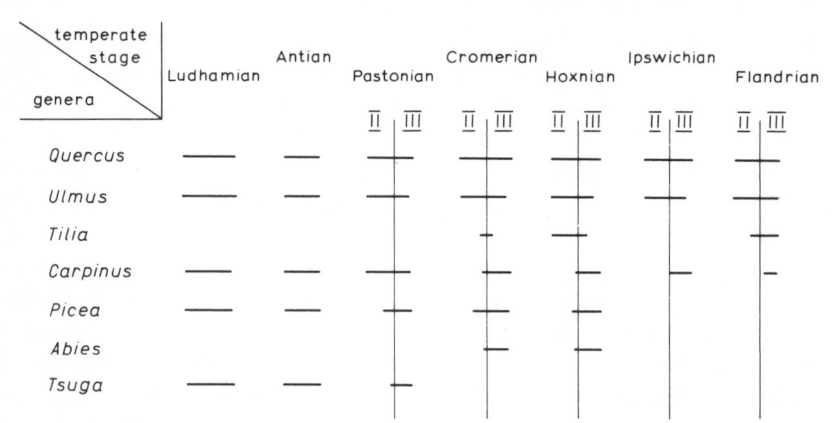

FIG. 6.19 The occurrence of some forest genera in East Anglia during zones II and III of the temperate stages of the Pleistocene (after Walker and West, 1970, fig. 2).

for herbaceous species, including northern and montane plants, during the cold stages must have led to the greater dispersion of these species, even though they became restricted in distribution during the temperate stages. Under such conditions it might be expected that some speciation of herbaceous species would have occurred. But such has not yet been demonstrated from studies of fossils.

We thus have little evidence of evolution and extinction during the Pleistocene. But we do have a good idea of local extinction of taxa in north-west Europe which have survived elsewhere.

The largest local extinction takes place at the beginning of the Pleistocene. In the Pliocene, the forests of north-west Europe were rich in genera now found in eastern Asia and/or North America, such as *Sequoia*, *Taxodium*, *Carya* and *Nyssa*, together with genera still surviving in north-west Europe. This 'Arcto-Tertiary' flora was decimated by the earliest cold stages of the Pleistocene. Clement and Eleanor Reid's explanation of this European extinction is as follows. Climatic deterioration forced temperate genera south. In Europe and western Asia there were barriers to this southward migration in the form of mountains, deserts and seas; in eastern Asia and North America there were no such continuous barriers. The result was that only the more hardy of the circumboreal Tertiary forest genera survived in Europe, while in eastern Asia and North America the present forest genera survived the Pleistocene climatic deteriorations and are much more closely related to their Tertiary precursors.

The extinction in north-west Europe was not entirely accomplished at the beginning of the Pleistocene. Certain genera, such as *Tsuga*, persisted into the Early Pleistocene, but by the beginning of the Middle Pleistocene the forest genera are those of modern north-west Europe, as are most of the shrub and herbaceous species. So, for example, the plant species (about 340 taxa) from the Cromerian temperate stage in East Anglia are nearly all found in the modern British flora. A few only (some 5 per cent) are exotic, e.g. *Picea abies*, *Trapa natans*, *Corema alba* and *Azolla filiculoides*.

In summary, we see in the Pleistocene great shifts in the range of species, little evidence of production of new species, and more evidence of extinction of species and great reductions in range. If we include plant communities in our consideration of evolution and extinction, we can also say that the great shifts of climate caused regrouping of species in combinations not now known, especially in the floras of the cold stages, an effect, in these stages, of cold conditions in lower latitudes than now normal, and also an effect of the apparent gradual reduction in biotype range of some species.

History of phytogeographical elements

The present British flora has been divided into a number of phytogeographical elements, each characterized by a common distribution area of the plants in the element. Thus the arctic–alpine element contains species with their main distribution in arctic or subarctic regions of the north and in high mountains at lower latitudes. The continental element contains species characteristic of central Europe. These divisions are useful in portraying the geographical affinities of our present flora. It may be tempting to say that each element has a distinct history. That the arctic–alpine element arrived during a glacial period and that the continental species arrived in the country during a period when continentality of climate extended into the country.

Table 6.1. *Some fossil occurrence of species in different phytogeographical elements of Matthews (1955)*

Oceanic southern	*Damasonium alisma*	Devensian; periglacial
	Ranunculus parviflorus	Ipswichian; temperate, zone Ip II
Oceanic West European	*Daboecia cantabrica*	Wolstonian; periglacial
	Erica mackaiana	Hoxnian (Gortian); temperate, zone Ho III
Continental southern	*Buxus sempervirens*	Hoxnian; temperate, zone Ho III
	Helianthemum canum	Devensian; periglacial
Continental	*Carpinus betulus*	Ipswichian; temperate, zone Ip III
	Ranunculus lingua	Devensian; periglacial
	Stratiotes aloides	Beestonian; periglacial
Continental northern	*Cicuta virosa*	Cromerian; temperate, zone Cr II
	Potamogeton praelòngus	Devensian; periglacial
Oceanic northern	*Armeria maritima*	Devensian; periglacial
	Najas flexilis	Flandrian; temperate, zone VI
		Devensian; periglacial
North American	*Eriocaulon septangulare*	Flandrian; temperate, zone VIIa
		Wolstonian; periglacial

However, when we look at the fossil record, we find that it is more useful to discuss the history of particular species, rather than the history of phytogeographical elements. Table 6.1 shows that species of the same element can be found under different climatic conditions.

We cannot generalize from the history of particular species to the history of particular elements. Some species occur in both periglacial and temperate environments, others either in periglacial or temperate environments only. Reconstruction of the past history by assembling fossil records is the most important key to present distribution. A most striking example has been the finds of species (e.g. *Erica mackaiana* and *Daboecia cantabrica*) of the so-called 'Lusitanian' element (Oceanic West European element of Matthews, 1955) in the Hoxnian (Gortian) interglacial of Ireland (pp. 162–3). Evidently these species have a long history in Ireland and further evidence may well demonstrate that some of these species were able to survive in the periglacial environment.

In conclusion, we can see that the study of the Ice Age flora and vegetation has added a new dimension to our knowledge of the Ice Age. The geological skeleton of ice advance and retreat has been clothed by a huge and recent increase in our knowledge of the Ice Age environment, revealed by the botanical studies we have discussed. Vegetational history now provides the background for ice advance and retreat and for the history of the fauna. It provides a clue to the detail of changing climate. And, of course, it also provides the means of interpreting the origin of the present British flora and vegetation, how these came to be present and how different factors have been effective in moulding them. We have considered many of these factors in this chapter. More discussion of man's effects on the natural vegetation of Britain will follow in Chapter 8.

7

The Ice Age Zoology of Britain

An interest in the animals of the Ice Age may have been aroused for different reasons. First, there is the non-scientist collector who has cabinets of specimens, many of them splendid examples of their sort but often without adequate recording of the exact localities and horizons from which they came. Furthermore, their very existence is often unknown, as scientists seeking to trace and examine the fossil animals from different Pleistocene deposits naturally turn to the known museums and the private collections in the hands of fellow scientists. Such collections, according to scientists, are a scandal and a disgrace, but how many Pleistocene malacologists have absolutely clear consciences that they have never picked up an odd mammoth tooth or bison jaw – only, of course, where there were plenty – to decorate the tops of their bookshelves? And how many Pleistocene botanists could swear that they have never yielded to the temptation to sequester a nice flint artefact? Seriously, although one cannot condone the hoarding of material of interest in private hands, it is too easy to raise the steam of righteous indignation against it. For all we know the presence of fine specimens in private hands may sometimes spark off an interest, which ultimately may result in a greater contribution to science than the loss of the few hoarded objects responsible for the initiation of that interest.

Second, there is the man who is scientifically interested in the history of the development and movement of a given group of animals in the Pleistocene. At the present time certain animals have curious distributions, for example the boreo-alpine distribution pattern in which the animals are confined to high latitudes in Scandinavia and to high altitudes in lower latitudes as in the Alps. The understanding of distributions such as these is made much easier when we have as complete a series of pictures as possible of the ways in which distributions have changed with climate and time. To be of value to scientists with these interests the animals concerned must be located and dated with the greatest possible precision and also be readily available for study; hence, scrupulously kept museum collections are vital. There is a tendency today to regard the museum curator merely as a cataloguer of material. Yet, without him many natural scientists would either have no material or useless material and so be unable to pursue their own purposes.

Those interested in the history of groups of animals are almost always specialists, because it is impossible to maintain knowledge on a wide enough front to cover all the various groups of animals adequately. We shall refer occasionally to various aspects of the history of certain animal groups in Britain, but will not follow this subject in detail. It would be possible to write on this topic in connection with the vertebrates, because most people are familiar with many of the species, but probably few, apart from specialists, would be able to sustain interest in the history of beetles and snails in Britain.

Third, there is the scientist who is interested in the animals of the Ice Age largely for what they tell him about the geographical or environmental conditions of the period. Probably the majority of Quaternary workers on animals and plants are interested in this aspect more than in the history of individual species and groups of animals. They sometimes call themselves palaeoecologists. Their work becomes most fascinating when the ecological conclusions derived from the study of different groups of creatures are compared. Full details of modern ecology are rarely known, so that when one is probing the past there is plenty of room for differing conclusions. There is most cause for congratulation, and we hope no grounds for the charge of collusion, when the botanist pronounces that the plants are typically those of a fen, the malacologist says that the snails are those that he would expect to find at the present time either in or on the marshy margins of fen drains, and the student of vertebrates produces a list of fishes which coarse fishermen in the Fenland still try to catch in the Old West River. Even before they have given their verdicts the study of the character of the sediment may have strongly indicated the nature of the deposit.

If all studies were so convergent one might be justified in smelling a rat. In other cases one or two lines of evidence may be ambivalent, and the conclusions from another line may be required to decide between the two. For example, the plants and the snails might be equally represented by those that live in fen drains and those that live on drier land at the margins. Are we then concerned with the deposits of a channel into which the drier land species have been washed, or are we concerned with an essentially land deposit to which periodic floods contributed freshwater species? If there are abundant fish remains then the first hypothesis would seem more likely to be the correct one.

However, the questions are often more involved than this. Most deposits, unless they have accumulated in an environment in which there is little waste or water movement, are curious mixtures obviously representing the sweepings from a number of environments. One has only to think of a small river, which itself contains many different niches in which different associations of plants and animals are found, and at the same time transports any material brought into it down its valley slopes or blown into it, to appreciate the complexity of the mixtures that might ensue. In these conditions if a hypothesis, that the assemblage of fossils could be explained by assuming them to represent, say, four environments, holds good for several different groups then the result is highly satisfactory. But different environments are more likely to be represented among the different fossils because of variations in the ease with which they may have been transported into a deposit. Thus, the wind will be completely unable to add to the fish fauna of a river, it is not likely to add materially to the number of snail environments represented, it probably will add fresh plant communities, or those parts of other communities which have light wind-borne seeds, while it will contribute pollen from the regional pattern of communities. In fact, the interpretation of the palaeoecology becomes a detective problem of assessing probabilities, requiring the collaboration of experts to arrive at the most likely solution.

To the student of palaeoecology absolute precision in identification may not be of such paramount importance as to the student interested in the history of species, for he considers general trends and there are always some unexplained anomalies in his interpretation. He relies on carefully stratified samples and the elimination of as much bias as possible in the assessment of the frequency of species horizon by horizon. These matters are considered in more detail later in the chapter. Nevertheless, he must be always on the look-out for different species and needs reference collections in museums for checking and comparison.

At the present time there are probably no groups of animals in the Ice Age studied in Britain by more than half a dozen experts with the exception of the larger vertebrates. Such experts rely on the observations by amateurs of interesting sections being exposed in trenches, road cuttings, boreholes, natural erosion of river banks and so on. The communication of details of these to any professional known to be likely to be interested is worthwhile as any of them may reveal information of vital importance in the study of the Ice Age. After all, it was a schoolboy who found the rhinoceros skeleton in the last interglacial deposit at Selsey which got into the national press.

General principles of the interpretation of Ice Age animal remains

Whatever group of animals one is using, one always interprets the past in the light of the present. It is assumed that the distribution of any animal is the same now as it was in the past, so that the general and climatic environment of the area where it formerly lived may be deduced. One can hardly do other than adopt this procedure; otherwise there would be no rational basis for a reconstruction of the past. When faced with species which are awkward to fit into the general pattern, one might be tempted to say that their environmental requirements had changed over the years. Alternatively one could regard them as contaminants, derived from older beds of different type or from the surface by slipping down rabbit holes. By such slick dismissals of incompatibles one would preclude any advance on present knowledge and the discovery of interesting and unusual associations of habitat. Yet, although it is impossible to work other than by the general rule of comparing present and past distributions to reconstruct the past, we must be prepared to face a number of questions.

(1) Is the present distribution climatically determined? It is most natural to make this basic assumption about the general distribution of a species, but not so easy to be precise about it. If the basic control is climate it could be either a temperature or humidity control. If a temperature control, it could involve the maximum summer temperature, the minimum winter temperature, or some such factor as the length of a season. Again, most animals may be influenced by climate through vegetation, if plants are their food, and it is not completely true that the sole control of vegetation is climate for soil enters into this balance. A recognition of the complexity of the matter is revealed when we talk of oceanic and continental species.

Even if the control is climatic, it is important to know at what level climate is effective. This was mentioned in Chapter 1, where it was stated that large animals and small animals are affected by different levels of climate. The climate in a marsupial's pouch is probably as even and monotonous as the macro-climate of equatorial, oceanic regions, although the climate experienced by the parent marsupial could be quite extreme. The micro-climate is very important in the case of small, non-flying animals.

If we assume we know the climatic requirements, it is still not always true that general distributions are climatically controlled. This is especially true of the larger vertebrates hunted by man. When they are near to extermination it may be that their present environment is not that which they like best, but that which man likes least and is least prepared to conquer. In long-settled, densely-peopled areas such as Europe it is very unlikely that the main control on vertebrates is climatic.

Finally, there are certain animals whose distributions are confined by their association with special environments. There are also plenty of plants with the same characteristics. Among the animals one can think of the house mouse, the brown rat and the cockroach, all of which are very closely connected with human activities. At a lower level are the dung beetles, which are known to occur as Late Pleistocene fossils in Britain.

An ancillary question in the same field is whether there has been time for the animal in question to have occupied the whole of the range climatically suitable for it. In general terms there probably has, especially when we remember that some plants, which are on the whole less mobile than animals, are in process of retreat in Scandinavia. It may be that certain very immobile animals have not occupied their full range, especially if they did not cross barriers when the barriers were crossable. In any case the same question applies to the interglacials. We do not know their length and so cannot know as yet whether there was time in them for the occupation of whole climatic ranges. The only thing that one can do is to assume that the two, postglacial and interglacial, are comparable.

(2) Can we be sure that the ecological and climatic requirements of the animal in question have not changed with time? The short answer to this is that we cannot be sure, but we dare not assume that they have changed unless we have sound and consistent evidence to this effect. The point can be illustrated from the non-marine Mollusca, the group of animals with which we are most familiar.

At the present time *Discus ruderatus* is a snail with a characteristic boreo-alpine distribution, yet it does not occur in cold Pleistocene deposits in Britain, as do certain other species which have similar modern distributions. Instead, it has been found in small numbers in the deposits of the last three interglacials, the Cromerian, Hoxnian and Ipswichian, and the postglacial. It seems to occur always at or near the climatic optimum of the period concerned. Its presence cannot be attributed to contamination and ignored, because it is too consistent. It cannot be attributed to cold micro-climates, which are far less likely to occur in interglacials, if they occur at all, than are warm micro-climates in cold periods. Nor can it be alleged that the deposits it is found in are cold deposits, because they have large and varied faunas including large numbers of species, a considerable number of them of southern type. With Mollusca the evidence of the majority must be accepted and we are forced to assume that the climatic tolerance of *Discus ruderatus* has changed, unless it is a species which cannot thrive in the presence of dense human populations and so is now confined to sparsely-peopled areas.

Another example, which is of a different type, is *Pisidium vincentianum*, a tiny freshwater bivalve shell. At the present time this, or something indistinguishable from it on the basis of shell characteristics, occurs westwards through Asia Minor to northern Pakistan. Yet, in Britain it is found associated with a restricted cold fauna in deposits laid down at some stage during the last glaciation in Britain. Again the evidence from the other Mollusca is overwhelming and must be accepted.

A third example is that of *Abida secale*, the postglacial history of which has been elucidated by Kerney (1962). The present distribution of this species in Britain is on dry hillsides and in quarries, usually on limestone, although it extends into open woods where there is plenty of warmth and presumably no excess moisture. In Europe its range is distinctly southern. Yet it occurs in deposits which are definitely late glacial in age in dry Chalk valleys in southern England. The important factors here seem to be the micro-climatic warmth and dryness and openness of Chalk hillsides irrespective of the regional climate. This is not so much an example of a species which has changed its requirements as one whose requirements were met in different ways in different general climatic conditions. But, again, the straightforward interpretation of the deposit containing it would have been very different had only its present distribution been considered and would have been in conflict with that derived from other Mollusca.

A fine modern example of a change in ecological requirements is provided by *Potamopyrgus jenkinsi*, which seems to have been a brackish water snail until the beginning of the last century when it started to spread into freshwater, which it did very widely and very rapidly.

Lastly, we must revert to the question of tolerance of the presence of man. It was suggested as an alternative explanation of the distribution of *Discus ruderatus* above. Certain snails appear in the southern and eastern parts of England in deposits which seem to date from the climatic optimum of the postglacial period, but they are now either very rare or extinct in those parts although still present more abundantly in the north and west of the country. Two such species are *Acanthinula lamellata* and *Lauria anglica*. The areas they now frequent are generally cooler and wetter than the south and east, and lime is not as ubiquitously distributed in the soils as it is in the south and east. Yet it would be very unwise to say that their ecological requirements have changed and that they have become more oceanic in type and less demanding of lime, especially as *Lauria anglica* is found in Portugal and Algeria. It seems much more likely that these species have failed to survive in the intensive farming conditions characteristic of much of the south and east: they are in fact anthropophobes. Thus, a characteristic which they may always have had, failed to affect their distribution until the density of human population became much higher than it was in the interglacials and early part of the postglacial. Yet not all anthropophobes have changed their distribution to a north-western one in Britain. *Ena montana*, for example, probably could not tolerate the acid soils of those parts and is instead confined to old beechwoods in limestone areas in the south and east. Perhaps its general Central European distribution means that it cannot stand the oceanic conditions of the north and west any more than their more acid soils.

(3) The third and last main question, or group of questions, which must be faced, concerns the relationship between the fossil assemblage and the living community. A number of effects come into play here.

The first is the question of selective preservation. The weaker the shell or bone the more readily it will be destroyed, but if all fossil assemblages have roughly the same composition to start with they should remain comparable with each other, but not with the living community from which they originated, except in general terms. The difficult case arises when one is trying to compare two different assemblages, each with two or three indicator species, but not the same species, characteristic, shall we say, of warm conditions. Then, although the deposits may really represent the same climate, differential destruction of the indicator species may make the deductions drawn from them differ, unless there is a checking technique, such as a pollen analysis, of the deposit available.

The other vital question is that of contamination. Most fossil assemblages, both of animals and plants, consist of a sweeping together of living communities from a number of niches. This can be readily realized in the case of the deposition of freshwater Mollusca by a river in flood. In addition to the species from a variety of freshwater habitats, there will be other species from any land habitats which happen to be adjacent to the course of a stream. Practically all deposits require skilled interpretation of the number of local environments which have contributed to their fossil assemblages. This is not normally treated as contamination, which is usually confined in meaning to the mixing of material of different ages.

Contamination can be of two sorts: obvious and subtle. The obvious sort is that which one encounters in working on deposits in the east of England, many of which are contaminated by micro-fauna from the Chalk (pl. 39), presumably released by freeze–thaw pulping of the Chalk to a large extent. Such micro-fossils, many of them Foraminifera, can usually be separated from Pleistocene fossils very readily by their form of preservation.

Subtle contamination occurs when, for example, a periglacial stream washes away an interglacial deposit and incorporates the fossils from it. Something comparable is going on at the present day on the northern shores of the Stour estuary in Suffolk, where *Corbicula fluminalis* is being washed out of the Stutton brickearth, which is Ipswichian interglacial in age, and incorporated into a modern estuarine gravel with inshore marine and brackish water species. It is sometimes said that one ought to be able to tell the two sets of fossils by their different degrees of wear. Possibly one can under ideal circumstances, but, if the interglacial deposit is fine and calcareous and saturated, its shells may be almost as fresh as those living in the periglacial stream. They may be even fresher than weathered land shells being delivered into the stream down the valley-side slopes. Again, if a very clayey deposit is dried out and becomes cracked, weathering agents will obviously work in the cracks and attack any shells found in the walls of the cracks, leaving those in the centres of the crack-bounded blocks unweathered. Thus, two modes of weathering can develop in a deposit without their having any significance in the question of contamination. If age were the only control of the degree of preservation everything would be so much simpler.

This is a really important question of the interpretation of mollusc faunas because we often find that faunas deposited in the cold conditions prior to a glaciation consist mainly of hardy species but with an admixture of small numbers of species which seem surprisingly southern for the context. The question naturally arises whether to consider the latter as indistinguishable contaminants or as survivals from the interglacial whose presence may throw light on the precise nature of the conditions leading up to the glacial. The question is always difficult to answer, and perhaps the only way to approach it is by allowing the evidence to accumulate so that we may know whether the condition is almost invariable or only occasional. If the former, contamination seems the less likely explanation.

(4) Certain difficulties also arise from the present state of knowledge of the distribution and identification of animals. There are far too few scientists interested in all the various groups of lower animals for our knowledge of them to be anything like complete even in Europe, let alone in the rest of the world. It is, therefore, completely unlikely that field collecting of such animals will have been done except in isolated localities. This will apply most strongly to the less accessible places, unless there is some particular interest in the faunas of such places, as there might be, for example, in the animals of the far north of Scandinavia. Thus, our knowledge of present distributions, on which we must base our interpretations of the past, is probably generalized from the study of relatively few localities and is always liable to be upset. This is less true of vertebrate animals and such things as mollusc shells, for the attraction of these for cabinet collectors has meant that knowledge about their distribution has been increased far beyond what it would probably otherwise have been.

It is also probably true that the more intensive the study of particular genera of animals, the smaller the number of species recognized and the greater the variation allowed to each individual species as a result. For example, some couple of hundred species of large freshwater mussels were recognized in France at the end of the last century, but only about half a dozen are recognized now. Of course, the exact status granted to a taxon (see p. 172) does not affect its ecological and climatic interpretation, provided that it is controlled by environmental conditions. Again, earlier records in Africa of the common pond snail, *Lymnaea*, listed a great variety of species, but there may well be only one 'species' composed of a whole variety of local races, apart from one or two introductions of European species. The contemporaneous state of the taxonomy of the group of animals concerned will be reflected in any records of Pleistocene animals.

Perhaps more serious is a different problem. If two animals which appear to be identical are found within, say, a couple of hundred metres of each other, few would hesitate to describe them as the same species. But, if those animals are separated by long distances or lengthy periods of time, it needs a bold man to give them the same name. For example, if specimens of the large freshwater snail, *Viviparus*, are found in Jurassic freshwater deposits and also in English rivers today and there seem to be no differences between their shells, does one have the nerve to give them the same name with all that this implies about the survival and immutability of a species over a long period of time? If we were just beginning to discover the freshwater Mollusca of Europe and our first two records of apparently identical species were from Crete and Lapland, it is unlikely that they would be given the same name. If knowledge of any particular group stays incomplete it is likely that the same animal may have been given different names in different places, so that, unless one is aware of this, the interpretation of a fossil occurrence will depend on which modern occurrence it is compared with. This is a real difficulty, especially when species which are alleged to be tightly controlled by climate or environment are under consideration. The opposite may also occur, two species alleged to be identical being in fact different, for the range of form within a species is not necessarily less than that between species.

Animals in the deposits of the Ice Age in Britain

As indicated earlier in this book, the general trend of Pleistocene research at the present time is towards the taking of series of related pollen and macro samples for detailed investigation in a laboratory. Very little collecting is done from open sections in the field, save of the larger vertebrate animals, and material selectively removed from open faces is usually recorded apart from that recovered from the series of stratified macro samples.

Such macro samples are usually taken at intercepts of 10 or 20 cm, of amounts between 1 and 2 kg. They are washed, sieved and sorted under varying degrees of magnification, so that material down to about 0·5 and occasionally 0·25 mm is usually looked at. Most of these sediments are derived from rivers, lakes and fens, so that it is hardly surprising that a great variety of fossil material is found (pl. 39). An exhaustive review of the various classes of animals recorded from Quaternary lake and bog sediments has been given by Frey (1964): this review covers America and Europe and includes many groups that have received scant attention in Britain.

Ideally one should pick out everything organic when sorting Pleistocene sediments. However, if one has 50 or 60 samples from one deposit and if they are all so rich that it takes a week to pick out the Mollusca and seeds and beetles from each – these are the fossils usually recovered – and if one is also approximately identifying them as they are picked out, so that perhaps the best part of 100 lots are being built up from each horizon, the thought of trying to extract even more is daunting. Furthermore, in the present state there are insufficient experts to make the number of identifications required for quantitative analysis and possibly insufficient knowledge of some of the animals for any detailed interpretation to be built up from their study.

Because of this we shall confine the discussion of animals in the British Pleistocene to a number of groups which have been particularly studied. These are the vertebrates, which were among the earliest studied and are certainly those which catch the public imagination to the greatest extent; the land and freshwater Mollusca, which have always had their followers and which have recently been used in quantitative studies designed to eliminate personal bias in selection and interpretation; the beetles and weevils, which may not have such a long and intensive history of study as the vertebrates and Mollusca, but which have recently been subjected to intensive studies like the Mollusca. In addition to these the marine Mollusca have a long history in Pleistocene studies but modern interpretations using them are only just beginning. Lastly, the Foraminifera are extremely useful in brackish and marine deposits and there are signs of their being used more systematically than before.

Apart from these the most common animal fossils are members of the Crustacea. Among these, the Cladocera, of which *Daphnia* is a well-known genus, have been studied in American and European deposits, and in Britain the Late-Devensian and postglacial deposit of Esthwaite Water has been studied by Goulden (1964). The Ostracoda, which are also Crustacea although they look like small bivalve shells, occur by the thousand in some deposits, but have not been studied in intensive detail yet.

VERTEBRATE ANIMALS

Although vertebrate remains are the most spectacular Pleistocene fossils, especially when whole skeletons of animals as large as elephant, rhinoceros or hippopotamus are discovered, they are not really the most useful. A single elephant may tell us little more than a single seed or a single snail, but it takes up a great deal more room and is far more difficult to stratify accurately. Yet, on the other hand, unstratified skeletal material in museums can sometimes be zoned by pollen analysis of fragments of the matrix in which it was embedded still found adhering to the bone. An early example of the use of scrapings from vertebrate skeletons was the identification of the food of a frozen Siberian woolly rhinoceros by picking the remains from its teeth: this was not an example of the wonders of twentieth-century Quaternary studies, for the analysis was made 125 years ago in Russia.

Much was done on the preparation of monographs about Pleistocene mammals in the last century and in the earlier part of the present century. These periods probably represented the heyday of collecting, because the greater use of hand labour meant that the location of many finds was more accurately known to someone than it would be now, even though careful records were not always kept. At present the concentration on a few pits of a much larger size means that the variety of material exposed is less, and the mechanical methods mean that many things may be badly damaged if not destroyed without being seen. A few mammoth teeth and bones found during screening gravel, which may have come from a pit of vast extent or, indeed, from one of a series of pits, are far from ideally stratified.

Most modern interpretations are based on percentage frequency studies, and not on the presence or absence of rarities, and the larger vertebrates are not suited to this type of study. Thus they have become in a sense incidental to the main purpose of Pleistocene studies, yet in many ways their careful study could prove to be very useful.

More than most animals they show a pattern of evolving faunas through the Pleistocene, much more so than do the non-marine Mollusca which are discussed below. The lowest parts of the Pleistocene, the Crag deposits of East Anglia, show the decline of forms inherited from the Pliocene and the emergence of ancestors of forms more common in the later Pleistocene. Among the former are the last of the mastodons, relatives of the elephants with long straight tusks, large heads and shorter limbs than the elephants proper. At the same time the elephants proper developed. The earliest of these was the southern elephant, *Elephas* (*Archidiskodon*) *meridionalis*, which, by the time of the Cromer Forest Bed, had begun to diversify into the ancestors of the two later forms, the straight-tusked form, *Elephas* (*Palaeoloxodon*) *antiquus*, and the form of the mammoth characteristic of the steppe environment, *Elephas* (*Mammuthus*) *trogontherii*. Again, there was only one form of rhinoceros, *Dicerorhinus etruscus*, in Britain until after the time of the Cromer Forest Bed. Among the voles it was the genus, *Mimomys*, which dominated the Pleistocene before the time of the Forest Bed, but this was later replaced by species of the genus *Arvicola*.

In later interglacials different forms of elephant and rhinoceros occur, while the rodent species show a number of changes. By the end of the last (Weichselian or Devensian) glaciation all the large Pleistocene mammals have disappeared and the only extinct elements in the British fauna are animals such as the beaver, *Castor fiber*, and the wolf, *Canis lupus*, which both became extinct in Britain comparatively recently.

It is also thought by some authorities that there are differences between interglacials not only in the forms of the species but also in the presence or absence of certain animals. Sutcliffe (1964), in particular, has maintained that there are significant differences between the mammal faunas of the Hoxnian and Ipswichian interglacials. The Hoxnian is characterized by:

(a) Two forms of rhinoceros, *Dicerorhinus kirchbergensis* (=*merckii*) and *Dicerorhinus hemitoechus*.

(b) The fallow deer, *Dama clactoniana*, was a larger race than the *Dama dama* of the Ipswichian.

(c) Horse is present.

(d) Both hippopotamus and hyaena seem to be absent.

Unfortunately, the chances of collecting a large and representative fossil mammal fauna depend on having large exposures under observation for a considerable number of years, whereas many interglacial deposits are studied by analysis of boreholes or temporary trench sections. Further, we do not have all that many Hoxnian and Ipswichian interglacial sites accurately stratified by pollen analysis. The possibility must be borne in mind that the next Hoxnian site might well yield hippopotamus. After our experience of the virtual absence of alder from the Ipswichian until it turned up abundantly at Wretton in Norfolk (see Chapter 10, p. 274), and also the discovery that the *Vertigo parcedentata – Pupilla muscorum – Columella columella* (pl. 42) snail fauna was a facies fauna characteristic of more than one period in the Pleistocene, one becomes chary of accepting the suggestion that no record till now should be construed as tantamount to absence. Hippopotamus (pl. 40) may well be absent from the Hoxnian, for it may have been unable to populate Britain in that period. On the other hand, one must beware of assuming that every deposit with hippopotamus, but not dated by pollen, must be Ipswichian. There are certain deposits, such as that at Barrington in Cambridgeshire, a famous site for hippopotamus, which might on this basis be called Ipswichian although certain local geological evidence suggests that it is much earlier. Such deposits ought to be put into suspense until a third line of evidence is found to decide the question. Conflict of evidence nearly always leads to rethinking and possible advance of knowledge.

In addition to the evolutionary changes in the mammal fauna, there are environmental changes reflecting the changes from interglacial climate and vegetation to glacial climate and vegetation. If one thinks only of present distributions, one might be inclined to believe that any interglacial with records of species such as lion, hyaena, rhinoceros, elephant and hippopotamus must have had a climate at least verging on the tropical. But the vegetation history shows that the climate was only a little warmer than it is now and that these animals must have lived in cooler climates in the past.

As Zeuner (1958, 1959) has suggested, the primary adaptation of some of these animals may have been to the vegetation and only secondarily to the climate, in that forest is more usually related to warmer climate than are open vegetation associations. A common view is that *Elephas antiquus* is the warm species and that *Elephas primigenius* is the cold species, whereas it may be that the former was adapted to woodland and the latter to open vegetation. There may well have been two forms of the latter, one adapted to the cold tundra, the last survivals being the frozen Siberian mammoths, and the other to the warmer loess-steppe. If this is so, it could explain why both of these two elephant species are sometimes found together. For example, in the Ipswichian it is known that the forest opened up in the *Carpinus* (hornbeam) zone at the end of the late temperate zone III. Under these conditions both species of elephant might have co-existed, so that there would be no need to argue that the presence of the mammoth demanded a cold climate or that it had been wrongly recorded. Other animals may also have different species adapted to different vegetation, for example rhinoceros, or different forms of the same species adapted to woodland and tundra, for example reindeer (*Rangifer tarandus*).

The various interstadials associated with the Devensian or Weichselian glaciation have a fairly distinctive cold mammal fauna, though similar faunas are known from the previous Wolstonian glaciation. The main elements are the mammoth (*Elephas primigenius*), the woolly rhinoceros (*Tichorhinus antiquitatis* = *Rhinoceros tichorhinus*), the bison (*Bison priscus*) (pl. 41), the reindeer (*Rangifer tarandus*), the giant deer (*Megaceros giganteus*), the musk ox (*Ovibos moschatus*), and occasionally the wolf (*Canis lupus*) and the Arctic fox (*Alopex lagopus*). In Scotland, because of its near total glaciation in the last glacial period, and in Ireland, which was nearly completely glaciated at the same time and also separated from Britain except at periods of low glacial sea levels, virtually the only Pleistocene mammal faunas are of this type, though not with all the elements present (the woolly rhinoceros is not known in Ireland, for example).

Although this assemblage probably indicates tundra conditions, certain of the mammals tolerate different environments as well. The mammoth has already been mentioned. In addition the musk ox probably inhabited both tundra and loess-steppe, while there are woodland forms of both *Rangifer* and *Bison*. The identification of the critical forms often requires the location of certain parts of the skeleton, usually the head and teeth. Isolated bones, especially when they are worn, cannot always be identified specifically.

It may be that there is hope in the future of the quantitative study of small vertebrate remains being integrated into the closely-stratified modern type of Pleistocene analysis. Reasons of size virtually forbid this in the case of most mammals, but in some deposits the macro samples, being searched primarily for seeds and snails, yield encouraging quantities of small vertebrate remains. The spines of the three-spined stickleback, *Gasterosteus aculeatus* (pl. 39, Q and R), are beautiful, readily-recognized objects, and such things as pike teeth and perch scales have been recorded and photographed. Fish teeth are fairly common in some deposits. In a preliminary manuscript report on small vertebrates found in the Ipswichian interglacial beds at Wretton, Mr J. N. Carreck lists a shrew, a lemming, several voles, a mouse, ?lizard, grass snake, viper, frog, pike, tench, chub, roach, rudd, ?eel and two species of stickleback. When it is remembered that these were all extracted incidentally by a non-expert in his search for seeds and snails, it will be realized that an expert on small vertebrates might recognize considerably more than these. Further, they were discovered at many different horizons so that a continuous picture of ecological conditions might be built up on them and compared with that derived from plants and invertebrates. Undoubtedly there will be difficulties at first through lack of ecological knowledge and also probably through lack of a wide range of reference material, but these difficulties have beset the investigation of every group of creatures. The amount of material recognized will certainly increase with experience.

NON-MARINE MOLLUSCA

The general methods used in dealing with the interpretation of groups of small animals in Pleistocene deposits can be well illustrated from the land and freshwater Mollusca. There is a long history of research into these: it probably stemmed originally from the cabinets of magnificently sculptured and coloured marine shells, via the handsome, modern non-marine Mollusca to the bleached, broken, corroded relics found in Pleistocene deposits in association with man's implements and the bones of the large mammals (pl. 42).

Most of the old studies were in terms of faunal lists collected from sections which probably represented very often long periods of time. They were often only records of presence with, at best, crude indications of frequency; two species may have been recorded as rare and common but it remains concealed whether this meant 50 of one and 500 of the other, or 1 of one and 10,000 of the other. Nevertheless, these old records are an invaluable basis especially in reconstructing broad-scale aspects of such things as the total faunas of interglacials.

Non-marine Mollusca have advantages and disadvantages as means of interpreting the Pleistocene. These may be either concerned with the technical aspects of preservation, recovery and identification or with the detective work of interpretation.

One of their main advantages is that they occur in a great variety of deposits but are usually best preserved in fine-grained sediments, such as alkaline peats, detritus mud, clay, silt and fine sand. They are well preserved in calcareous sediments, irrespective of whether these are preserved in anaerobic (waterlogged) conditions or not. On the other hand, they are not common in acid environments and they are quickly leached out of non-calcareous sediments. Thus their distribution commonly overlaps that of plant remains, but plants without snails may occur in acid conditions and snails without plants in oxidizing conditions. There is enough overlap to check the interpretations provided by the two groups against each other, so that one can then more confidently use interpretations based on one group where the other is missing.

When Mollusca occur they are found usually in great numbers, probably greater than those of the other mainly used groups, so that they lend themselves to quantitative work.

The shells are fragile, but as they do not bend or stretch appreciably there are no distorted specimens. Although only broken shells may be found, they have their original shapes and, unless badly corroded, their original ornamentation, or at least enough of it to be diagnostic.

The identification is facilitated by the fact that there are manuals of the mollusc faunas of most European countries often with good illustrations, for example the works of A. E. Ellis in Britain, and plenty of reference material in museum collections, in addition to a number of specialized journals dealing with the animals (for example, in England we have the *Journal of Conchology* and the *Proceedings of the Malacological Society of London*).

Samples for the extraction of Mollusca should be of the order of 1–2 kg in weight and ideally from intercepts of not more than 10 cm either in open section or as parts of a core. They should preferably be the same samples that are used for the recovery of macroscopic plant remains and other animals.

It is usual to soak most samples in a bowl or basin of water and allow the sediment to disintegrate, if it will of its own accord or else with gentle stirring. Most sediments in the fine sand to silt range will do this readily, especially if they have been dried out first. Most purely organic deposits can be disintegrated by stirring. The clays are more difficult because of the way in which the particles are flocculated. Dispersal can often be aided by solutions of sodium or potassium hydroxide or calgon, though even with these fairly vigorous agitation may be needed. After this the sediment is washed in a sink through a sieve: for non-marine Mollusca it is usual to sort down to a mesh size of 0·5 mm, though material smaller than this, down to 0·25 mm, should be retained for inspection, especially if other groups of fossils are being recovered as well. It is always difficult to decide how vigorously the material should be swirled in the water, particularly in the case of very clayey sediments. Undoubtedly, the rougher the handling the greater the number of broken specimens, but in fine sediments many specimens are already crushed and the vast mass of fragments one gets is largely made up of these. One has to balance one's ability to identify fragments against the desirability of having the largest possible number washed out.

For Mollusca the material is then gently dried: we use an aluminium sheet over a box containing a soil-heating element. Gentle drying is necessary so that the seeds in the samples do not shrivel and shrink more than is necessary, as they would if oven-dried. It may take two or three days for a fine-grained, organic residue to dry out. When the material is dry it is sieved through 2 mm, 1 mm, $\frac{1}{2}$ mm and $\frac{1}{4}$ mm sieves, as it is difficult to sort material of very varying grain size. The coarsest fraction is inspected through a hand lens on a black glass plate. The other fractions are sorted systematically under a low-power binocular microscope by sprinkling the dried material thinly on a flat glass dish ruled into strips the width of the microscope field. This enables an almost total recovery of identifiable material to be made.

This method of sorting is ideal for Mollusca. It suits some investigators of seeds: others like to sort their seeds wet. It is bad for delicate plant material, such as mosses, which tend to become shredded and destroyed. It would do adequately for marine Mollusca, but not for Foraminifera, which are smaller in size and in the study of which it is customary to go down to about 125 μm for sorting purposes. The range of sieve size suitable for non-marine Mollusca is also suitable for beetles and weevils though the various parts of individuals are separated by this method. If they can be identified and statistical studies are being used this probably does not matter. The other method in felted organic sediments is to split the material along the bedding planes and pick out the individuals, though this has the disadvantage of introducing bias into the tally of the specimens recovered.

Precautions have to be taken not to count the same individual twice. This means that the counted part must include a characteristic feature of the shell, usually the apex, but in the case of some species the aperture. One should count one individual per two valves of the bivalve shells, but these are so easily crushed and rendered unrecognizable that we record the number of valves as individuals. It does not really matter because the aim is comparability between samples. It is probably a vain hope to try to estimate the original composition of the population, simply because of the fact that preservation is so variable.

For a start, there are those Mollusca, the slugs, which have virtually no shell or merely very degenerate shells. The common genus, *Arion*, has merely a few calcareous granules. There are plenty of records of *Arion*, but whether the granules so named are really from the slug is doubtful in view of their astonishing frequency at times. With most other slugs the shells are often recovered and can sometimes be approximately identified. Snails with thin shells composed of a few whorls, such as *Succinea* (pl. 42C) and *Vitrina*, are readily destroyed, as are the thinner bivalves, *Anodonta*, *Sphaerium* and some species of *Pisidium*. Big shells are more readily crushed in sediment than little shells. Very elongated shells are more often broken than small compact shells. In fact the thicker the shell material and the more nearly spherical the shell, or the more nearly spherical the cross-section of the whorls, the better the preservation. So, what one achieves is comparability between samples sorted by one operator, and near-comparability between samples sorted by different operators with comparable degrees of expertise.

But this process of selection, in spite of its defects, is much better than

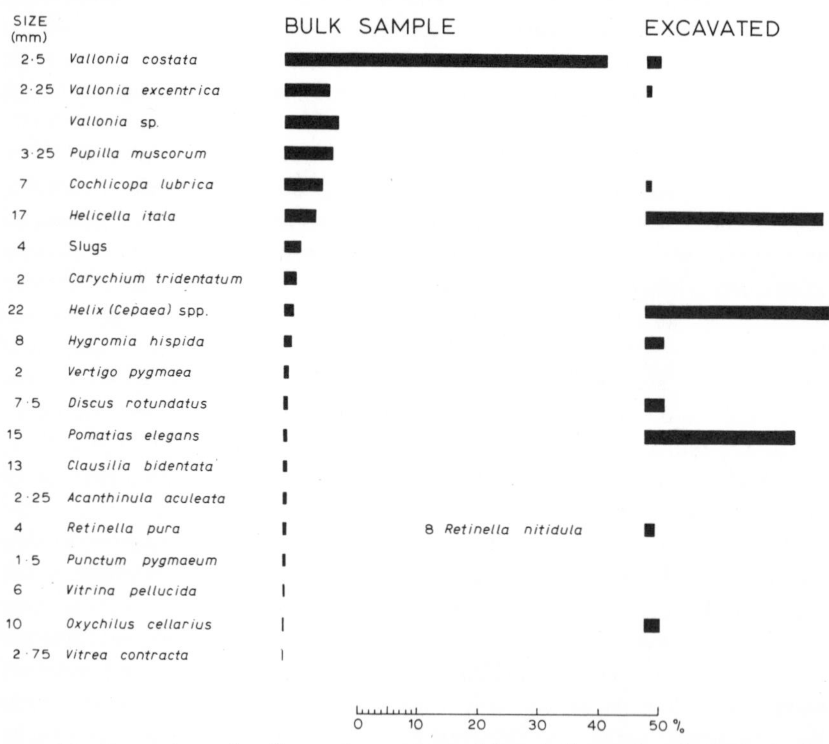

FIG. 7.1 Comparison of molluscan faunas obtained from bulk sample and visual selection from excavation, Bronze Age barrow, Arrton Down, Isle of Wight (after Sparks, 1961, fig. 1).

visual selection of snails from sections in the field, a procedure which leads to gross distortion of the fauna present. An example of the differences between bulk sample sorting and visual selection is shown in fig. 7.1. The importance of size in visual selection is overwhelming. The only apparently large species under-represented in visual selection is ***Clausilia bidentata***, but this is a long, spindly shell usually found in fragments. Further, it is dark brown in colour when alive and weathers buff, the same colour as the loam. Colour probably also explains the complete absence of ***Pupilla muscorum*** from the selected sample, for this species weathers to a similar colour.

The usual form of diagram drawn up from the counting of identified specimens is a percentage frequency histogram, although at times it is interesting to use absolute frequency diagrams in conjunction with percentage frequency ones. For example, if Mollusca are reoccupying an area we can assume in an ideal example that perhaps 50 of one species only are recovered from an earlier part of the deposit, while at the next level there are again 50 of that species and perhaps 150 of another species which occupied different parts of the environment when conditions suited it. On a percentage frequency diagram species A would obviously show a decline from 100 per cent to 25 per cent and the fact that it had not decreased absolutely would be obscured. Again absolute frequency, provided that the samples are of the same size, reflects climate: milder climates suit Mollusca and milder climates mean more complete vegetation and hence less soil movement and slower rates of accumulation. More Mollusca are associated with slower accumulation, thus leading to peaks in curves. There are, of course, other factors to be considered, so that each section needs to be interpreted on its own merits.

Three questions arise in the interpretation of any group of fossils:

(1) Are they useful indicators of age?
(2) Are they useful indicators of climatic conditions?
(3) Are they useful aids in interpreting the local environment?

They are less than perfect as indicators of age. This problem can be approached in a number of ways. The first of these is an extension of the method originally used to divide up the Tertiary period, namely by defining the interglacials in terms of the percentage of their species which are now extinct in Britain. On this basis, the Cromerian has about 21 per cent extinct, the Hoxnian about 17 per cent, the Ipswichian about 15 per cent and the early part of the Postglacial about 5 per cent. But any single deposit may show values for extinct species varying from nil up to figures above the values quoted, so that a Hoxnian deposit, for example the type site at Hoxne, may show a smaller percentage of extinct species than several good, rich postglacial sites.

We may also find that certain species are confined to particular interglacials. For instance, *Nematurella runtoniana* is apparently confined to the Cromerian and *Viviparus diluvianus* to the Hoxnian. But the number of deposits known from each interglacial is small and the hypothesis can only be held tentatively and may have to be abandoned in the face of stronger evidence of age of another type.

The same is also probably true of faunas of particular type. The best-known example is the loess type fauna with *Pupilla muscorum*, *Vertigo parcedentata* and *Columella columella* (pl. 42G). But this has been found under the same conditions associated with more than one glaciation, although for years it was regarded as a typical last glaciation cold fauna in Britain. It is known from this country, the Netherlands, Germany and Czechoslovakia in beds which are not Devensian (Weichselian) in age. There is another fauna found early in interglacials in which the freshwater species, *Planorbis laevis* and *Planorbis crista*, are dominant, but this has been found already in beds which are Hoxnian at Hoxne and Hatfield, Hertfordshire, and Ipswichian at Selsey and Wortwell, Norfolk.

These comments do not necessarily apply to short-distance correlations. For example, it is possible sometimes to zone in isolated parts of a section by using the spectrum of Mollusca, for example the lateral samples at the Hatfield sections (fig. 7.2). It may even be possible within small river basins to guess which beds are related, but very often here one has a position in a terrace sequence as evidence as well as the Mollusca. This can be done fairly readily, for example, with the Barnwell Terrace of the river Cam. It can also be done with a series of very similar deposits in similar geographical situations, such as the chalky washes studied by Kerney (1963) on the North Downs.

State of preservation is virtually useless as a potential age detector. Cromerian shells preserved in anaerobic organic deposits may appear a lot fresher than bleached shells on a limestone hillside that are a few decades old at most. Some of the latter may even still be alive, as we once found when we washed some dried, bleached shells from a hillside in Jordan only to find them crawling about and eating the label pot in with them, apparently in protest at a wrong provisional identification.

As indicators of climate Mollusca are far from ideal. In the first place it is difficult to establish the climate from freshwater faunas because this involves inferring the climate from what can be inferred about water temperatures. Secondly, many of the species have very wide climatic tolerances, for example from Sicily or even North Africa to well north in Scandinavia. If one uses single indicator species for climatic deductions, an ecological change eliminating that species might easily be interpreted as a climatic change. It has proved possible to use climatic groupings

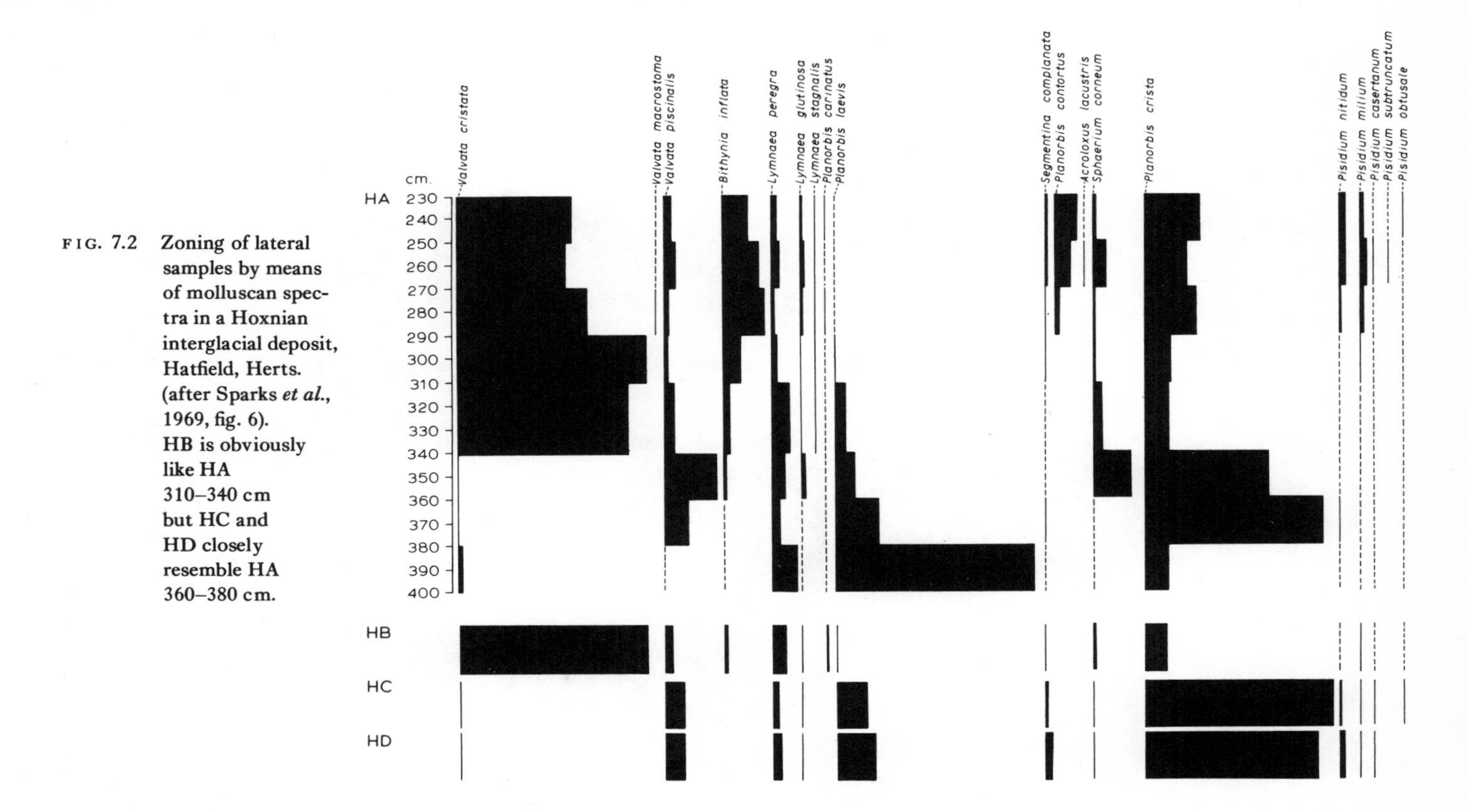

FIG. 7.2 Zoning of lateral samples by means of molluscan spectra in a Hoxnian interglacial deposit, Hatfield, Herts. (after Sparks *et al.*, 1969, fig. 6). HB is obviously like HA 310–340 cm but HC and HD closely resemble HA 360–380 cm.

based on the northern limits in Scandinavia. Four such groups have been used:

(A) Species reaching to, or almost to, the Arctic Circle.

(B) Species reaching approximately 63°N.

(C) Species reaching 60–61°N., i.e. approximately the northern limit of the oak.

(D) Species whose northern limits are outside, or in the very south of, Scandinavia.

Histograms of these groups have been used through the second part of the Ipswichian interglacial at Histon Road, Cambridge and Stutton, Suffolk (fig. 7.3). It will be noticed that the really tolerant species, group A, are well-represented throughout, thus demonstrating the comparatively small nature of the climatic difference between interglacials and the present, while the successively less tolerant groups are increasingly confined to the milder, middle parts of the interglacial. There are, of course, irregularities in detail, so that general trends must be studied.

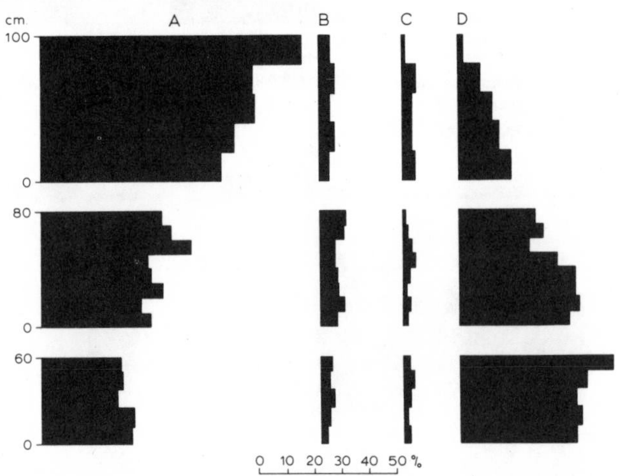

FIG. 7.3 Analysis of distribution groups of non-marine Mollusca in an Ipswichian interglacial deposit at Stutton, Suffolk (after Sparks, 1964, fig. 1). For explanation, see text.

One must beware of placing too much climatic interpretation on single specimens of warm species, even very southern species such as *Belgrandia marginata* (pl. 42F) and *Corbicula fluminalis*, because these are known to occur in quite cold conditions accompanying the onset of glaciation, whether as survivals or as contaminants is not yet certain.

Similarly, it would be unwise to build up an ideal pattern of molluscan response to climate in some such terms as the following. In the early part of an interglacial the mollusc faunas will lag behind the plants in reflecting the improving climate because their likely rate of colonization is slower. So much depends on local conditions as Kerney's work (1963) on the North Downs has shown. Whereas the succession of mollusc faunas in the early part of an interglacial represents immigration rate, the succession in the second part represents survival potential. Thus, it might be theoretically argued that the mollusc fauna curve should lag behind the true climate curve. When we get to the glaciation it would be nice to be able to believe that before the first phase of the glaciation the fauna would be dominated by hardy species, but that less hardy ones should survive in small numbers. After the first phase of the glaciation, the interstadial should be characterized by a very hardy, restricted fauna with none of the less hardy species surviving so that an interstadial fauna ought to be distinguishable from an Early Glacial fauna. Unfortunately, this simple model is not yet proven.

Although few ideal schemes seem to last long in the face of geological evidence, certain climatic inferences clearly emerge from the study of non-marine Mollusca. Interglacial faunas, rich in number of species and number of individuals, are usually readily distinguishable from restricted cold faunas dominated by large numbers of very few species. Even here, it must be remembered that adverse climatic conditions can produce faunas very similar to those produced in adverse local conditions, such as poorly-oxygenated swamps or arid hillsides, so that all chance of confusion is not eliminated. Finally, by the use of climatic or distributional groups the general trend of an interglacial section may be revealed.

Mollusca are probably most revealing in the interpretation of local environmental conditions. Again, one is faced with a wide range of tolerance among many species, usually among those with the widest climatic tolerance. On the other hand, some species are very restricted, for example *Pomatias elegans* to dry, friable, calcareous soils in which it burrows, and *Ancylus fluviatilis* to moving water preferably with a stony bottom.

For the determination of local environments the Mollusca can either be treated as single species or as groups. Wherever possible, it is probably better to treat them as single species in the construction of frequency diagrams. This is only really practicable where the general environment remains approximately homogeneous for a reasonable period of deposition. In these conditions one is dealing with pretty well the same species throughout and not with greatly changing patterns. One must be able to recognize the likely homogeneity of environment from other evidence, e.g. the position of the deposit in the present landscape. Thus, in Kerney's work (1963) on the Late-Devensian and postglacial deposits present in dry valleys in the North Downs, it is apparent from the lithology and attitude of the deposits that their general conditions of deposition must have been as hillside washes in dry Chalk valleys. With this knowledge it is instructive to graph each species separately, because one can then see which species react in the same way to environmental change, whereas when one groups species one merely hopes that one includes in the groups creatures which react in similar fashion. The use of single species can also be adopted sometimes in interglacial deposits, for example in the deposits of the Hoxnian pond investigated near Hatfield (fig. 7.2). The interpretation of the changes through this deposit may be made by considering changes which occur together at given levels.

Yet with many interglacial deposits the total form of the deposit is unknown, especially when it has been revealed by boring or coring. In the case of the large, very mixed fauna often found in such deposits, one's interpretation is correspondingly coarser and the use of mollusc groups necessary. In interglacial deposits one is usually dealing primarily with freshwater molluscs and these are grouped as follows:

(1) Slum species, which can stand very poor conditions of periodic drying, poor aeration and probably large temperature change typical of small dirty bodies of water. Most of these species will of course tolerate better conditions as well.

(2) Catholic species, which are found in almost all freshwater environments except real 'slums'.

(3) Species characteristic of slowly-moving, clean, plant-rich streams, originally called ditch species.

(4) Moving water species; the water movements are not necessarily rapid and may be caused either by winds or currents. On the whole larger ponds and streams are involved than for group (3) species.

Of course, there is often considerable overlap, because all environments may be represented by dead Mollusca deposited by a stream and the study of the groups really gives only a generalized picture.

At the same time the land Mollusca may also be grouped. Because they are much rarer in interglacial deposits, it is not always possible to do this satisfactorily and different groups may be separated for specific purposes, for example, marsh and associated species, dry land or xerophilous species, open ground species which can be either dry land or marsh, and 'woodland' species, a rather vague and ill-defined group.

An example of the use of such groups for the continuous interpretation of a deposit is provided by the Ipswichian interglacial deposit at Histon Road, Cambridge (fig. 7.4). The groups of freshwater Mollusca are numbered as they are above while groups 1 and 2 of the land Mollusca are respectively marsh and associated species, and dry land species.

This deposit can be divided into several sections. At the very bottom of the section illustrated there is a dominance of slum freshwater species and marsh land species, a typical combination probably representing a riverside marsh. From A to B the groups of freshwater molluscs reach peaks successively until there is a dominance of the moving water group at B. This seems to indicate progressive changes towards river conditions. From B to C the catholic and moving water species dominate the freshwater molluscs. This is a common and logical combination. They are accompanied by dry land species rather than marsh species, again a common phenomenon, because,

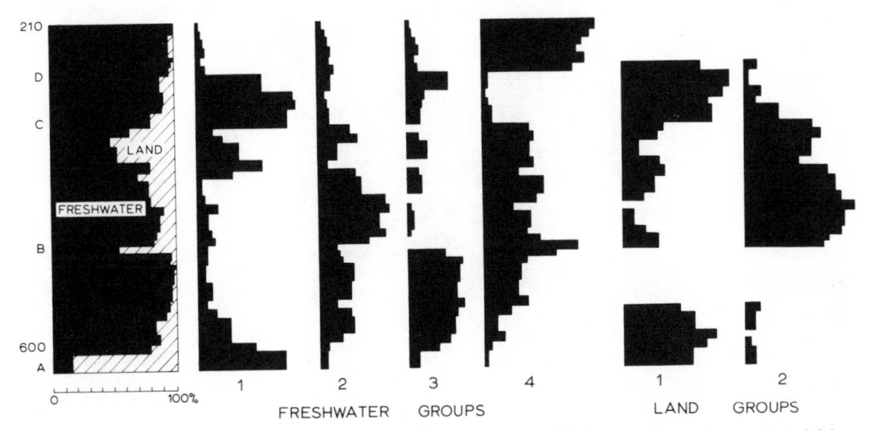

FIG. 7.4 Analysis of ecological groups of non-marine Mollusca through an Ipswichian interglacial deposit at Histon Road, Cambridge (after Sparks, 1961, fig. 5). For explanation, see text.

when stream conditions are fully established, they will transport dry land Mollusca, which soil creep carries into them higher in their courses, as readily as the marsh species from the sides of the stream. Starting at C and lasting to D, there is an abrupt reversion to marsh conditions, shown very clearly by both the slum group of freshwater species and the marsh group of land species. At D there is an equally dramatic reversion to the moving water group. All this is consistent with deposition in an aggrading floodplain consisting of a variety of riverside and stream environments. If one graphs the open ground molluscs of this deposit, chiefly species of *Vallonia*, they confirm the openness of the environment demonstrated by a study of the plants.

Another very useful purpose served by Mollusca is the detection of brackish influences and the changes to general brackish conditions. This is usually done by using *Pseudamnicola confusa*, *Hydrobia ventrosa* and *Hydrobia ulvae*, here ranged in their apparent order of tolerance of salt. The complete change to brackish conditions in the Ipswichian interglacial at Selsey, Sussex, is quite clearly shown (fig. 7.5) by the tremendous rise of *Hydrobia ventrosa* (pl. 42A) followed by an increasing importance of *Hydrobia ulvae*.

Temporary slight manifestations of brackish conditions are revealed in the Ipswichian interglacial deposit at Wretton, Norfolk (fig. 5.14). Such incursions are very important when we are trying to discover where sea level stood at different times during interglacials. Unfortunately, they are very rare.

FIG. 7.5 Brackish Mollusca in an Ipswichian interglacial deposit at Selsey, Sussex (after Sparks, 1961, fig. 2).

One could continue to illustrate the interpretations of climatic and local conditions made possible by the study of non-marine Mollusca. But we have perhaps already devoted to them more than their fair share of space. There are reasons for this: they are the group of animals most familiar to us, they have been studied for a long time and they reveal a surprising amount of information. They also illustrate fairly well the general procedure in dealing with groups of small Pleistocene animals.

MARINE MOLLUSCA

Many marine molluscs are large and noticeable fossils, so that they have been recorded from Pleistocene sections over a long period, but, as with their non-marine relatives, many of the early records were of presence and absence or in rough measures of frequency from whole sections or major divisions of sections. The problem here was similar to that with non-marine Mollusca, namely that a much more detailed stratigraphy and recording had to be adopted to enable the study of these animals to be integrated with botanical work.

Most of the Middle and Upper Pleistocene deposits, from the time of the Cromerian Upper Freshwater Bed onwards, are either freshwater or, much more rarely, brackish. Below these beds lie the East Anglian Crags, originally classed as Pliocene and extensively studied in the latter half of the last century and the beginning of this in general terms. These beds are largely marine so that the study of both marine Mollusca and Foraminifera (see below) comes into its own as an adjunct to botanical studies. It is most useful having alternative sources of information, because, under marine conditions, certain distortions tend to appear in pollen diagrams.

The re-investigation of these deposits using modern techniques started yielding results in the early 1960s and a deep boring at Ludham in Norfolk, which passed through 55 m of Crag deposits, was financed by the Royal Society. The vegetation of these deposits was analysed by West (1961), who showed the existence of two cold and three temperate phases. The same cores were used by Norton (1967) in an analysis of the marine Mollusca present.

The problems involved were similar to those involved in using non-marine molluscs from cores. The amount of sediment was limited and much of the material was fragmentary. This involved precautions in counting and also difficulties in identification. In work such as this there is usually far more difficulty with the first study or two because the broken material that one wants for comparison is not readily available. Museums in the past have kept the best specimens, which were in many cases the only ones recovered and which their curators are usually unwilling to have broken so that one can compare pieces of shells. It is not satisfactory to compare fragments with the corresponding area on an unbroken shell, and, indeed, with fairly tightly coiled gasteropods, it cannot be done. So that until the investigator has built up a reference collection of broken and juvenile specimens or accumulated enough fully-grown ones for him to break as necessary, progress must be slow and tedious.

It proved possible at Ludham to characterize the depositional environment mainly in terms of different littoral zones, related primarily to water depth, nature of the bottom and the approximate salinity. The interpretation of the climate from the marine Mollusca was less satisfactory, just as it often is with the non-marine Mollusca. A climatic interpretation was attempted by analogy with the known times of immigration of marine molluscs into the Oslo Fjord in southern Norway in the postglacial. It was found that at all levels there were species which had only immigrated into the Oslo fjord in the climatic optimum of postglacial time. On this basis it would be necessary to conclude that the climate never really dropped below that degree of mildness, a conclusion at variance with those drawn from both the plants and the Foraminifera.

Norton (1967) pointed out that there were some peculiarities in that the Ludham borehole horizons contain species which do not overlap in their distributions today, so that the meaning of the associations is obscure. Perhaps there was a much greater seasonal range of sea temperature at the time, or perhaps ecotypes of widely distributed species were produced which were able to withstand temperature conditions not tolerated by the present forms of the species. Meanwhile, the more widely distributed Foraminifera and the plants were seen as the better climatic indicators, with the marine Mollusca contributing a lot to the understanding of local conditions.

The conclusions drawn by Norton from the other Lower Pleistocene deposits he studied are consistent with those drawn from the Ludham borehole, which represents the longest sequence.

Certain later interglacial deposits also contain marine horizons, for example the Hoxnian deposits at Clacton, Essex, and in the Nar valley in south-west Norfolk, and the Ipswichian deposit at Selsey, Sussex, and the presumed Ipswichian March gravels in the Fenland. Careful analysis of the marine Mollusca of these deposits might yield information about the depth zone of the Mollusca and hence about the likely sea level. Certain glacial deposits contain marine Mollusca, for example some of those of the Cheshire Plain, but these must represent either sections of the sea floor ploughed up by the ice sheets or disturbed interglacial deposits, because the sea level must have been low in glacial periods. One of the most problematic of these deposits is the Corton Beds which occur just north of Lowestoft in Suffolk; it is still doubtful whether the marine molluscs in these sands are in place or have been deposited secondarily by an ice sheet (see Chapter 5, p. 146).

FORAMINIFERA

Foraminifera have been mentioned in dealing with marine Mollusca. These small invertebrate animals have been recorded from time to time in British Pleistocene deposits, for example by Macfadyen (1932) and Funnell. Their main use up to the present has been in the analysis of deep-sea cores, which may be the only deposits in the world retaining a full Pleistocene sequence.

They are potentially very useful, because their range of habitat is brackish to fully marine and they are fairly sensitive to temperature changes. Thus they may tell us something about the marine and brackish horizons where they occur, for example in the Hoxnian deposits at Clacton and in the Ipswichian deposits at Stone, Hampshire, and in the March gravels, and so lead along yet another line of evidence to that very important topic, contemporary sea level.

The most intensive use of Foraminifera so far has been in the study of the Lower Pleistocene of East Anglia by Funnell (1961, 1962), who has used them quantitatively. The technique of investigation is in essence similar to that used for Mollusca of all types, but, because of their smaller size, Foraminifera are usually recovered from smaller samples. The standard sample size is 100 g, i.e. one-twentieth of that of a standard mollusc sample. Thus it should be possible to analyse the Foraminifera in shorter core sections, in thinner cores and in less rich samples than is possible with Mollusca. The material, if clayey, requires dispersion: boiling in hot sodium carbonate or the use of dilute hydrogen peroxide is recommended. The material is sieved basically through a 240 mesh sieve (the openings in this are 65 μm), because most Foraminifera fall in the size range 125–1000 μm.

By studying percentage frequency of the species Funnell was able to show virtually the same succession of colder and warmer periods at Ludham as those confirmed by the pollen analyses from the later Royal Society borehole. The analysis of the Foraminifera was based upon another bore, located very near the Royal Society borehole and termed the Ludham Pilot Boring. Later, Foraminifera were used on another Lower Pleistocene site at Easton Bavents and on a probable Hoxnian deposit from the Inner Silver Pit off the Lincolnshire coast. In all cases the Foraminifera stand out as temperature indicators, and are probably less useful both for local environmental conditions and for assessing the age of the deposits, though the latter may be possible in broad terms.

Thus these groups of animals so far discussed supplement each other. The non-marine Mollusca are very abundant and widespread, very good for assessing local conditions but probably less reliable for assessing climate; marine Mollusca seem very good for the determination of local conditions where other lines of evidence are much inferior; while Foraminifera are very useful for the establishment of climatic sequences in marine deposits and can also be used to differentiate low salinity and inshore deposits, as at Stone, Hampshire.

INSECTS

Practically all the work of significance on Pleistocene insects so far has been concerned with Coleoptera, the beetles and weevils. They have been recorded from Pleistocene deposits for a century or more, rather like the non-marine molluscs though not in such large numbers, but the last fifteen years has seen their study put on a firm and detailed stratigraphical and quantitative basis. In this sphere, the Geology Department of Birmingham University under Professor Shotton has been outstanding, with Coope (1961, 1962, 1965 and 1967) contributing a whole series of important and detailed papers on the Coleoptera of various deposits.

The technique of sampling and analysis is not all that different from that used for the recovery of non-marine Mollusca as the size range is very similar for both. The various parts of beetles tend to become separated after death and especially after washing and sieving, so that one finds separate heads, thoraces and wing cases (pl. 39V). This does not matter provided that the species are identifiable from these fragments and provided that suitable precautions are taken to record the minimum number of individuals that the fragments represent.

As an alternative it is possible to sort through such sediments as felted or laminated peats by peeling layers off and running over exposed surfaces with a low-power binocular microscope. By this method it is possible to recover more related fragments of individuals, probably at the expense of introducing a strong bias into the sample. The worker interested in historical biogeography would probably not mind the bias; the Pleistocene worker interested in reconstructing the past from the assemblage of beetles would probably not be prepared to tolerate it unless he already had an analysis through the same deposit of frequency on as objective a basis as possible.

Beetles possess advantages and disadvantages in Pleistocene studies. The first advantage is that they are fairly common and that they also occur in a variety of deposits as they are well-adapted to preservation. They are probably not as common as seeds and fruits or as non-marine Mollusca: they are far less common than ostracods can be, though very much more useful.

Like the non-marine Mollusca there is little sign of species evolution in the Ice Age. In this they are very different from the vertebrates. This is a disadvantage in that it makes beetles unsuitable for zoning the Pleistocene, unless they should prove to have different patterns of occurrence in different interglacials, which is the basis of correlation by pollen zones. There is no evidence yet that this is a real possibility, and in this the beetles are again rather like the molluscs. The lack of evolution means, on the other hand, that deductions about past environments can be made much more safely because in every case we have a modern distribution on which to base such conclusions. Where the number of extinct species is large, it becomes correspondingly difficult to deduce the conditions under which they lived: it has to be done by analogy with presumed near relatives.

Many species of Coleoptera have well-marked preferences for particular environments and this characteristic is ideal in interpreting past conditions. In this respect beetles seem much superior to snails. These controls vary considerably. In some species it seems that certain thermal factors, such as whether the summer is long and hot enough for the species to complete a life cycle or arrive at a state in which it can hibernate, are paramount. Moisture factors may also be very important, not only in differentiating those species which spend a considerable part of their life cycle in water but also in separating the hygrophiles from the xerophiles among the land beetles. Chemical effects, such as pH values of acidity or the availability of minute quantities of salt, may have significance. In addition to these there are species of beetle whose location is governed by food factors: they may feed on certain types of plant matter; they may be predators on specific lower animals; there are carcass beetles and dung beetles. The presence of such specialized species is very useful in deducing environment.

In the elucidation of climate they again possess many advantages. A number of species seem to have very limited ranges at the present time: most useful are those with restricted ranges in north Scandinavia, for their presence in deposits is a clear indication of the climate which is presumed to have prevailed then. Fortunately, the Scandinavian beetles have been monographed by Lindroth and their distribution analysed in terms of the vegetation zones of the Scandinavian highlands. This is most useful in interpreting British fossil faunas.

It has also been claimed that, because they are flying insects, they are capable of colonizing new areas as soon as the climate becomes suitable and the beetle curve, as it were, will bear the closest resemblance to the climatic curve. It may, in fact, reflect the climate more closely than the vegetation curve. This is unproven, of course, and will probably remain so. It is more likely to be true with sudden changes of climate and can only involve beetles not dependent in any way on particular plants. With slower changes of climate it seems unlikely that plant colonization will not be able to keep pace with the rate of climatic change. It could be true if the series of vegetation zones has been caused not only by climatic changes but also by the slower development of successional communities.

To illustrate the possibilities of beetle research, two sites will be taken, one glacial and the other interglacial. A section of deposits dating from the middle of the Ipswichian interglacial was exposed in building foundations in Trafalgar Square and in the material Coope (1967) discovered a large beetle fauna. This included a very high proportion of dung beetles, the habitats of which were probably provided by the hippopotamus, elephants and lions, whose bones were found in the same deposit. Further, these dung beetles are scarab beetles and not species of the genus, ***Aphodius***, which dominates British dung beetles at the present day. Scarab beetles are typical dung beetles of central and southern Europe. So, this one group of beetles ties in closely with the climate and ecology suggested by other evidence, vertebrate, plant and molluscan.

At the opposite end of the scale is the Full Glacial deposit at Fladbury in Worcestershire. For a beetle fauna, the Fladbury fauna is not very large, 47 species, but it was derived from one block of peaty material with a dry weight of only about 1 kg. The gravel pit from which the material was derived has yielded bones of the mammoth, woolly rhinoceros and possibly bison, and the beetle fauna was fully consonant with this vertebrate fauna. Of the 47 species of beetles only one is not found north of the Arctic Circle at the present time, though many of them are, of course, found farther south as well. Eleven of the 47 species are now extinct in Britain and a further 8 are confined to the north of Britain. All of the extinct species are either characteristic of northern Scandinavia and Russia at the present time, or characteristic of these regions and the high alpine regions of central Europe as well. These facts speak for themselves.

It appears then that beetles differentiate the climate and certain aspects of the local environment more definitely than do the Mollusca. So far most of the detailed work on British Ice Age beetles has been on deposits of Devensian age especially in the Midlands, but interglacial faunas are known and no doubt work will proceed on them.

Although we may expect advances in knowledge of Ice Age zoology, the work is so time-consuming that the rate of advance may well be slow. Further, there are not all that many fossiliferous deposits known from some parts of the Early and Middle Pleistocene, and the discovery of new ones is often a matter of chance.

8

Man in the Ice Age in Britain

The evolution of man as a member of the world's fauna took place against the background of the changing environment we have already described. This evolution is shown by the finding of skeletal remains as fossils, and of cultural remains (pl. 43) which have become incorporated into sediments. It is also shown by the effects of man's activities on the environment, effects such as forest clearance and domestication of animals, which leave their trace in the fossil record.

These anthropological and archaeological aspects of the evolution of man are intimately tied to the geological record, which will therefore give a chronological record of evolution of man and his culture and a record of the environments in which this evolution occurred. In this chapter we shall be concerned with the relationship between archaeology and environmental history, especially as it is seen in Britain. The development of this subject in the last hundred years or so is of great interest in showing the evolution of man's thought about his own evolution.

In pre-Darwinian times, in the first half of the nineteenth century, man was considered to be of recent origin. Thus, according to Archbishop Ussher's chronology man had been created in 4004 B.C. Two developments in the middle of the last century led to the displacement of such chronologies as Ussher's. One was Darwin's exposition of the origin of species in geological time by natural selection. The second was the finding of stone tools, which, it was egotistically announced, must have been fabricated by man, in deposits which were in their geological relationships evidently very old, and in deposits where the remains of extinct large mammals, such as mammoths, were also found. Stone tools had in fact been found long before this. Thus, at Hoxne in Suffolk, John Frere in 1797 recorded flint hand-axes which he said had been 'used by a people who had not the use of metals', who, on geological evidence, belonged 'to a very ancient period indeed, even before that of the present world'. However, in the climate of thought at that time such ideas were not acceptable, and it was not until the 1860s that Frere was proved right by further investigations by the geologist Joseph Prestwich and the archaeologist John Evans at Hoxne and in the Somme valley in north-west France.

At the same time the glacial theory of the Pleistocene origin of the drift was becoming accepted and controversy then set in regarding the age of man in relation to the glacial periods. Was man preglacial, interglacial or post-glacial in origin? With the development of the details of Pleistocene stratigraphy in the last half of the last century, it became recognized that man was at least interglacial in origin. There were then considerable efforts made to demonstrate that man was much more ancient, and had an origin in the Pliocene, by the finding of flint tools in Pliocene deposits in south-east England. Although many if not all of these tools (eoliths, fig. 8.6B) are now recognized as being the result of natural processes (such as frost-cracking) rather than human flaking, there is now no doubt that the earliest stone tools from Africa are of an age (*c.* 1·75 million years) which includes many of the deposits in which the controversial eoliths were found. Further, since that time the Plio-Pleistocene boundary has been redefined at a much lower level to take in cold climatic stages which occurred in preglacial times, so that the protagonists of 'Pliocene' man have been proved right, even though their evidence appears faulty.

With the development of detailed knowledge of Pleistocene stratigraphy and of dating methods in the last thirty years it has become possible to provide a far better chronological scale for the evolution of skeletal and cultural remains; and the development of techniques such as pollen and mollusc analysis has led to greater knowledge of the details of vegetational, faunal and climatic changes associated with this evolution.

In Britain there is a very considerable record of cultural remains through the Middle and Upper Pleistocene, but skeletal remains are very rare until late Flandrian time. Although we are concerned with Britain it will be necessary to refer to the more complete sequences known from continental Europe and Africa.

But before considering these aspects further, we should examine the stratigraphical contexts in which the remains are usually found. These vary very much. Artefacts and other cultural remains may be in their original place in habitation sites or they may be reworked into other sediments. Thus in Britain heavy stone tools such as hand-axes often occur in coarse gravels, and must have been redistributed from their place of origin. Their minimum age is then given by the age of the river gravel. On the other hand, if artefacts are found in fine lake muds or peats, they are almost certainly contemporaneous in origin with the organic deposits, having been introduced into the sediment during the habitation activites which produced them. Such a geological context for a culture is very satisfactory, for it offers a good basis for dating, through radiocarbon assay of the organic components associated with the habitation site or organic deposits, as well as the means of studying the environment through the fossils contained in the organic deposits.

In the later Flandrian, where habitation sites and man's activities become more complex, there are increased opportunities for the survival of remains in hearths, storage pits and burial monuments. Thus with a barrow, the old soil of the time of building may be preserved, as well as the cultural remains associated with burial. The monument itself may be covered by or be resting on terrestrial peat, again a stratigraphical context which will allow dating and environmental studies.

Open habitation sites such as those considered above contrast with cave sites, where the stratigraphy is often far more complex and difficult, for the reason that the environment of deposition varies very much locally and so does the source of sediment, in the form of stalagmite, sediment from outside the cave and from the cave rocks. The regional environment is more difficult to study under these conditions, especially in the absence of well-preserved fossils of the kind we find associated with open sites and organic deposition. Nevertheless a large part of the record of the Palaeolithic is from cave sites where good shelter was obtainable at the time, so that cave sites are an essential study.

Remains of early man

Britain boasts few Pleistocene skeletal remains of early man, but these must be put into the context of man's evolution as known from other regions. Difficulties of interpreting hominid evolution arise from the fragmentary nature of the fossils and from the difficulties of age determination of finds from widely separated sites. The fragmentary finds hinder the development of a clear taxonomy. Each new find of a skull or skull fragment may indicate to some a new species or genus, to others a variation of subspecific rank. Hominids are distinguished from apes by their bipedalism (thought to be related to the change in habitat from forest to open country) and larger brain case (fig. 8.1). It is usual to distinguish two genera, *Australopithecus* (fig. 8.1B) and *Homo* (figs. 8.1C, D, E, F). The former show an early form of bipedalism, with very small brains (*c.* 600 cc); possibly they used tools. The latter show advanced bipedalism, a gradual enlargement of the brain, the development of a power, and then a precision grip with the hand, and the manufacture of tools.

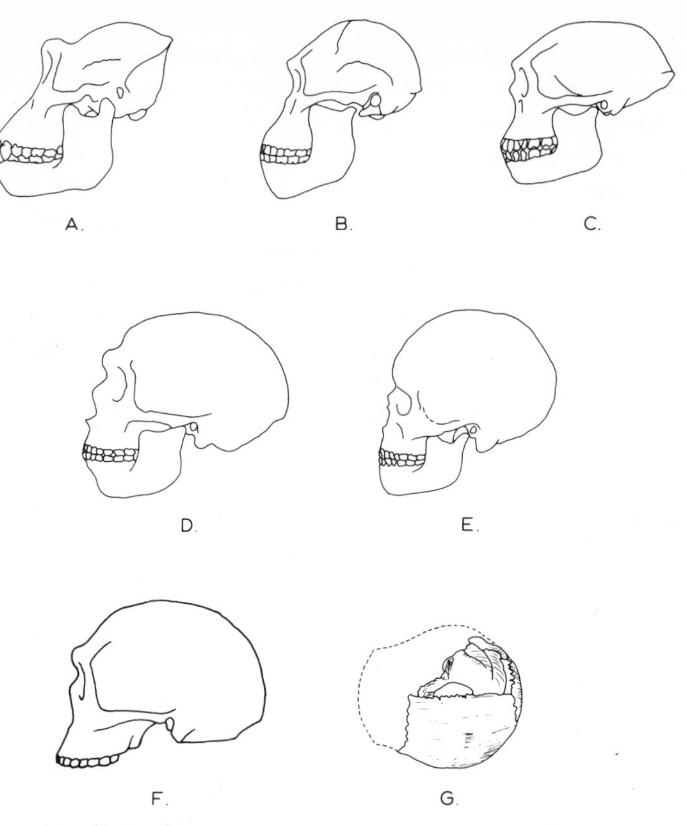

FIG. 8.1 Skulls of fossil man, man and ape. Approximately $\frac{1}{7}$ natural size. A, female gorilla (after W. E. Le Gros Clark, *The Fossil Evidence for Human Evolution*, Chicago, Ill., University of Chicago Press, 1964, fig. 13). B, *Australopithecus*; C, *Homo erectus*; D, *Homo sapiens neanderthalensis*; E, *Homo sapiens* (from J. K. Brierley, *A Natural History of Man*, Heinemann, 1970, after A. Hardy, *The Living Stream*, Collins, 1965). F, The Steinheim skull: the Swanscombe skull when complete was probably almost indistinguishable from this skull; G, Swanscombe skull bones, seen from above (after K. P. Oakley, in the *Proceedings of the Geological Association*, 1952, Vol. 63, figs. 8 and 9).

Nearly all the *Australopithecus* remains come from the Lower Pleistocene of South and East Africa, though they are also found in the Middle Pleistocene and in south-east Asia. The number of species is few, and the specific distinctions are a matter of opinion.

The genus *Homo* likewise is thought to contain few species, but with variation from site to site suggesting a strong element of subspecific differentiation. The oldest fossils believed to belong to this genus come from the

Lower Pleistocene of East Africa in Olduvai Gorge, at a similar level to where *Australopithecus* remains were found, together with pebble tools. A potassium–argon date of 1·75 million years has been obtained from this lower part of the sequence at Olduvai Gorge. The remains, designated *Homo habilis*, include skull fragments, a jaw, and parts of hands and a foot. The last suggest the possession of a power grip and bipedalism. A second species, *Homo erectus* (fig. 8.1C), comprises several finds from Asia and Africa and these are Middle Pleistocene in age. The skulls show an increase in brain size (950–1050 cc) over *Australopithecus*, but they also show a number of characteristics which make them distinct from our own species, *Homo sapiens* (fig. 8.1E). The skull is thicker and smaller with a low vault and frontal flattening, with a prominent occipital crest and brow ridges. *Homo erectus* remains have been found in China (Pekin Man), Java, North Africa and East Africa, where a skull was found at Olduvai Gorge associated with a hand-axe industry. The third species, *Homo sapiens*, comprises several subspecies from Europe, Asia and Africa, which show a considerably larger brain size (1300–1700 cc). The oldest find thought to be within the species is the Swanscombe skull (fig. 8.1G) from the Middle Pleistocene of the Thames valley, associated with a Lower Palaeolithic hand-axe industry. Our own subspecies, *Homo sapiens sapiens*, with the earliest occurrence in Europe at about 35,000 years ago, appears to have replaced at that time an earlier subspecies, Neanderthal Man, *Homo sapiens neanderthalensis* (fig. 8.1D). Neanderthal Man is characterized by the absence of a chin, by prominent brow-ridges, and other differences so that some taxonomists consider he should have specific rank. Remains of Neanderthal Man have been associated with Middle Palaeolithic flint industries.

Very little is known of the origin of the present races of man. Our own subspecies appeared in Europe some 35,000 years ago, and is associated with the Upper Palaeolithic, spanning the last half of the Devensian or Weichselian (last) glaciation. It appears that during this time, especially in the temperate regions freed from ice by climatic amelioration, man spread widely into new regions and from the Old to the New World. He was aided in this by cultural advances as well. In more equatorial regions where climatic change may not have been so marked, and the climate may have been more hospitable over a long period, *Homo sapiens sapiens* may have had a higher antiquity. The difficulty of investigating the origin of races from the geological point of view is that skeletal remains hardly show any of the characteristics associated with race.

Table 8.1 summarizes the geological relations of the genera and species we have discussed.

As far as Britain is concerned we are in a marginal position in relation to the centre of man's evolution in Africa. The earliest hominid remains and

Table 8.1. *Geological age of various human skeletal remains*

		Homo sapiens sapiens			Upper Palaeolithic
	35,000	——			
Upper Pleistocene		*Homo sapiens neanderthalensis*			Middle Palaeolithic
Middle Pleistocene	*c.* 200,000	*Homo erectus* (Pekin, Java Man, etc.)	*Homo sapiens* (Swanscombe Man)	*Australopithecus*	Lower Palaeolithic
Lower Pleistocene	1·75 million years	*Homo habilis*		*Australopithecus*	

certain stone artefacts are of Middle Pleistocene age, though as we shall see, there are controversial Lower Pleistocene artefacts.

We shall now go on to consider first the skeletal remains and then the cultural remains in the Pleistocene of Britain.

Skeletal remains

The oldest skeletal remains found in the British Pleistocene are the skull bones (fig. 8.1G) found in the Middle gravels of the 100-ft Boyn Hill Terrace of the Lower Thames at Swanscombe near Gravesend, Kent (fig. 8.2). Their age is Hoxnian, in the Middle Pleistocene, with the dating based on fauna and stratigraphy; further, on the basis of the vertebrate fauna it has been suggested that the gravels in which the fragments were found are late Hoxnian. The occipital and two parietal bones were recovered from the same site over a period of twenty years, 1935–55. They are of a young individual, 20–25 years, and probably female. The skull remains are similar to those of the Steinheim skull (fig. 8.1F), and they have been attributed to ***Homo sapiens steinheimensis.***

The find is important for several reasons. It is certainly dated to a parti-

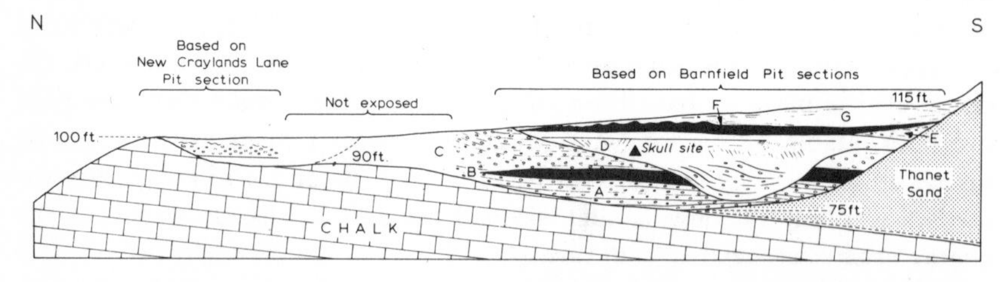

FIG. 8.2 Diagrammatic section across the 100-ft (30 m) Terrace at Swanscombe, Kent. A, Lower Gravel; B, Lower Loam; C, Lower Middle Gravel; D, Upper Middle Gravel (and sand); E, solifluxion wedge; F, Upper Loam; G, Upper Gravel (and clay). (After K. P. Oakley, in the *Proceedings of the Geological Association*, 1952, Vol. 63, fig. 4.)

cular stage in the Pleistocene, the Hoxnian, suggesting the antiquity of the *Homo sapiens* lineage. The skull bones are accompanied by hand-axes of Middle Acheulian type (fig. 8.6D), thus relating this culture in this instance to *Homo sapiens*, and lastly they are associated with a fauna of large vertebrates, including woodland species such as *Elephas antiquus* and grassland genera such as ox and horse. Unfortunately pollen has not been found in the Swanscombe sequence, so it is not possible to make any vegetational reconstruction. Thirteen teeth ascribed to Neanderthal Man (*Homo sapiens neanderthalensis*) were found during 1910–11 in a cave sequence at La Cotte de St Brelade in Jersey. No Neanderthal remains have been found in Britain, though the cultures associated with Neanderthal Man on the Continent are found. The remaining skeletal remains in our area have all been referred to *Homo sapiens sapiens*. At seven sites, all cave sites, remains of last glaciation age have been found, associated with Upper Palaeolithic industries. A radiocarbon date of 18,460 years B.P. of the 'Red Lady' of Paviland Cave near Swansea, associated with mammoth and an Upper Palaeolithic (Late Proto-Solutrean) industry, placed the age of the Red Lady near the beginning of the glacial advance in the later part of the Devensian cold stage. Further finds of Mesolithic age have been made, for example at Thatcham, Berks (9290 years B.P.) and in Argyllshire. With the development of more sophisticated burial habits after the Mesolithic, skeleton remains become increasingly abundant.

Before finishing this note on skeletal remains of *Homo sapiens sapiens*, we should refer to the many finds in Britain of *Homo* remains once believed to be of great antiquity but now known to be much younger. They include skeletons at Tilbury, Essex, at Galley Hill, near Swanscombe, and skull fragments in London, Bury St Edmunds, all thought to be Palaeolithic but now known by fluorine and nitrogen analyses and radiocarbon dating to be Flandrian. The other skeletal remains to mention are the famous finds from gravels of the 50-ft terrace of the river Ouse at Piltdown, Sussex. The reconstruction of the Piltdown skull after finds reported in 1912 always posed a serious problem for students of man's ancestry, for the finds showed cranial characters of recent *Homo sapiens sapiens*, with the jaw of an ape, all of reported Early Pleistocene age. The problem was solved by a detailed study of the bones in the 1950s, which showed beyond doubt that the cranial bones had been artificially stained and that the jaw and canine tooth, also stained, were that of an orang-utan. Moreover, a skull fragment and the jaw have been dated by radiocarbon to ages of 620 and 500 years respectively. The ingenious hoaxer or hoaxers were the cause of a long scientific controversy over their significance with a concomitant waste of time, even though the hoax may have stimulated thought on the problems of human evolution.

Cultural remains

Man has been defined as the tool-maker. It is the ability to produce a material culture, through conceptual thought and communication, that distinguishes man from ape. The earliest artefacts, that is to say implements or other objects made by man, are pebble tools associated with remains of *Australopithecus* and *Homo habilis* in the basal layers of Olduvai Gorge in East Africa at an age of some 1·75 million years. From that time to the present day, there has been a continual evolution, very slow at first, of man's tool-making abilities and his cultural complexity. Of this 1·75 million years, all except the last 3000–4000 years in Western Europe is occupied by the Stone Age (see fig. 8.4). This is the first of the periods of the so-called Three Age System; it was followed by the Bronze Age and the Iron Age, bringing us in Britain to the time of Roman occupation at the beginning of the Christian era.

The Three Age System was developed in the last century during the preliminary sorting of artefacts into a time sequence. Since that time the progress of knowledge has made this simple classification seem too simple. What we have in reality is a series of cultures, distinguished from one another in some ways, similar in other ways, and characterized by the kind of artefacts and their relative frequency. Refined methods of dating allow ages to be ascribed to some of these cultures, and then their connections in space and time with each other can be determined. The difficulty of reconstruction of the sequence of cultures and of the factors which cause their evolution lies first in their dating. Even the application of radiocarbon dating, though it has proved immensely useful, is difficult because of the changing relation between radiocarbon and sidereal years (see Chapter 9). The second difficulty lies in segregating the effects of diffusion of cultural traits and of migration of peoples from studies of excavated material culture.

These complicating factors recognized, we now have to give a brief account of the sequence of archaeological cultures known in Britain during the Pleistocene. A culture is composed of a number of industries. An industry is a group of artefacts of varying types made at the same time evidently by the same people. Thus a particular industry will be found at a particular level in an excavation. If a number of industries appear to belong together in type, time and space, they may then be considered a culture. Cultures are named after a type site, e.g. the Acheulian (hand-axe) culture, after St Acheul, a suburb of Amiens, or after a particular artefact type such as the Beaker culture, or after a characteristic type of structure. An industry will be described from a particular site and will be named after it.

In the following account we have only taken account of a few representative well-dated sites. The increasing richness of the archaeological record in the Flandrian cannot be dealt with in detail; an outline is presented to show the relations between archaeology, geology and climatic change. Radiocarbon dating forms the basis for the sequence younger than about 50,000 years B.P. For industries older than this the methods of geological correlation and relative dating have to be used. There are few Middle and Lower Palaeolithic sites well-integrated into the record of environmental change because only too often the finds are not accompanied by abundant biological evidence which can form the basis for environmental reconstruction and relative dating. Many of the finds of Lower Palaeolithic implements are in gravels, where they might well be derived from earlier deposits which have suffered reworking in a fluviatile environment or by solifluxion.

THE STONE AGE

The Stone Age is conventionally divided into three parts, the Palaeolithic, Mesolithic and Neolithic. The boundaries between them are not so clear as the names suggest, for reasons already discussed. The ages of these divisions is shown in Table 8.2. The Palaeolithic is divided into three parts, Lower,

Table 8.2. *An outline of the archaeological sequence in Britain*

Stage	^{14}C *age* B.P.	*Culture, industry or stage*	*Division of Stone Age*
	3,000–1,000	Iron Age	
	2,000–2,600	Late Bronze Age/Early Iron Age	
	2,700–3,900	Middle Bronze Age	
	3,000–3,800	Early Bronze Age	
Flandrian	3,500–4,000	Beaker	Neolithic
	3,500–5,300	Windmill Hill	
	6,000–7,000	Sauveterrian	Mesolithic
	9,550	Maglemosian	
	6,000–10,000	Creswellian	
Devensian	20,000–30,000	Proto-Solutrean	Upper Palaeolithic
		Mousterian	
Ipswichian		Levalloisian	Middle Palaeolithic
Wolstonian		Levalloisian	
Hoxnian		Acheulian (Swanscombe Man) (zone Ho II) Clactonian (zone Ho II)	Lower Palaeolithic

Middle and Upper, related to advancing techniques of tool manufacture and to increasing richness of culture.

THE PALAEOLITHIC

The beginnings of tool manufacture are shrouded in doubt by the similarity of primitive tools to naturally-occurring flaked pebbles. The earliest dated tools identified are found in Africa (Lower Pleistocene, 1·75 million years) and are of the so-called chopper tool or pebble tool type, made by striking a few flakes from the side of a pebble in one or two directions (see fig. 8.6A). Such an industry has been associated with *Homo habilis* and *Homo erectus*. In Britain such Lower Pleistocene industries have not been found. But early in this century many flints from the Lower Pleistocene Crags were described as being artefacts, such as the flints, some flaked bifacially, in the Red Crag near Ipswich (fig. 8.6B), and the so-called rostrocarinates from the base of the Norwich Crag near Norwich. All are now thought to be natural products. They do not satisfy the requirements for identification as a tool, namely, that the object conforms to a set and regular pattern, that it is found in a geologically possible habitation site, preferably with other signs of man's activities (e.g. chipping, killing or burial site), and that it shows signs of flaking from two or three directions at right angles.

There are abundant Lower Palaeolithic remains in south and east Britain (fig. 8.3). This distribution is perhaps the first expression of the economic advantages of living in the south-east. Over 3000 sites are known which have yielded Lower Palaeolithic artefacts. The earliest accepted culture found in Britain is the Clactonian (fig. 8.6C), a primitive industry producing flakes with a wide angle between the bulbar face and the striking platform (see fig. 8.5). This culture may be a descendent of a pebble tool culture. At the type site, Clacton-on-Sea, Essex, the industry is situate in a freshwater deposit of the Hoxnian (partly in zone Ho IIb) temperate stage, a zone of deciduous forest in the climatic optimum part of the interglacial. Presumably the Clactonian industrial activity took place on the shores of an ancient river, perhaps part of the equivalent of the Thames at that time. At Swanscombe, on the south bank of the Thames near Gravesend, a similar Clactonian industry is found on the Lower Gravel of the 100-ft terrace and is correlated with zone Ho II or III of the Hoxnian.

In the Middle gravels at Swanscombe we find a rich Acheulian hand-axe industry (fig. 8.6D), associated with the Swanscombe skull bones (*Homo sapiens*). The Acheulian hand-axe represents a new stage of flint tool manufacture, with two-way flaking on both sides of the implement, a development from the simple pebble tools. Hand-axes have been found in association with *Homo sapiens* and *Homo erectus*. The Acheulian hand-axe culture in south-east Britain has been firmly dated to the Hoxnian temperate stage by the find of hand-axes and waste flakes in interglacial lake sediments containing pollen at the type site, Hoxne, in Suffolk. At Hoxne, the industry was located in the sequence close to a horizon where it was shown by pollen analysis that forest cover diminished in favour of herbaceous vegetation. Whether the encampments of Acheulian man were a causal factor in this is unknown. It seems

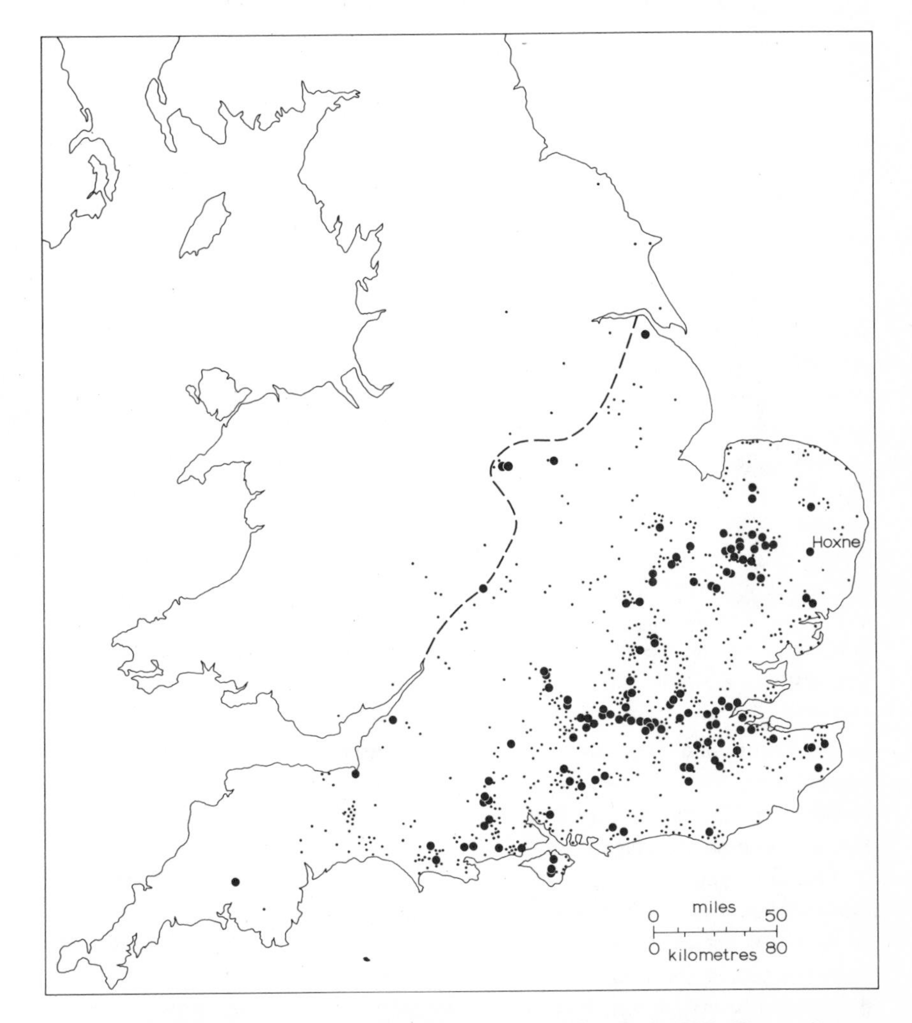

FIG. 8.3 Distribution of Lower Palaeolithic artefacts (after Roe, 1970). The small spots represent single finds or groups of less than ten. The large spots represent more than ten, occasionally up to a thousand. The dashed line indicates the approximate limit of Pleistocene deposits with flint common. (From F. W. Shotton, in the *Proceedings of the Geological Association*, 1968, Vol. 79, fig. 1.)

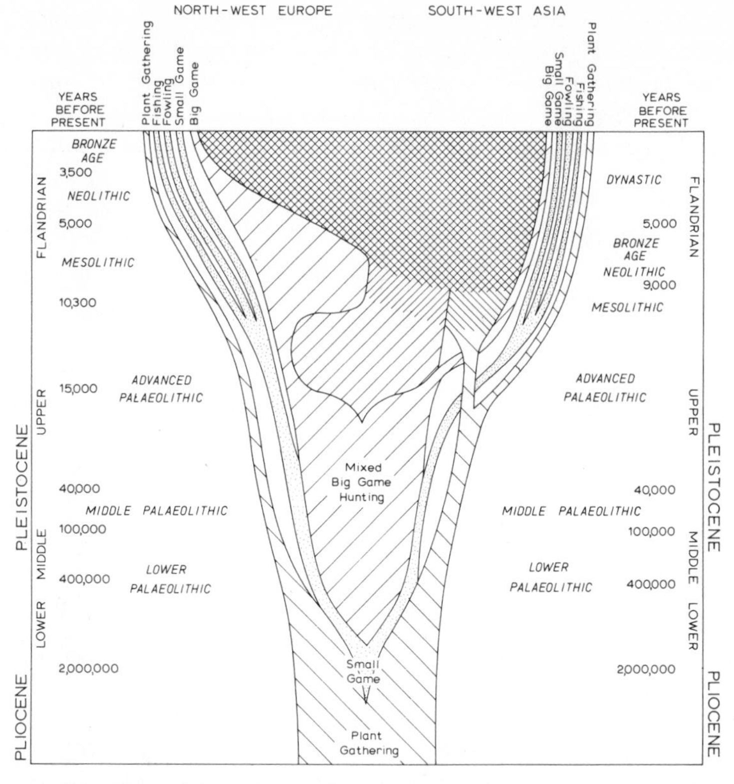

FIG. 8.4 Chart of the evolution of the food quest by Primates. The transition from plant gathering to farming in south-west Asia is shown on the right, with the later spread of farming to north-west Europe. (After Clark, 1967, fig. 123.)

perhaps more likely that they moved into the area in a period of retreat of the forest cover.

Both the Clactonian and Acheulian cultures are, then, associated with the Hoxnian temperate stage.

The next youngest industry is associated with a new type of tool production, using a core prepared in such a way that the form of a flake struck from it was accurately determined. This technique is known as Levalloisian and resulted in the formation of the so-called tortoise core, from which a flake could be struck which would show many flake scars on one side, and on the other side the bulbar surface of the struck flake (fig. 8.6E, F). The Levalloisian technique is associated with the Middle Palaeolithic. In Britain rich finds of tortoise cores and flakes have been found beneath a chalk solifluxion deposit at Baker's Hole, Northfleet, near Gravesend, at a level

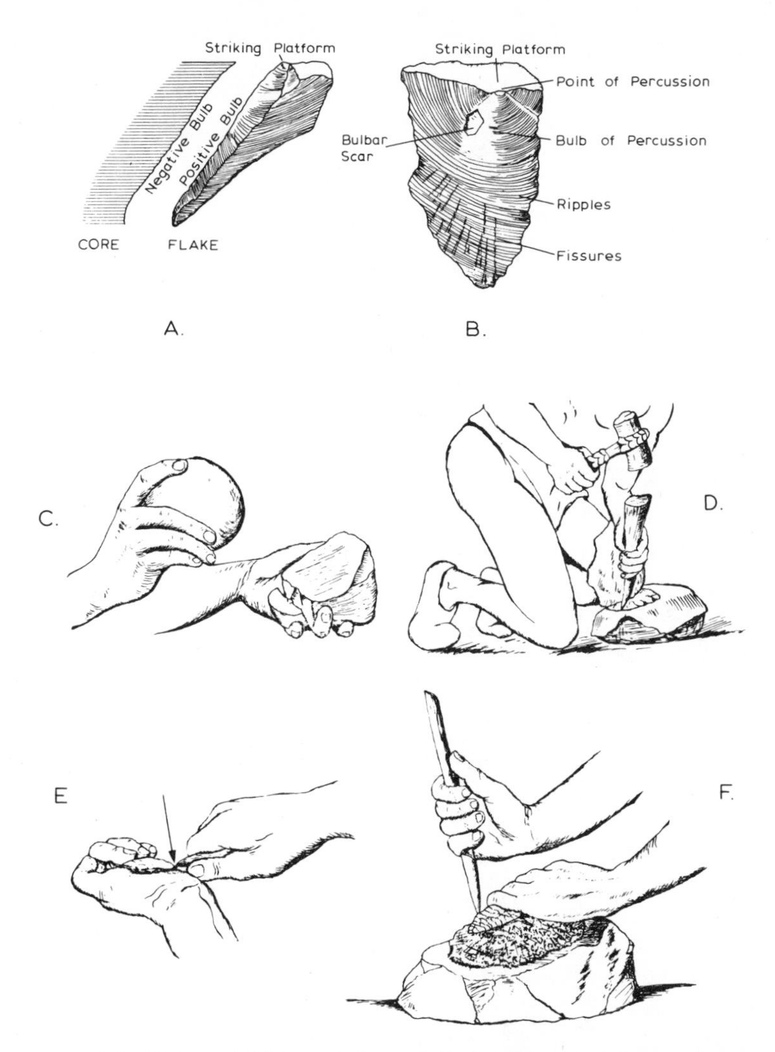

FIG. 8.5 Flakes and flaking techniques. A, B, flint flake struck by man, in two views. Note the bulb below the point of percussion; C, producing flakes by direct percussion with a hammer stone; D, producing flakes by indirect percussion, with a bone or wooden punch; E, F, two methods of dressing the edge of a tool by pressure flaking. (After Oakley, 1967, figs. 4 and 11.)

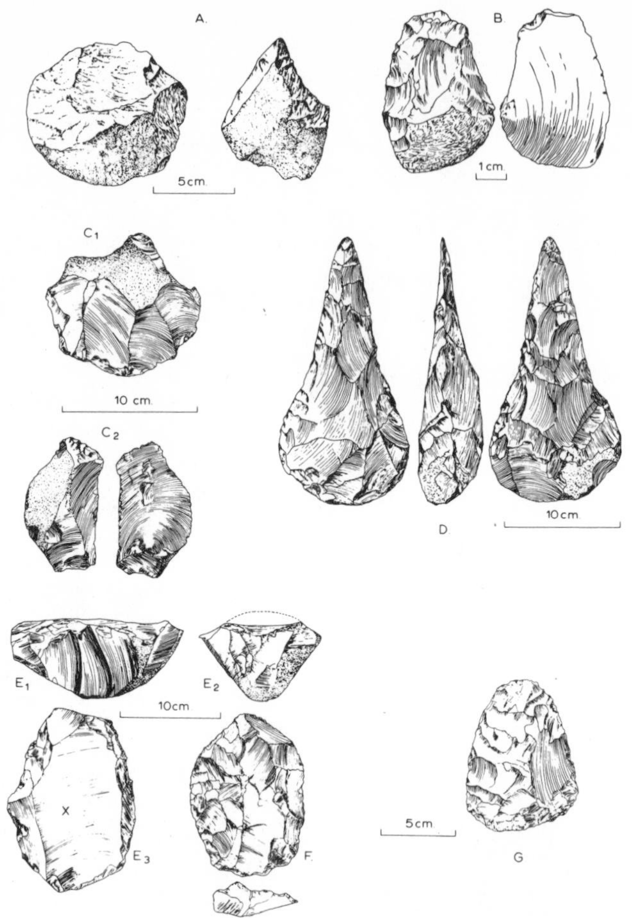

FIG. 8.6 Palaeolithic tools and a Crag eolith. A, pebble tool from Olduvai gorge, Tanzania (after Roe, 1970, fig. 4); B, so-called flint side-scraper from the Lower Pleistocene Red Crag at Bramford, Ipswich (after Reid Moir, ***Antiquity of Man in East Anglia***, London, Cambridge University Press, 1927, fig. 7); C^1, Clactonian flint core; C^2, Clactonian flake tools, both from Lower Gravel, Swanscombe, Kent (after Oakley, 1967, fig. 21); D, Middle Acheulian hand-axe from Lower Middle Gravel at Swanscombe, Kent (after C. D. Ovey, ***The Swanscombe Skull***, Royal Anthropological Institute of Great Britain and Ireland); E^1, E^2 and E^3, three views of a Levalloisian tortoise core from Northfleet, Kent; F, flake struck from tortoise core, Northfleet, Kent, similar to the flake struck from E^3 at X; G, small hand-axe of Mousterian type from Kent's Cavern, Torquay (after Oakley, 1967, figs. 22 and 23).

lower than the 100-ft terrace of Swanscombe, and probably to be dated to the Wolstonian cold stage. Another industry of Levallois type has been found in fluviatile deposits at Brundon, Suffolk, in a horizon dated to the Ipswichian temperate stage.

A second Middle Palaeolithic culture is the Mousterian, named after a cave site in the Dordogne. This culture made prominent use of flake tools with well-made flake points and scrapers predominating, and the Levallois technique still playing a part. Tools were also made from small discoid cores. The Mousterian culture belongs to Neanderthal Man, an association made possible by the Mousterian practice of burying their dead in caves. The Mousterian in Britain has been found in a number of cave sites, in north Wales, Somerset, South Devon, Derbyshire and Kent, with probable ages of 35,000–40,000 years B.P. On the Continent the latest Mousterian is found at about 30,000 years. The culture is thus associated with the middle part of the last glaciation. It gave way to the Upper Palaeolithic at about 33,000 years, at a time when Neanderthal Man was replaced by *Homo sapiens sapiens*.

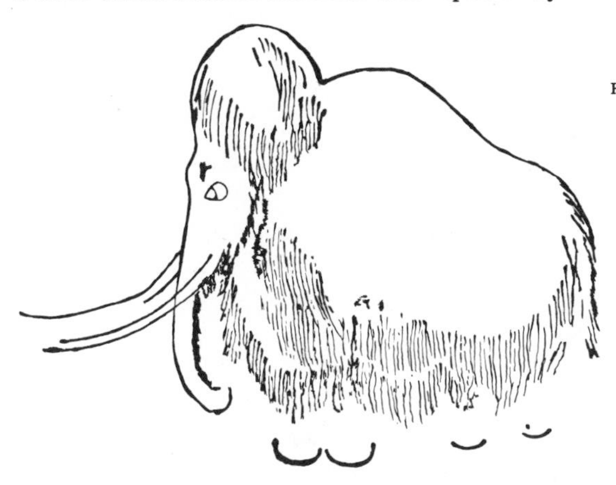

FIG. 8.7 Mammoth engraved on rock in the cave of Font-de-Gaume, Dordogne, France (after Breuil). Greatly reduced. (From W. B. Wright, *Tools and the Man*, London, Bell, 1939, fig. 84.)

The Upper Palaeolithic is characterized by the widespread use of rather narrow flint blades with parallel flake scars (fig. 8.8A–D), giving a prismatic appearance to the flakes. Such flakes were produced by indirect percussion, producing punch-struck flakes (see fig. 8.5). The flakes also often show a nearly vertical retouch, and could be used as knives and projectile points. Another common tool was the graver or burin, a flake with a strong chisel-like working edge. The Upper Palaeolithic shows a great increase in industrial evolution with many cultures appearing over the comparatively short time of some 33,000–10,000 years. Bone work became highly developed and so did art (fig. 8.7).

In France there is a well-known succession of cultures through the Upper Palaeolithic, known from cave sites, with distinction between them made on typology of flint tools and the development of bone artefacts and cave art. In Britain such a complete succession is not seen. Two general phases are evident. An early phase, Proto-Solutrean, has leaf-shaped tools with extensive retouch (fig. 8.8A). The age is thought to be in the period 20,000–30,000 years, but the dating evidence is not precise. Finds have been made in a few caves in north and south Wales (Paviland Cave, Swansea), south Devon, Somerset and Derbyshire. The material is scantier than that of the Mousterian and perhaps represents seasonal occupation. The later phase contains no leaf points; they are replaced by backed blades and comprise the Creswellian industries (after the caves of Creswell Crags, Derbyshire) (fig. 8.8B–D), allied to the Magdalenian of the Continent. Backed blades have one edge blunted for holding and hefting and the other sharper.

Some sixteen sites in Britain show the Creswellian. Most are cave sites, but a few are open. The industries are much richer than those of the earlier phase and very variable in tool composition. Radiocarbon dates indicate that the age of the Creswellian spans a period of a few thousand years at the beginning of the Flandrian, at the time of the early postglacial afforestation with birch and pine.

The Creswellian can be considered as intermediate between the Upper Palaeolithic and the Mesolithic. The latter contains flint industries, e.g. Tardenoisian, with very small flakes, microliths, which are 2 or 3 cm long or less (fig. 8.8E). The Mesolithic peoples had to contend with a forest environment in contrast to the Weichselian periglacial environments of the Upper Palaeolithic. A number of Mesolithic cultures have been distinguished in Britain. The Maglemosians used assemblages of microliths to tip, barb and arm their projectiles (fig. 8.8F). We see the appearance of the flint adze, probably in response to the necessity of clearing forest. Abstract art, engraved on bone, is also characteristic. A Maglemosian site at Star Carr, near Scarborough, has proved to be one of the most rewarding sites excavated in Britain. The site was an encampment of a hunter-fisher people on a lake edge, and there was a very good record of the environment of the time in the biological evidence from the lake muds. A clearing had been made in birch forest and a rich collection of flints and worked bone, including harpoons, was found on a platform at the lake edge.

A second culture is the Sauveterrian, a microlithic culture without adzes. Sites have been found on the Pennines, East Anglia, south-west England and Wales. The Larnian–Obanian culture is a later variant with heavy axe tools found on the north-west coast of Britain and in northern Ireland.

THE NEOLITHIC AND AFTER

The Neolithic under the Three Age System came to be associated with the advent of pottery, the occurrence of polished stone axes, and with the

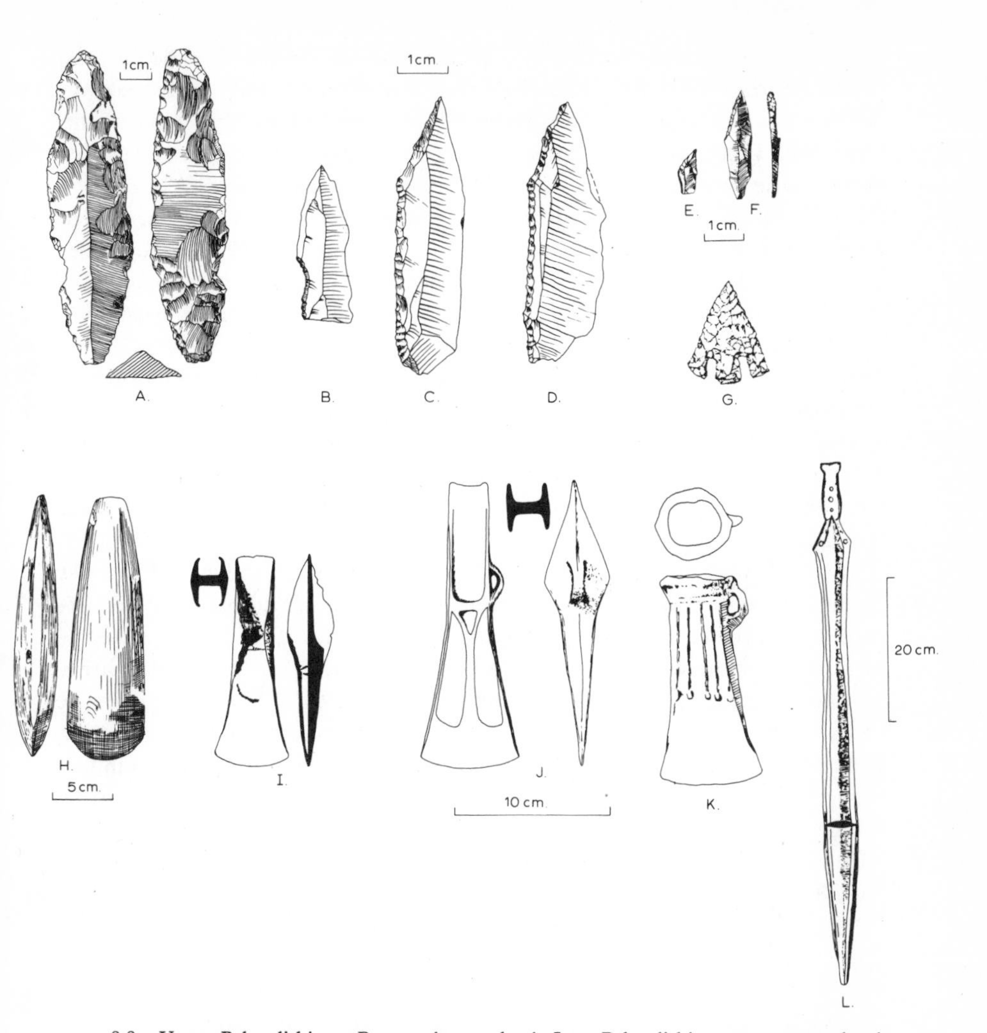

FIG. 8.8 Upper Palaeolithic to Bronze Age tools. A, Late Palaeolithic stone weapon head: 'Proto-Solutrean' point from Vale of Clwyd (after Oakley, 1967, fig. 26a); B, C, D, Creswellian flints from Anston Cave, Yorkshire: B, small burin, C, D, angle-backed blades (after P. A. Mellars, ***Antiquity***, 1970, fig. 1); E, Tardenoisian point, Ham Common, Surrey; F, Maglemosian point, Kelling Heath, Norfolk; G, Bronze Age barbed and tanged arrowhead, Lakenheath, Suffolk; H, polished stone axe, Kelso, Scotland (all after Oakley, 1967, figs. 12e, 13e and 29g, h); I, Middle Bronze Age wing-flanged axe, Trillick, Co. Tyrone; J, Middle Bronze Age palstave, Wilton, Wilts. (from Roe, 1970, fig. 99); K, Late Bronze Age socketed axe, Llynfawr, Glamorgan; L, Middle Bronze Age sword, Penrhyndeudraeth, Merioneth (from Roe, 1970, figs. 100 and 104).

appearance of farming settlements, in contrast to the hunting, fishing and food-gathering economies of earlier times. But in recent years, with the great development of archaeological excavations in the Middle East and south-west Asia, much more has come to be known about the origins of settled communities practising farming and it is clear that the processes of cultivation and domestication of plants and animals occupied a long period of several thousand years before the making of pottery, itself first recorded in that area in about 6000 B.C. These traditions were initiated and developed by Mesolithic people. Thus in Shanidar Cave, in mountainous north-east Iraq, in the 9th millennium B.C., there is evidence from faunal remains, stone querns and reaping knives that sheep were herded and that cereals such as wild barley and wheat were harvested. The following few thousand years covers the development of more sophisticated animal husbandry and cereal cultivation, with the establishment of complex settlements, and a subsistence economy of the type which forms the basis of our present civilization (see fig. 8.4).

In north-west Europe, in the same period, we see at the end of the last glaciation the utilization of reindeer herds as a specialized basis for subsistence, but in the following few thousand years, instead of the further development of the kind seen in the Middle East and south-west Asia, a Mesolithic hunting, fishing, food-gathering economy persisted. The encroachment of the Flandrian forests, replacing the late-glacial tundras, must have been a great shock to Upper Palaeolithic man, and the rather specialized reindeer economy was replaced by the more widely-based economy with hunting, fishing and food-gathering. It is not until *c.* 3500 B.C. that farming systems penetrated to Britain (fig. 8.4), with pottery and polished stone axes (figs. 8.8 and 8.9).

The cultural sequence in Britain in and after the Neolithic becomes progressively more complicated. With the increased mobility of people, and the rapid development of industrial techniques, with avenues of migration for people and techniques into Britain from all along the western seaboard of continental Europe, it is hardly surprising that the distinctions between cultures in space and time become difficult to maintain.

In general, it is possible to distinguish in Britain an earlier Windmill Hill Culture, named after a causewayed camp in north Wiltshire, which is represented by coarse, rather plain and sparsely decorated pottery with round-bottomed bowls (figs. 8.9A, B), and a later secondary Neolithic which has increasingly decorated pottery (figs. 8.9C, D). The form of graves undergoes change, the long barrows with multiple burials and few grave goods of the Windmill Hill Culture giving way to the megalithic tombs and structures prominent by the end of the Neolithic. In the Neolithic, there was increasing discrimination in the use of stone for axes and several axe factory sites are

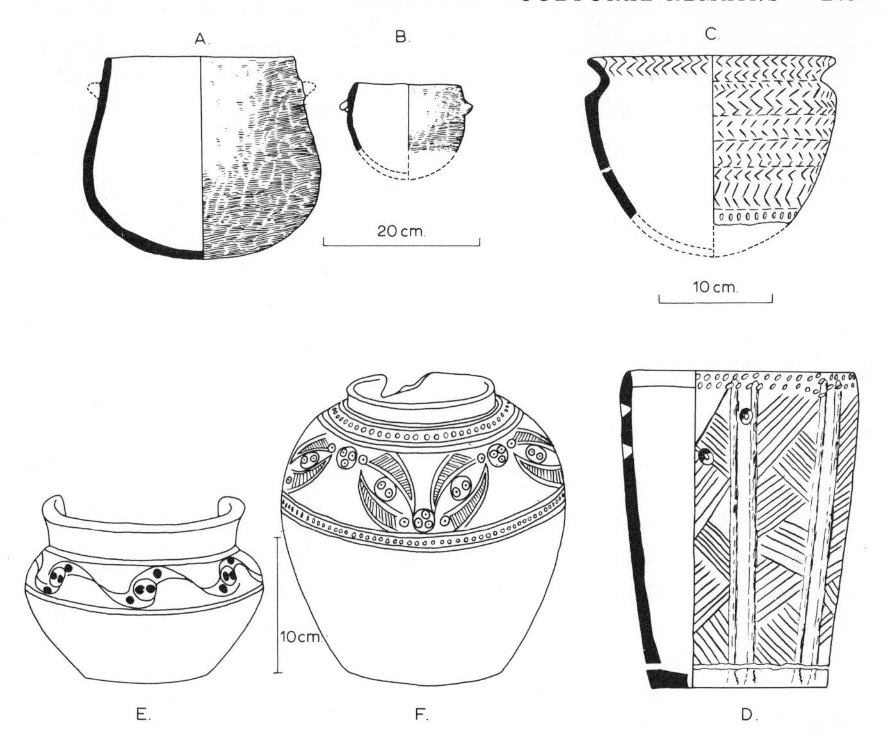

FIG. 8.9 Pots. A, B, plain round-bottomed pots of the earlier Neolithic from Maiden Castle, Dorset (after S. Piggott, *Neolithic Cultures of the British Isles*, Cambridge University Press, 1970); C, D, decorated pottery of the Neolithic: C, 'Peterborough' bowl with cord impressions, from West Kennet long barrow, Avebury, Wilts.; D, Rinyo-Clacton pot with impressed and grooved decoration and applied vertical cordons from a barrow at Wilsford, Wilts. (from Roe, 1970, figs. 75a, b); E, F, decorated Iron Age pottery from Glastonbury, Somerset (from Roe, 1970, fig. 124).

now known, from which axes were exported. Examples are seen at Great Langdale, in the Lake District, and Graig Lwyd in Caernarvonshire.

The Beaker Culture is involved in a phase transitional to the Bronze Age. The name refers to the decorated, well-fired pottery drinking vessels characteristic of this culture. The pottery was accompanied by high-quality flintwork, and there was a tradition of working in copper and gold, so bringing metal working to this country. The Beaker Culture overlaps the secondary Neolithic and is associated with some of the henge monuments. Thus the first phase of stone erection at Stonehenge is associated with this culture.

The Bronze Age can be divided into three indistinct phases, Early, Middle and Late. The Early Bronze Age Wessex Culture lasted from *c.* 1600–1400

B.C. The round barrows of the chalk uplands of southern Britain, rich in grave goods, are associated with this period, as are certain henge monuments such as Stonehenge in the form now preserved. Weapons include stone battle axes and metal daggers. In the Middle Bronze Age there is an increased use of metal with rapiers, slashing swords (fig. 8.8L), and spearheads, and farming was much more extensively practised. In the Late Bronze Age, a further sophistication of metalwork occurs with beaten sheet bronze. There are fewer sites and monuments of the Late Bronze Age, but many hoards of bronze objects, either finished or of scrap, have been unearthed. The evolution of the bronze axe provides a valuable indication of age at this time (figs. 8.8I–K), with flanged axes in the early period, palstaves in the middle, and socketed axes in the Late Bronze Age. By the end of the Bronze Age high standards of metal working had been established, and there was a decline in the use of stone tools. At about 500 B.C. iron working was introduced into the country from the Continent with the Celtic Hallstatt Culture. Within the Iron Age, again three periods have been distinguished in the past, A, B and C. Iron Age 'A' includes the Late Hallstatt Culture and the beginning of the La Tène Culture, 'B' the middle La Tène period, and 'C' the late La Tène period with the immigration of Belgic tribes into southern England. The Hallstatt Culture, with its long iron swords, built graves with rich goods and timber structures. The La Tène culture shows a very rich decorative art applied to metal objects and pottery. The earliest use of coins in Britain, and the use of the two-wheeled chariot, are also associated with this culture. The Iron Age was a great time for the building of hill forts, especially in the period between the earliest Iron Age settlements and those of the Belgic invaders. The fortifications such as Maiden Castle in Dorset, were successively protective against a series of invasions from the Continent starting in La Tène times and continuing into the Roman period. The Iron Age 'C' Belgic invasions are mentioned by Caesar, and this then brings us to the Roman period and the beginning of recorded history in England, though the Iron Age lasted longer in the northern areas which were not subdued by the Romans.

Man's activities and the history of the vegetation

The effects of man's activities on the environment have recently become obvious in industrial societies. Talk of clean air, river pollution, marine pollution and conservation of the environment takes a prominent place, quite rightly, in everyday life. Such effects have become more and more obvious in the accelerated technological developments of the last fifty years. They are being recorded for posterity in the resultant chemical and physical characters of sediments now being laid down in lakes and seas.

In the same way, there is a recorded history of man's exploitation of the environment in Pleistocene sediments. The effects of exploitation are seen in changes in the fossil assemblages and kinds of sediment. As with the doubts expressed over the identity of the earliest tools, there are problems in interpreting the earliest examples of man's effects on the environment. Changes in vegetation associated with archaeological settlements of the Neolithic and later give clear evidence of the effects of agricultural practice and land clearance on the natural vegetation. But in earlier times, in the Mesolithic and Palaeolithic, it is possible that certain practices affected the vegetation in a way difficult to distinguish in the fossil record from natural events. Thus at Hoxne, Suffolk, the Acheulian horizon contained charcoal and was related to a short period when forest retreated to give place to grassland. Is such a deforestation the result of a natural fire, Acheulian man using the clearings as habitation sites, or was the clearance intentional? The same difficulty occurs in the Mesolithic, where charcoal has been found associated with weeds common later in Neolithic and later clearances. Thus there is some evidence of forest clearance in pre-Neolithic times. Perhaps the hunting practice of driving game involved the use of fire, which would destroy the natural vegetation, or perhaps local clearings accompanied even short-lived settlements.

The use of fire has a long history. The earliest evidence for it is seen in the caves at Choukoutien, occupied by Pekin man, at a time thought to be 200,000–300,000 years ago. In Europe, evidence of fire first occurs associated with the Acheulian hand-axe culture. Charcoal has been found in the Acheulian horizon at Hoxne, Suffolk, and fire-crazed flints in the Middle Acheulian workshop debris of Swanscombe Man. In the Lower Palaeolithic evidence of the use of fire is not so widespread in archaeological contexts as in the Middle and Upper Palaeolithic, and it is possible that the Lower Palaeolithic users were fire collectors rather than fire producers, relying on natural conflagrations, collecting fire, and conserving it. On the other hand, the regularity with which hearths are found associated with the Middle and Upper Palaeolithic sites leaves little doubt that Neanderthal Man and his successors were capable of fire production. If this was so, then it seems likely that they would have used fire to exploit the environment, and driving game or clearing forest would be obvious uses. Thus the effects on the vegetation seen in the Mesolithic may well have been intentional and related to the use of natural resources.

The deforestation phase at Hoxne and the clearances of the Mesolithic and Neolithic are seen in pollen diagrams as decreases of tree pollen (AP) frequency and rises of herbaceous pollen (NAP). The herbaceous element contains pollen of weed species associated with human habitation and land clearance such as grasses, *Plantago*, *Rumex*, *Polygonum*, Chenopodiaceae and *Urtica*.

The widespread introduction in the Neolithic, at about 3000–3400 B.C. in Britain, of clearance associated with crop cultivation resulted in a very obvious horizon in pollen diagrams covering that period. The increase in the pollen of grasses, *Plantago lanceolata* (see fig. 6.4), spores of *Pteridium*, the occurrence of cereal pollen grains, the decrease in tree pollen frequencies, especially that of *Ulmus*, are all characteristic of the time. In these early clearance periods three phases may be distinguished, of clearance, of farming,

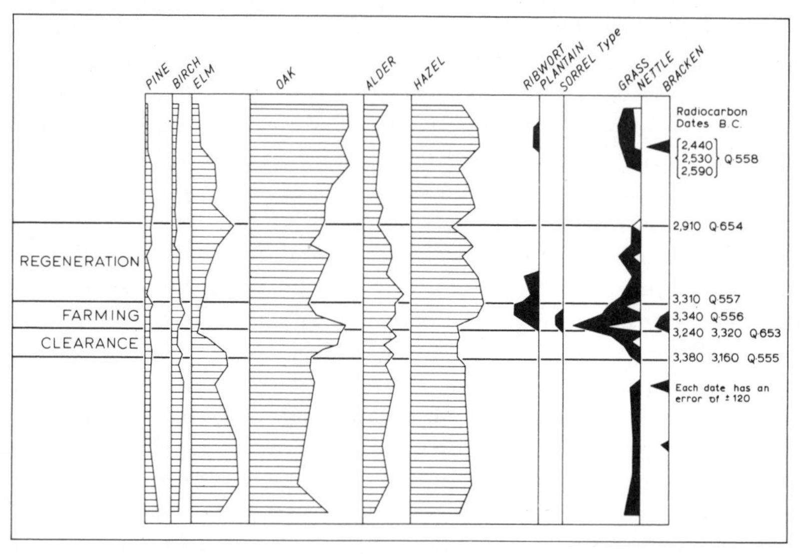

FIG. 8.10 Pollen diagram from Fallahogy, Co. Londonderry, showing clearance episodes and their radiocarbon ages. The pollen types indicating clearance are on the right, with their curves blacked in. (After N. Stephens and R. E. Glasscock, eds, *Irish Geographical Studies*, Department of Geography, Queens University, Belfast, 1970, fig. 5.9.)

and of regeneration of the forest. Fig. 8.10 shows a pollen diagram from Northern Ireland which demonstrates these phases. The clearances were probably made by fire, in a similar way to the fire clearances still practised by aboriginal farmers in various parts of the world. The settlements would be short-lived, for soil exhaustion might soon ensue, and a new clearance, with soil enriched by ash, would be preferable to continued cultivation of one area. Such shifting cultivation occurs with primitive farmers at the present time, and it would help to account for the extensive nature of the Neolithic clearances in Britain.

On the most fertile soils, regeneration of forest might follow abandonment of an area. Such regeneration is seen in pollen diagrams in the return to dominance of the mixed oak forest (fig. 8.10). But in areas of poorer soil, clearance and cultivation may have changed the soil properties for good, and the clearances may have had a lasting effect on the forest cover.

The decline of the frequencies of the pollen of *Ulmus*, mentioned above as accompanying clearance, is seen around 3000 B.C. in all British lowland pollen diagrams, and in north-west Europe as well. The decline has been attributed both to climatic change and to use of elm as a fodder plant for cattle. There is evidence for both views. Even in the recent past, dried elm foliage has been used as winter fodder in Scandinavia, and the foliage has been found fossil in Neolithic settlements in Switzerland. Moreover, at its northern limit in Scandinavia in the present century, the growth of elm is known to be very sensitive to climatic change. Whatever the reason, the elm decline marks the beginning of the pauperization of the British native mixed oak forests.

In the Neolithic, then, we see the first certain occurrences of small temporary clearances, with cereal growing, both wheat and barley, and cattle browsing. The clearances are marked by increased frequencies of pollen of *Plantago* and grasses and spores of *Pteridium*. In the Bronze Age, a contrast of farming intensity develops between south-east England and the rest of the country to the north and west. Pollen diagrams from the south-east show continuous high frequencies of herbs and weeds characteristic of mixed farming, whereas in the rest of the country there are either continuous low frequencies of herbs or successions of small temporary clearances. The contrast must be associated with the increase in population in the south-east and the development of more permanent agricultural systems there. In the Iron Age, clearance was made easier by the use of the iron axe and extensive clearance in the north and west started at about this time. In Romano-British times new plant introductions, including rye, were made. In the north and west, clearance intensity varied with political stability. Thus pollen diagrams from southern Scotland show a marked clearance phase at *c.* A.D. 400 after a period of instability during the Brigantian wars.

The nature of agricultural practice can be deduced from studies of recent pollen rain in areas of different agricultural practice, and the results can then be applied to pollen diagrams. Arable areas show low frequencies of *Plantago* pollen and *Pteridium* spores, and high frequencies of pollen of *Artemisia*, *Rumex*, Chenopodiaceae and other arable weeds. Pastoral areas show the reverse. Changes in the arable–pastoral proportion can be seen in pollen diagrams. Fig. 8.11 shows clearance types and arable–pastoral relations at various sites in western and northern Britain. The later development of clearances in the north is shown, leading to the present situation of almost complete deforestation.

In the south-east, a pollen diagram from Old Buckenham Mere, fig. 8.12, shows a change in the emphasis from pastoral to arable farming which

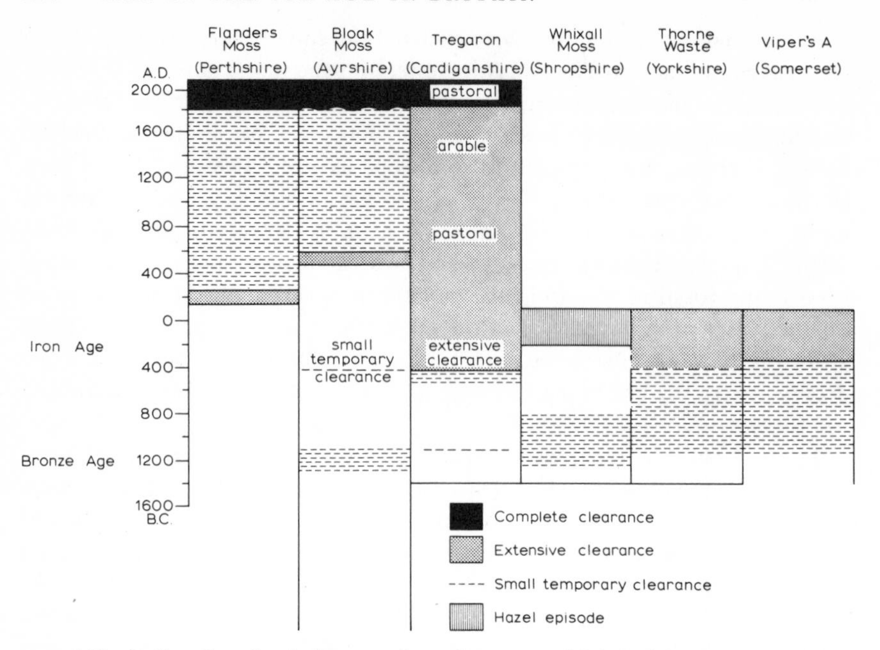

FIG. 8.11 Radiocarbon dated clearances in various parts of Britain (after J. Turner, in the *Proceedings of the Royal Society*, 1965, **B**, **161**, fig. 28).

accompanied Anglo-Saxon settlement, with its multiple ox-teams capable of ploughing heavier soils than those which Iron Age farmers could cultivate. This pollen diagram is also interesting in showing the cultivation of certain crops. The curve for *Secale* pollen starts in Roman times, and those for hemp (*Cannabis*) and flax (*Linum*) in Anglo-Saxon times.

In Britain we are distant from the areas in south-west Asia (the Fertile Crescent) where wild plant cultivation first occurred and domestication by selection followed. Evolved crops came into use in Britain with the settlement of Neolithic and later people, long after the period of domestication in south-west Asia (see fig. 8.4). The fossil evidence does not allow us to draw a similar conclusion about domestication of animals and their introduction to Britain. For example, domestication of cattle may well have been attempted in many different areas in Mesolithic times as an adjunct to hunting.

The destruction of Britain's natural forests in the last 5000 years is the principal effect of man in that time. The process, as well as clearing forest, promoted soil changes and may well have been responsible for the replacement of forest by heathland in many areas. Such a change of vegetation also accompanied the natural replacement of forests in the uplands by blanket bog, as a result of climatic change.

The consequence is that our natural forest vegetation is very limited in

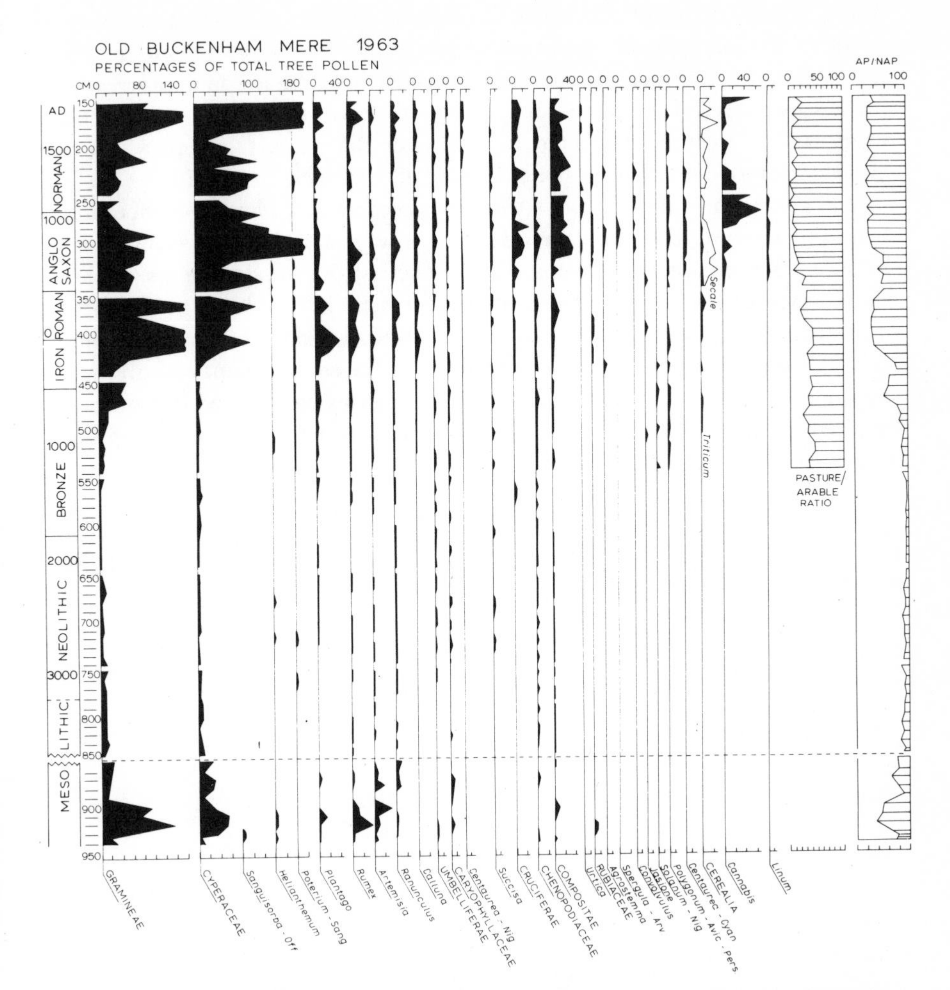

FIG. 8.12 Pollen diagram from Old Buckenham Mere, Norfolk, showing the relative frequency of herbaceous plants in the Flandrian. The basal sediments are of Late-Weichselian (late-glacial) age, and show high frequencies of herb pollen (NAP). There follows a period of forest dominance in the Flandrian, with low NAP frequencies. The NAP frequencies recover towards the top of the diagram, related to clearance and the spread of agricultural activities following the arrival of Neolithic man. Pollen grains of many cultivated species are recorded. Certain taxa (to the left) are associated with pasture, others (to the right) with arable, so giving the ratio of pasture/arable pollen. (After H. Godwin, in the *New Phytologist*, 1968, Vol. 67, figs. 6 and 7.)

extent. Small areas of oakwood in western Britain and birchwoods in northern Britain probably approach native forest. The woodlands of south-east Britain have in many areas been managed for timber and coppice production over several centuries, and though they are in a sense descendants from native forest, management policies over the years, together with the introduction of exotic species of tree, have altered their composition.

The agricultural development which proceeded over some five thousand years and affected the natural environment in ways described above, has been followed by the industrial development of the last one hundred years. In this short space of time, combustion of natural fuels by products of chemical manufacturing processes, advances in the use of agricultural chemicals and the explosion of nuclear weapons have affected the environment each in their own way. Certain of these changes affect the processes involved in radiocarbon dating, on which we rely for dating past events, as we shall see in the next chapter.

9

Dating Ice Age Events

Every geological section tells a story of events in the past. The sediment type, its structure, and contained fossils all give information on environment at the time of deposition. We may thus have evidence of climatic, vegetational or faunal history, of ice advances and retreats, of episodes of solifluxion and permafrost development, and of episodes of sea level change. To obtain an overall synthesis of the Ice Age, it is necessary to fit all these observed events into a framework of time, so that we can see first what series of events occur at one place during the passage of time, and secondly, how events in different places can be related in time. The first process involves dating, the second correlation of geological sequences in different places. Both these aspects of synthesis of events into a framework of time and space are discussed in this chapter.

Dating can be of two kinds. Events can sometimes be dated to years, so-called absolute or chronometric dating, or events at one place can be dated in terms of events at another place, with no reference to a particular scale of time, so-called relative dating. For example, the sequence of climatic fluctuations in the Ice Age makes it possible to correlate, and thus relatively to date, the last interglacial deposits in various parts of Europe, all separated from the present temperate stage by deposits indicating cold or glacial conditions of the last major glaciation. We shall first consider the various kinds of absolute dating, and then go on to discuss relative dating and other aids to dating and correlation.

Absolute dating

A scale of time to which events may be referred can be obtained from two sources. One is by counting a sequence which results from the annual layering caused by natural rhythmic processes, such as varved clay deposition and tree ring formation. The second involves the measurement of geochemical changes in radioactive substances which take place with the passage of time. In Britain, neither varved clay studies nor tree ring analyses have been carried very far. It is radiocarbon dating that has provided the most satisfactory method of dating. We shall briefly consider other methods of absolute dating before discussing the radiocarbon dating method, its applications and results.

VARVES AND OTHER LAMINATED SEDIMENTS

In certain sediments an annual lamination may result from deposition of a particular sediment type at one time of the year and another sediment type at another time of year. Thus varved clays (pl. 35) form in lakes receiving drainage from melting glaciers because in the summer melt both sand and some clay are deposited, forming a light-coloured layer, but later, in the winter, as meltwater input decreases, the supply of sand and clay is diminished and clays, already in suspension, sediment and form a darker layer. By counting the annual pairs of light and dark layers, the length of time a varve sequence took to deposit can be measured. In the Baltic region varve sequences are found in lakes outside the successively more northern limits of the receding last glacial ice sheet. The sections can be correlated over a distance, relying on patterns of change of varve thickness which are regionally determined by meltwater characteristics. By this correlation varved sediments have been used as a basis for dating the successive margins of the retreating ice during the last 15,000 years. This is the period covered by ice retreat which gradually left Scandinavia ice-free except in the mountains. Though varved clays of last glacial age occur in northern Britain, the exposures are isolated and there has been no synthesis of an ice retreat calendar.

Other laminated lake sediments form under temperate conditions in deep water, where seasonal differentiation is preserved in the lack of a bottom fauna to disturb it. Such lamination can be annual. For example, in the interglacial lake at Marks Tey, Essex, in each annual period of deposition there is first a layer rich in diatoms, probably formed in the spring, following a rapid increase in the diatom population of the lake, then a darker more organic layer formed later in the year as organic detritus settles to the bottom of the lake. By counting these pairs of layers, which occur throughout most of the interglacial, it has been estimated that the interglacial period lasted some 30,000–50,000 years. The pollen diagram of the Marks Tey lake has been time-calibrated by relating the pollen sequence to the lamination sequence. It is clear that lakes with laminated sediments are very useful in dating environmental history. Unfortunately such lakes are very scarce in Britain.

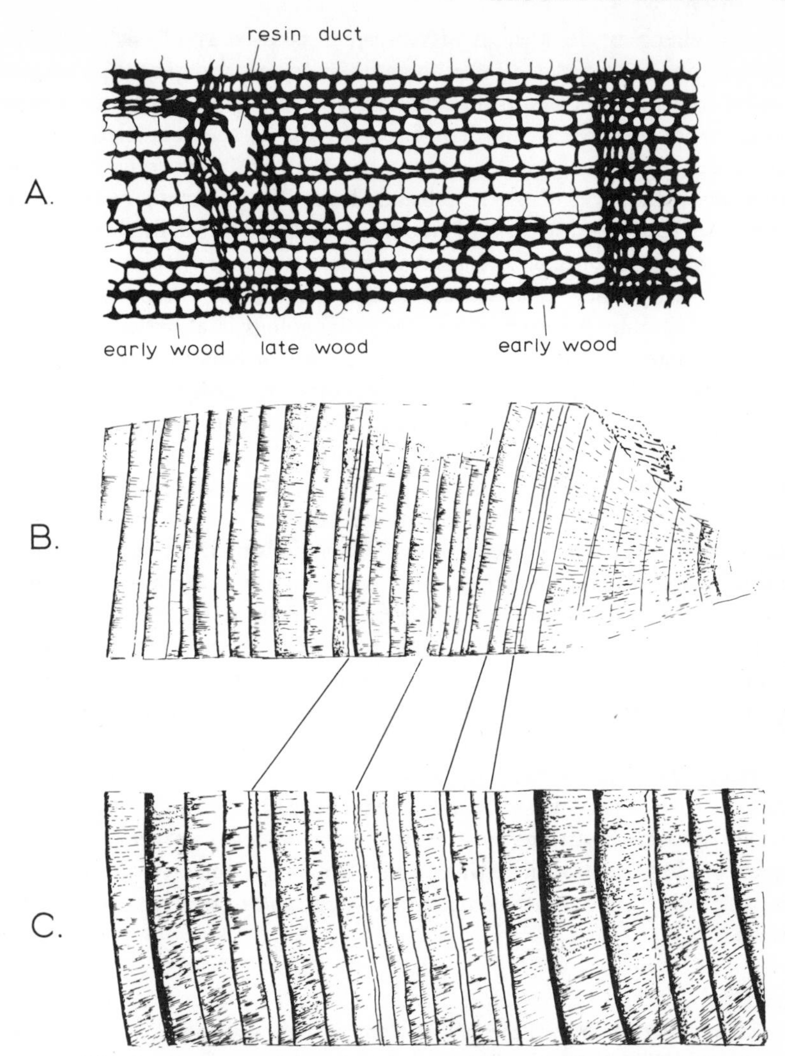

FIG. 9.1 A, transverse section through a conifer wood, ***Pinus strobus***, showing annual rings (× 30) (after West, 1968, fig. 9.2A). B, C, transverse sections through trunks of two conifers showing the correlation of their annual rings (after M. A. Stokes and T. A. Smiley, ***An Introduction to Tree Ring Dating***, Chicago, Ill., University of Chicago Press, 1968, fig. 6).

Tree rings provide another basis for chronology. Seasonal changes in the cells produced in the course of wood paths are expressed as annual rings (fig. 9.1). By counting annual rings of a tree trunk its age may be measured. By such measurements soils or moraines on which trees grow may be dated or the time a montane tree limit was established may be determined.

Just as changing thicknesses of varves can be used as a basis for correlation from one varved clay section to another, so can changing thicknesses in tree ring series from trees of different ages be used as a basis for correlation of rings from one tree to another (fig. 9.1B, C). By such correlations a tree ring or dendrochronological sequence has been built up going back some 7000 years with wood of the Bristlecone Pine, *Pinus aristata*, of south-west North America. The thickness of tree rings is related to the rate of growth of the tree, so that favourable and unfavourable times of growth are recorded in a tree ring series. So, to some extent, climatic history is recorded in these sequences over a long period. A tree ring chronology in a particular area may also be used to correlate the ages of logs used in buildings and excavated from archaeological sites. The Bristlecone Pine chronology has proved very useful in assessing the relation between radiocarbon chronology and ages in calendar years. By radiocarbon dating wood of known age, discrepancies emerge between radiocarbon dating results and the scale of calendar years. We shall come back to this important point later.

Long tree ring chronologies have only been built up in a few areas, such as the south-west of North America. They are most useful for studying climatic history where it can be shown that tree ring thickness can be related to a particular climatic factor such as winter cold or the annual drought season. Such places may be in dry areas marginal to deserts or at high altitudes near the tree limit. In areas of deciduous woodland such as occur in north-west European lowlands it is more difficult to build up a chronology by correlation from one tree to another, because the varying ecological conditions from tree to tree may mask trends in sequences of tree rings. But studies of this sort are now beginning in Britain and north-west Europe, so that a tree ring chronology useful for dating timber may soon emerge. The method is being applied to timbers in mediaeval buildings and to tree stumps preserved in bogs.

RADIOMETRIC DATING

A number of geochemical methods of dating can be applied to the Ice Age. They are based on the measurements of amounts of elements which in the process of time are either formed by or are subject to radioactive decay. The rate of decay is known, so that the time interval may be determined between the present and the time at which the parent element was fixed and its decay began. In the Ice Age radiocarbon dating is the most important of these methods, though others have been applied. These other methods include potassium–argon dating, relying on the radioactive decay of ^{40}K to ^{40}A in crystalline rocks, such as those of volcanic origin. This method can be used in the older parts of the Ice Age and has provided the first good indication of the date of the beginning of the Ice Age, in the bracket 1·5–3·5 million years, depending on the definition of the boundary used. The long-lived Uranium isotopes ^{238}U and ^{235}U decay through short-lived isotopes to lead. Some of the intermediates have half-lives suitable for Ice Age dating. Dates have been obtained on marine carbonates back to some 200,000 years with this method, dates which are important in determining the chronology of marine deposits of the Ice Age.

Potassium–argon dating and uranium-series dating can be applied to sediments of 100,000 years old and older. Radiocarbon dating has its use in the more recent part of the time scale. Its half-life is taken to be 5568 years, and it is effective to ages of about 40,000 with suitable apparatus. Dates of greater age can be obtained by very careful measurement of the radioactive decay. Thus the limit of radiocarbon dating is normally up to some 40,000, with possibilities to about 60,000 years. The method itself is based on measuring the radiocarbon activity of biogenic material such as wood, peat, shells and bones. Radiocarbon, ^{14}C, is produced in the upper atmosphere by the interaction between nitrogen and cosmic ray neutrons (fig. 9.2). The radiocarbon so produced is oxidized to carbon dioxide and mixes with the normal (^{12}C) atmospheric carbon dioxide. It is then taken up by plants and animals and also becomes dissolved in sea and fresh water. Here it will be taken up by living organisms and exchange reactions will also occur between dissolved carbon dioxide, carbonates and bicarbonates to form inorganic sediments. Thus radiocarbon is incorporated into the biosphere and its concentration reaches an equilibrium with the atmospheric, freshwater and ocean carbon reservoir. On the death of the organism, the radiocarbon is trapped, there is no exchange with the reservoir, and it will decay with the passage of time. The rate of decay is given by the half-life, taken to be 5568 years. The concentration of radiocarbon falls to half of its value every 5568 years. By measuring the radioactivity of fossil biogenic material containing carbon, the date at which death took place and equilibrium by exchange ceased, can thus be determined.

The proportion of radiocarbon in the carbon reservoir is extremely low, about one atom in a million million carbon atoms. Very sensitive measuring equipment (pl. 44) is required to detect this concentration. Detection of disintegrations can be made in a gas proportional counter, where carbon is introduced as carbon dioxide, methane or acetylene gas, or in a liquid scintillation counter with carbon introduced as benzene. The chemical preparation techniques for producing gas or liquid from the sample are lengthy and complex. The materials used for dating are the natural carbon-containing biogenic substances. The richer they are in carbon the better. Thus charcoal is best, then wood, peat, bone and shell in that order. Bone can be utilized, even though its carbon content is low, for the collagen can be extracted and its radioactivity measured.

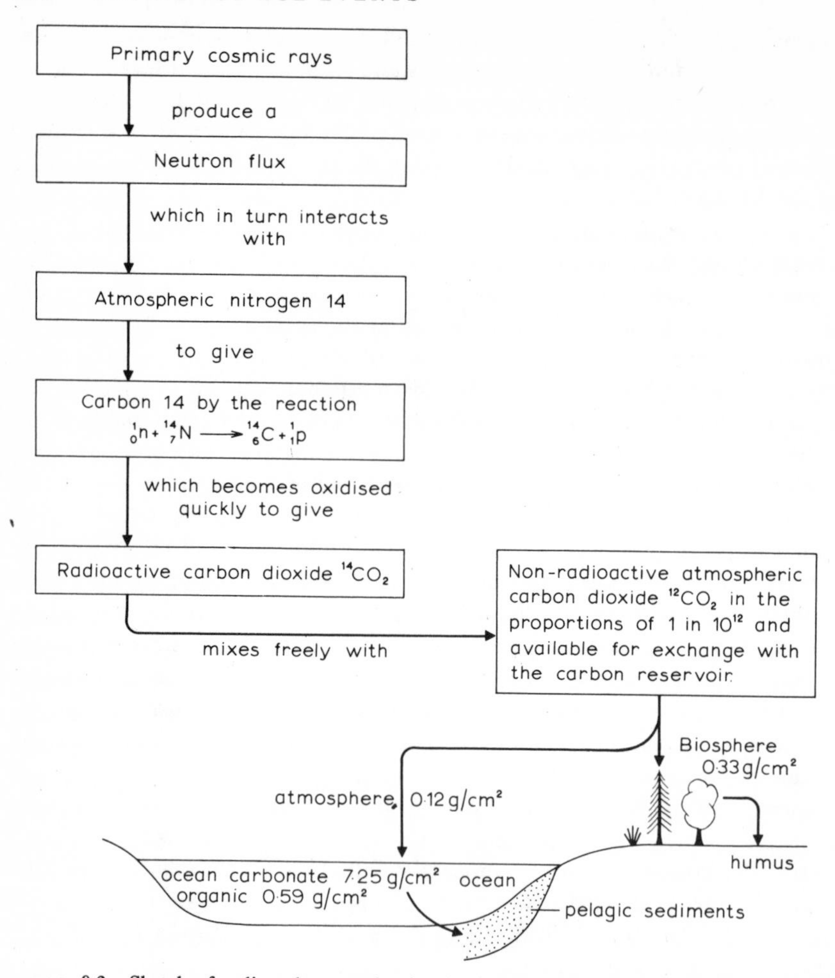

FIG. 9.2 Sketch of radiocarbon production in the atmosphere and its fixation in the biosphere and other carbon reservoirs. The amount of carbon (^{12}C and ^{14}C) in different parts of the exchange reservoir is shown, expressed as grams of carbon in exchange equilibrium with the atmospheric carbon dioxide for each square centimetre of the earth's surface. The main reservoir is the ocean carbonate.

The frequency of disintegrations measured is compared with that of a modern standard, and the difference will give the age of the sample in radiocarbon years. The measurement of the specific activity of a sample is very much complicated by the presence of background radiation, much stronger than that of the sample, which will also be measured by the detecting apparatus. Precautions have to be taken to reduce this unwanted radiation to as little as possible. The counting apparatus may be shielded in a lead housing, which attenuates soft gamma radiation derived from the surroundings. Cosmic ray mesons which penetrate to the counter may be cancelled electronically using an anticoincidence shield in the form of a ring of Geiger counters or a plastic scintillation shield. A pulse produced by a cosmic ray meson in the anticoincidence shield will then automatically prevent the recording of a pulse produced by the same particle in the detector itself.

The radiocarbon age is calculated as follows:

$$t = \frac{T}{\log_e 2} \times \log_e \frac{(S - b)}{(S_o - b)}$$

where t = age of sample in radiocarbon years
T = half-life of radiocarbon
S = count of fossil sample
b = background count
S_o = contemporary count

Thus the counters measure the radioactivity of the gas or liquid from the fossil sample (S), of the background (b) and of the modern standard (S_o). The common standard for recent radioactivity of living material is taken as 95 per cent of the activity of standard oxalic acid provided by the U.S. Bureau of Standards. The radiocarbon date is stated in terms of measured probabilities because of the random nature of radioactive decay. A date is given as a time interval within which the true age will lie with a certain probability. The error given for each determination is calculated on the counting statistics and refers to one standard deviation, within which there is a 68 per cent probability that the real date lies. The results of the assays are given as the radiocarbon age B.P., before the present, taken as 1950, or as a date A.D./B.C.

The use of the oxalic acid standard avoids the danger of different laboratories using different standards for modern activity, some of which may suffer from increased radioactivity as a result of nuclear weapon tests, and others from decreased activity as a result of infusion of dead carbon into the atmosphere by industrial processes using fossil fuel, all of which is of such an age that it contains no radiocarbon. The 95 per cent activity of the standard oxalic acid is close to the activity of wood which grew prior to the effects of the industrial revolution. This activity is then taken to give the modern radiocarbon activity of the natural carbon reservoir of the biosphere and atmosphere.

The limit of dating is set by the half-life and by the ability of detectors to measure very low activities. With gas counting the limit is near to seven half-lives, 30,000–40,000 years, unless a process of isotopic enrichment has been carried out. Using the latter process dates of up to 64,000 years have been obtained. If liquid counting is used, with a large volume of liquid, say 150 ml, dates have been obtained around 60,000 years and perhaps even greater ages are determinable.

We have briefly considered the process of obtaining a radiocarbon date. Let us now consider the meaning of this date. The radioactivity of the sample has been measured. Can we be sure that this remanent radioactivity is the product solely of the organisms which gave rise to the organic carbon? There are sources of error here. Old carbon may have been introduced; for example, by the presence of coal fragments or in submerged aquatic plants by photosynthetic absorption of old carbon derived from surrounding ancient rocks. These will cause the age to be older than the true age (Table 9.1). Then there is the reverse error caused by the introduction of more

Table 9.1

(After Kim, Ruch and Kempton, 1969)

A. Effect of contamination by infinitely old carbon on true age

% contamination	5	10	20	30	40	50
Years older than true age	400	830	1800	2650	4100	5570

B. Effect of contamination by modern carbon (oxalic acid standard) on true age

True age	Apparent age given by		
	1% contamination	5% contamination	10% contamination
600	540	160	modern
1,000	910	545	160
5,000	4,870	4,230	3,630
10,000	9,730	8,710	7,620
25,000	23,400	19,000	15,500
40,000	32,800	23,200	18,000
60,000	36,600	24,000	18,400

recent carbon by, say, the intrusion of recent rocks into the sample, the deposition of younger humic material, or the exchange of carbonate between shells and percolating ground water. These will both reduce the age of samples. This type of error may be especially effective in reducing the age of samples near the limit of radiocarbon dating (Table 9.1). Even the introduction of very small amounts of radiocarbon into a sample of great age will give an apparent age to a sample too old to have any measurable activity of its own.

Another source of error is what may be termed statigraphical. Can we be sure that the object dated can be certainly related in age to the geological sequence it is found in. For example, logs in a raised beach may not be the same age as the beach, or charcoal in an archaeological site may result from the burning of wood much older than the time of occupation the charcoal is supposed to represent.

By careful collection of samples and subsequent microscopic examination of samples it is possible to minimize the effects of these errors, but they should always be borne in mind in considering the interpretation of radiocarbon dates.

Apart from these sources of error, other considerations must be taken into account in interpreting radiocarbon dates. Certain assumptions have been made in establishing the method of radiocarbon dating. These assumptions are that carbon atoms in the reservoir are in equilibrium; that the cosmic ray flux, which establishes the radiocarbon, has remained constant over a long period of time; that the atmospheric source of radiocarbon is geographically constant in its specific activity, so that dates from different regions are comparable; and that the half-life of radiocarbon has been accurately determined. These are all very considerable assumptions. Perhaps the most critical one is the matter of the cosmic ray flux, which will determine whether or not the concentration of radiocarbon in the atmosphere in the past has remained constant. In recent years, many radiocarbon dates have been obtained from tree rings of trees up to 7000 years old. The Bristlecone Pine, mentioned previously in connection with dendrochronology, offers the possibility of relating sidereal years, with a chronology based on annual tree rings, to radiocarbon chronology. Fig. 9.3 shows the relation between the two. It will be seen that there are substantial discrepancies, up to 900 years nearly 5000 years ago, between the two. Short-term fluctuations are imposed on a longer-term trend.

The points in fig. 9.3, based on large numbers of radiocarbon dates of tree rings, have been used to draw up a complicated curve relating the radiocarbon and tree ring chronologies. The curve fluctuates so much that a single radiocarbon age may mean more than one true age. If it can be shown that the curve follows a similar course in different continents, we may then have a secure basis for converting radiocarbon to true age, but at present there is insufficient evidence to use any one curve as a basis of the conversion.

Changes in the intensity of the earth's magnetic field may be responsible for the alterations in the cosmic ray flux which have produced the long-term trend of discrepancy, while other factors, such as sun-spot cycles, may be affecting the cosmic ray flux to produce short-term cycles.

In spite of these difficulties in interpreting radiocarbon ages, there is no doubt that radiocarbon dating has provided a secure chronological basis for events of the last 50,000 years. Its great advantage is that it is applicable to organic material all over the world, using biogenic carbon arising in the

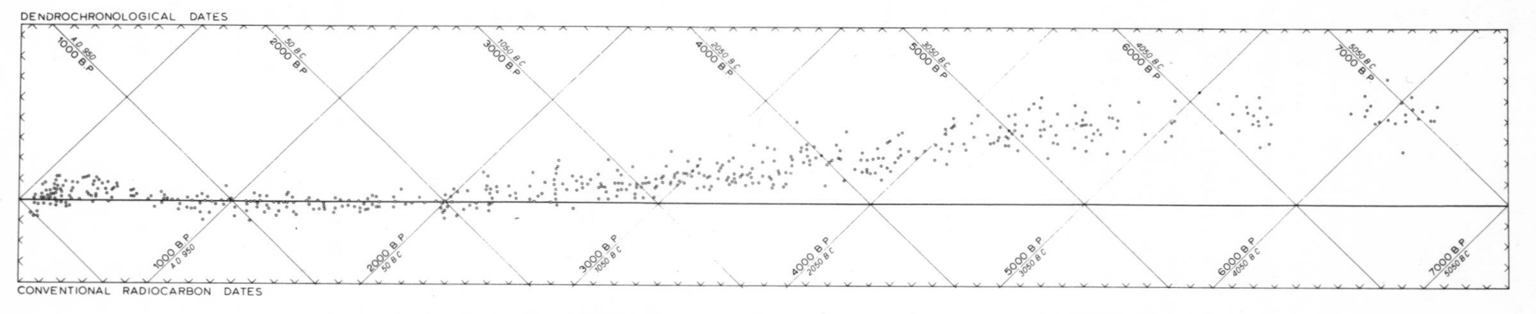

FIG. 9.3 Tree ring (including *Sequoia*, and species of pine) calibration of conventional radiocarbon ages. The radiocarbon dates are from wood in the northern hemisphere, with radiocarbon ages determined by laboratories in Arizona, Gröningen, Philadelphia, Cambridge and La Jolla. (After Olsson, 1970, pls. I, II and IV.)

terrestrial, freshwater and marine environments, in contexts which are significant for climatic change, plant and animal life, archaeology and sea level. This means that correlations are possible on a scale that was never before possible. Before the introduction of the method by Libby in the late 1940s, detailed correlations between distant places were not possible. The varve chronology could be applied to a few regions, but not Britain, and there was no widely applicable dating method. The introduction of radiocarbon dating has brought about a chronological revolution.

Relative dating of events and other methods of correlation

The methods of dating now to be discussed concern the correlation of events discerned in stratigraphic sequences. Such events may be related to faunal or vegetational changes, to sea level change, soil formation and weathering, volcanic eruptions, or climatic changes. From such correlations a sequence of geological events is worked out, which may in turn be related to an absolute chronology. An example is the building up of the East Anglian sequence which we have already considered (pp. 142–51). The vegetational history of each interglacial stage is different from that of the others. Interglacial deposits can be grouped according to their vegetational histories, and a stratigraphical sequence built up to which interglacial deposits of unknown age can be correlated.

Changes in the composition of faunas and floras have been used as a basis for relative dating in the Ice Age as in earlier geological epochs. But the effects of the rapid climatic changes in fauna and flora tend to mask the evolutionary changes significant for the stratigraphy of pre-Ice Age rocks. Some genera of animals, e.g. *Elephas, Rhinoceros*, and some species of Foraminifera do show evolutionary change and stages of their evolution have been used as a basis for correlation (see Chapter 7). A few species of plants became extinct during the Ice Age in Britain, and their presence or absence may be used as a basis for correlation. Examples of these European extinctions and first appearances are shown in fig. 9.4.

Because of the widespread occurrence of pollen in Ice Age deposits, pollen analysis has become an important method of relative dating. The development of pollen analysis of Flandrian peat and other sediments in north-west Europe has led to the subdivision of Flandrian time into a number of pollen-analytical zones, each characterized by a particular assemblage of forest trees. This succession has become a basis for relative dating on the assumption that pollen zones are to a certain extent synchronous. This method of relative dating has been much used in north-west Europe, in relation to vegetational history, climatic episodes and archaeology. But the basis for the subdivision of Flandrian time has now passed to radiocarbon

Stage \ Species	*Azolla filiculoides*	*Tsuga* (Hemlock Spruce)	*Abies alba* (Silver Fir)	*Mammuthus primigenius* (Mammoth)	*Euctenoceros ctenoides* (Tegelen Deer)	*Dicerorhinus kirchbergensis* (Merck's Rhinoceros)
FLANDRIAN				—		
DEVENSIAN				•		—
IPSWICHIAN				•		•
WOLSTONIAN	—		—	•		
HOXNIAN	•		•			•
ANGLIAN					—	
CROMERIAN	•		•		•	
BEESTONIAN						
PASTONIAN	•				•	
BAVENTIAN		—				
ANTIAN		•				
THURNIAN						
LUDHAMIAN		•				

FIG. 9.4 Occurrence of various plants and animals in the stages of the Pleistocene.

dating, though in many instances pollen analysis is still a useful method of relative dating. With deposits beyond the range of radiocarbon dating, e.g. interglacial deposits, pollen analysis is still of prime importance in revealing the vegetational histories characteristic of each interglacial and of the glacials, and in this field it remains an important method of relative dating.

The fluorine method is useful for the relative dating of bones and teeth found in sand and gravel. These remains contain a proportion of hydroxy-apatite which traps fluorine in the ground water. The fluorine to phosphate ratio is in part a measure of the time the fossils have been able to trap fluorine, so that bones in the same deposit which have suffered the same ground water conditions should show similarities in fluorine content. Thus a mixture of bones of different age in the same deposit may be revealed by the fluorine–phosphate ratio. Other chemical methods, such as nitrogen and uranium analyses, have been used to determine the relative ages of bones and teeth. The fluorine and other analytical methods have proved very useful in Britain for checking whether human remains are really as old as they have been thought to be. They were used in the investigation of the Piltdown skull and its associated bones, which led to the exposure of these forgeries.

Chemical and textural changes which accompany weathering and characterize soil formation have been used to identify fossil soil horizons. Correlation of soil horizons from one section to another has been used as a basis for chronology, especially in sections of loess, where intervals between the times of deposition of loess blankets are masked by soil formation. Few fossil soils are yet known in Britain, and their scarcity renders any correlation impossible at present.

Ash from volcanic eruptions settles over a wide area, and may form a useful stratigraphic marker, especially if the ash falls are dated. The Laacher See ash of the Eifel is widely distributed in north-west Germany. This ash fall has been dated by radiocarbon of the underlying and overlying organic sediments to the Allerød interstadial of Late-Weichselian times. In Britain, no ash falls of this sort have yet been found, though perhaps there is a possibility that the ash of Icelandic eruptions may have drifted to the extreme north and north-west of Britain.

Another means of correlation is based on changes in the earth's magnetic field, in direction over short intervals of time, and in showing reversed polarity over longer periods. The known change of direction of magnetization of particles in lake sediments has been used as a basis for correlation of recent lake sediments. Certain rocks, including igneous rocks and some sedimentary rocks, become magnetized by the earth's magnetic field at the time they are formed, so that by measuring the direction of their magnetization, the earth's field may be determined at the time of rock formation. If volcanic rocks are used to determine palaeomagnetic fields, samples of the same origin can be used for potassium–argon dating. Thus a chronology can be built up showing polarity epochs of normal and reversed magnetism, dated by the potassium–argon method. The chronology is applicable on a world-wide scale and to it can be correlated palaeomagnetic results from other kinds of rocks which are not dateable by potassium–argon. Fig. 9.5 shows the polarity epochs and palaeomagnetic events now known, calibrated by potassium–argon measurements. The method has been used to some extent in the Middle and Lower Pleistocene of East Anglia, but the detail is not yet so well known as to allow correlation with the more detailed palaeomagnetic sequence on the Continent.

Climatic change forms one of the main bases for correlation in the Ice Age. We have already mentioned the correlation of interglacials and glacials. Climatic episodes can also be characterized by physical measurements. Thus the relative abundance of oxygen isotopes $^{16}0$ and $^{18}0$ in biogenic carbonates (tests of Foraminifera, mollusc shells) secreted by animals living in water is related to the abundance of these isotopes at the time the carbonate was formed. The ratio in water will depend on the water temperature among other factors. Therefore by measuring the $^{18/16}0$ ratios in calcium carbonate skeletons, say tests of Foraminifera found at successive levels in a deep sea core, it is thought possible to deduce past temperature changes which occurred while the sediment was forming. The ratio of the isotopes in the carbonate is measured by a mass-spectrometer. The ratio changes are very

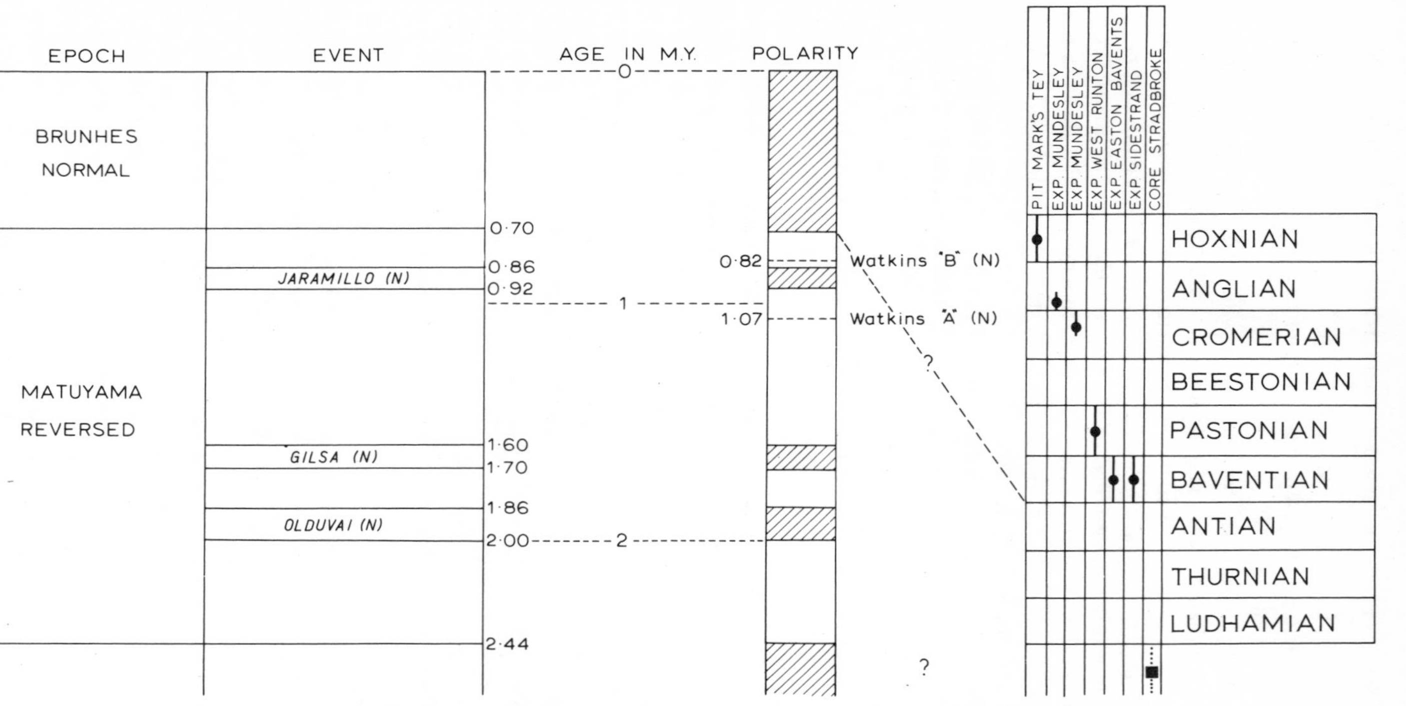

FIG. 9.5 Magnetic polarity changes in the last 2·5 million years. N, normal polarity within a reversed epoch. Possible correlations of the polarity stratigraphy with East Anglia stratigraphy are shown on the right. In the diagram on the right, the black circles indicate normal polarity of samples from pits, the black square from a borehole sample. Vertical unbroken lines through the circles give the range in time in which a polarity determination must be placed; the vertical dotted line gives the range in time in which a determination possibly must be placed. (After H. M. von Montfrans and J. Hospers, in *Geologie en Mignbouw*, 1969, Vol. 48, fig. 1, and Montfrans, 1971, fig. 31.)

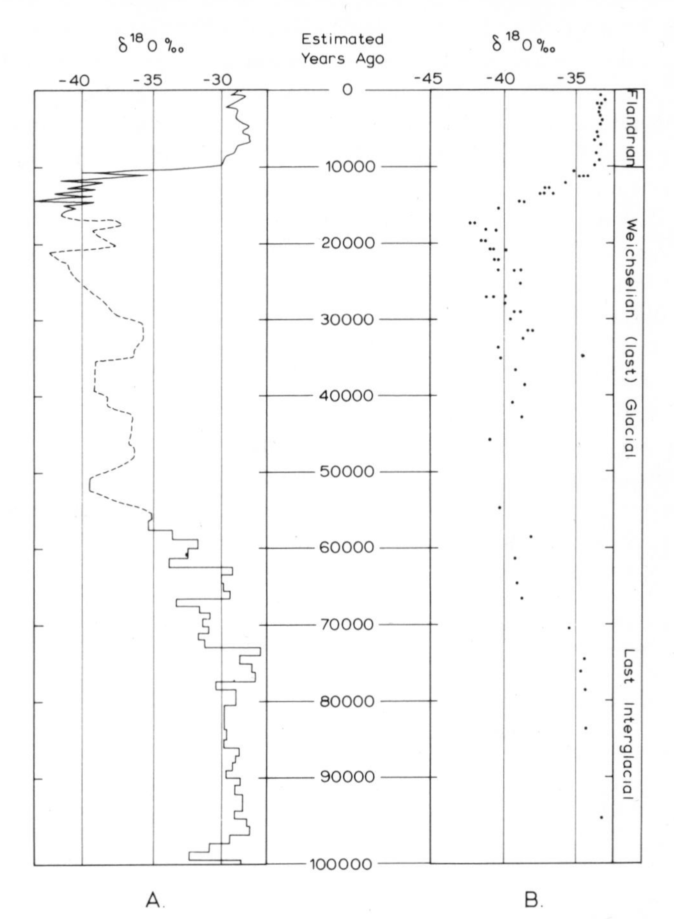

FIG. 9.6 Climatic variations in the last 100,000 years, expressed as changing $^{18}O/^{16}O$ ratios. δ is the per mille deviation of $^{18}O/^{16}O$ ratio in an ice sample from that of Standard Mean Ocean Water. Colder temperatures are indicated by move of the curves to the left (decreasing δ values). A, variations in the Camp Century ice core, Greenland (after W. Dansgaard *et al.*, in *Science*, 1969, Vol. 166, fig. 5). B, variations in the Byrd Station ice core, Antarctica (after Epstein *et al.*, 1970, fig. 1).

small; ^{18}O enrichment in respect to ^{16}O increases 0.02 per cent per 1°C temperature fall, and the instrument used has to be very sensitive. Other factors, such as the dilution of sea water, also affect the $^{18/16}O$ ratio, and the validity of the measurements as indicators of temperature change needs further investigation. It is possible that the changes in isotope ratio are effected less by temperature than by ocean dilution on melting, and concentration on forming, of the world's ice sheets. Nevertheless, episodes of differing isotope ratio are apparent in deep sea cores based on such measurements, and the cycles of change are a very useful basis for correlation of marine sediments, regardless of the significance of the results for temperature change.

$^{18/16}O$ ratios have also been measured in cores of ice from the Greenland and Antarctic ice sheets (fig. 9.6). The age–depth relations of cores are estimated from measured accumulation rates and calculations of thinning through-flow. The changes in isotope ratio are thought to reflect climatic change, and the curves from both poles have a certain similarity. This is a most important point, as it argues for a similar pattern of climatic change in both hemispheres. The curves indicate that the last cold stage or last glaciation started some 75,000 years ago and ended some 11,000 years ago. These dates are close to those obtained from other evidence.

Let us finally consider how all these methods of dating and correlation

Range of method	Time scale 10^3 years (not linear)		^{14}C NW Europe	K/A Europe	K/A Africa	Th^{230}/U^{234} (coral, shell, oolite near or above present sea level i-g) Mediterranean, Morocco
	10					
^{14}C	20	Weichselian	Flandrian 10250 B.P.			
	30					
	40					
	50					
	60		Chelford i-s			
	70					
	80		Eemian/Ipswichian			Ouljian
	90		i-g			(5-8m) sl
	100					
						Tyrrhenian ? sl
K/A	125					
	150			Holstein (= Hoxnian) i-g		
	200			Mindel g		Anfatian sl
	250			(Rhine middle and old		(25 - 34m)
	300			Middle Terrace)		
	350			Cromer i-g		
	400			Günz g		
	450			(Rhine High Terrace)		
	500					
	550					
	600					
	1000					
				Villafranchian f (S France)		
	2000				Olduvai (Villafranchian f)	
	3000					

Features dated: f = fauna; g = glacial; i-g = interglacial; i-s = interstadial; sl = sea level

FIG. 9.7 An outline chronology of events of the Pleistocene. The range of dating methods is shown on the left.

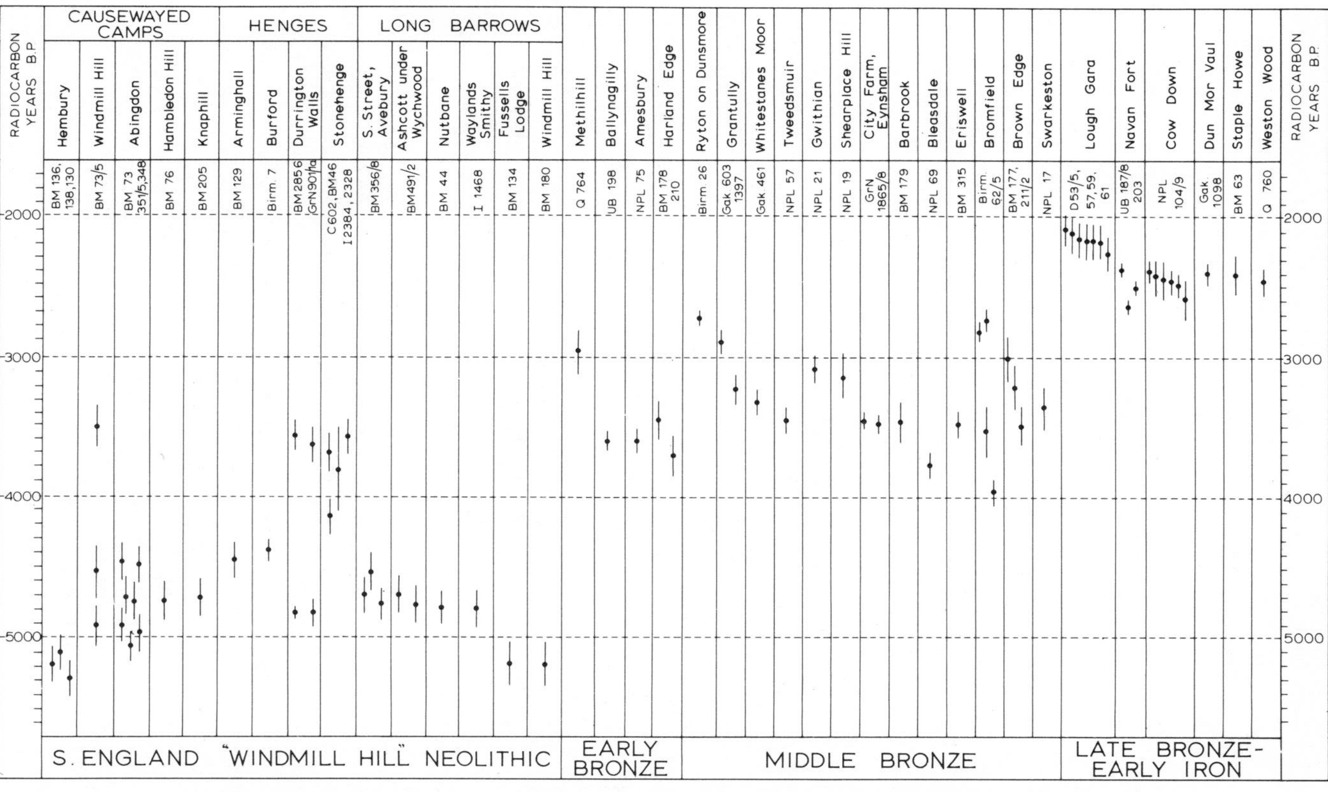

FIG. 9.8 Radiocarbon ages of various Neolithic and post-Neolithic sites in the British Isles. The vertical line through each date gives the quoted probable error of the determination (1 standard deviation). The code letters and figures below the site names refer to the laboratory code for sample dated. (After Godwin, 1970, figs. 1 and 5.)

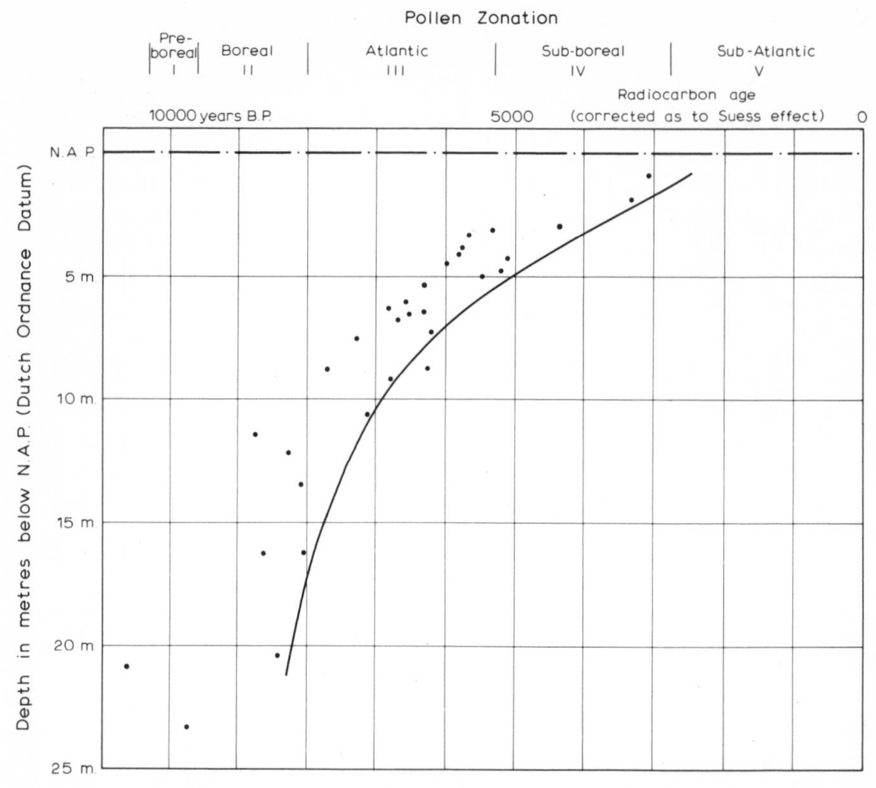

FIG. 9.9 Time-depth graph showing radiocarbon ages of peats at depth in the Netherlands. The peats dated are the oldest at various sites and formed in response to rising sea level. The curve is drawn through the lowermost points at any given age on the assumption that peat growth may have started at different heights above high tide level. (After Jelgersma, 1961, fig. 22.)

have contributed to the Ice Age chronology of Britain. Radiocarbon dating has provided a good chronological calibration of the stratigraphy of deposits of the last 40,000 years (fig. 9.7). This period covers the latter half of the last glaciation (the Devensian stage) and the Flandrian stage, which started some 10,000 years ago. Because of the wide application of radiocarbon dating, already mentioned, we have a very good idea of the archaeological chronology (fig. 9.8), of the chronology of the rise of sea level after the last glaciation (fig. 9.9), of the chronology of the great vegetational and faunal changes of the time, and of the ice advances and retreats of the latter part of the last glaciation. In this period of the last 40,000 years, then, we owe much to radiocarbon dating; and we owe a certain amount to relative dating by pollen analysis.

In those periods of the Ice Age which are beyond the range of radiocarbon dating, about 96 per cent of the length of the Ice Age, we have to revert to other methods of dating. Having built up a local British stratigraphical sequence based on temperate and cold stages, we have to try and correlate this sequence to a sequence which can be dated in some other part of the world, using another dating method which can carry us back 2 or 3 million years to the beginning of the Ice Age. This can be done to some extent by correlating our interglacial stages with those of the Continent, which themselves can be related to some extent to potassium–argon dates of European volcanic rocks. In the Middle and Lower Pleistocene it is also just beginning to be possible to correlate the climatic stages with palaeomagnetic polarity epochs (fig. 9.5). This, allied to potassium–argon dating, promises to be the key for an absolute chronology of the older parts of the Pleistocene. Thus we may hope in time to have climatic stages, and substages, identified on terrestrial or marine fauna, flora, geology, and perhaps oxygen isotopes, dated by correlation with the palaeomagnetic sequence and potassium–argon dating.

10

The Physical Geography of the Last Interglacial in Britain

In earlier chapters we have discussed various aspects of the Ice Age in Britain. We cannot make an integrated reconstruction of the geography of the whole of the Ice Age simply because we have not enough evidence ranging through the period. The easiest reconstruction would be of the Flandrian (postglacial) period in Britain, but we feel that an attempt should be made to present a coherent picture of a more remote period. This will show how difficult it is to fill in more than the outlines of the picture. The choice of the last (Ipswichian) interglacial period is dictated by two facts: the first is that there is more evidence available than for the other interglacials, and the second that we have been directly involved ourselves with palaeo-ecological studies of Ipswichian interglacial deposits over a number of years.

In the old-fashioned type of geographical account the physical geography was usually presented in the order relief, climate and vegetation, but in reconstructing the past it is preferable to reverse the order. Vegetation is the only feature, apart from the animals, of which we have direct evidence in the form of fossils, and hence we are more sure about this than about the climate and the relief. The climate can to some extent be inferred from the fossils, though there are problems in such interpretations which we have discussed in earlier chapters. The relief can only really be reconstructed from the attitude and nature of deposits, and the position of sea level at any given time by lithological, floristic and faunistic aspects of depositional sequences. In attempting this reconstruction, which cuts across the systematic treatment of earlier chapters, certain facts and ideas are bound to be reiterated if the account is to retain coherence.

Vegetation

In 1928, Jessen and Milthers published a comprehensive account of the last interglacial in Denmark and north-western Germany, in which they introduced a scheme of pollen zones. This, with the exception of a climatic oscillation at the end, has stood the test of time and has proved to be largely applicable to British last interglacial deposits, with the exception that the *Picea* (spruce) zone, *h*, has not yet been found in Britain. British Ipswichian interglacial sites have been studied botanically from about 1950 and those for which pollen evidence is available are as follows (the references quoted are the latest ones containing the evidence used in the reconstruction that follows):

(1) Histon Road, Cambridge: deposits revealed in sewer trenches when civilization spread from Cambridge to the nearest villages about 1950. The site is on the Barnwell Terrace of the river Cam (Hollingworth, Allison and Godwin, 1950; Walker, 1953; Sparks and West, 1959).

(2) Bobbitshole, Belstead, Ipswich: deposits exposed in the construction of sewage works in a minor valley near Ipswich (Davis, 1955; West, 1957; Sparks, 1957; Sparks and West, 1967).

(3) Trafalgar Square, London: Thames deposits revealed in the foundation of Uganda Government building (Franks, 1960).

(4) Selsey, Sussex: foreshore deposits exposed at low spring tides and revealed by severe coastal erosion on the east side of Selsey Bill (West and Sparks, 1960).

(5) Stone, Hampshire: foreshore deposits exposed at low tide and sampled by coring (West and Sparks, 1960).

(6) Stutton, Suffolk: brickearth exposed on the shore and in the cliffs on the north side of the Stour estuary south of Ipswich (Sparks and West, 1964).

(7) Ilford, Essex, a cored section through a famous Thames deposit kindly provided by George Wimpey and Co. Ltd (West, Lambert and Sparks, 1964).

(8) Aveley, Essex: deposits exposed in a pit in association with elephant skeletons (West, 1969).

(9) Wortwell, Norfolk: a deposit revealed during sewer construction (Sparks and West, 1968).

(10) Wretton, Norfolk: large sections of low terrace sediments exposed in the construction of the marginal relief channel around the eastern side of the Fens (Sparks and West, 1970).

The location of these deposits and their ranges in the interglacial are shown in fig. 10.1. They are all in the south-east of England and this limits the applicability of the geographical construction we shall base upon them.

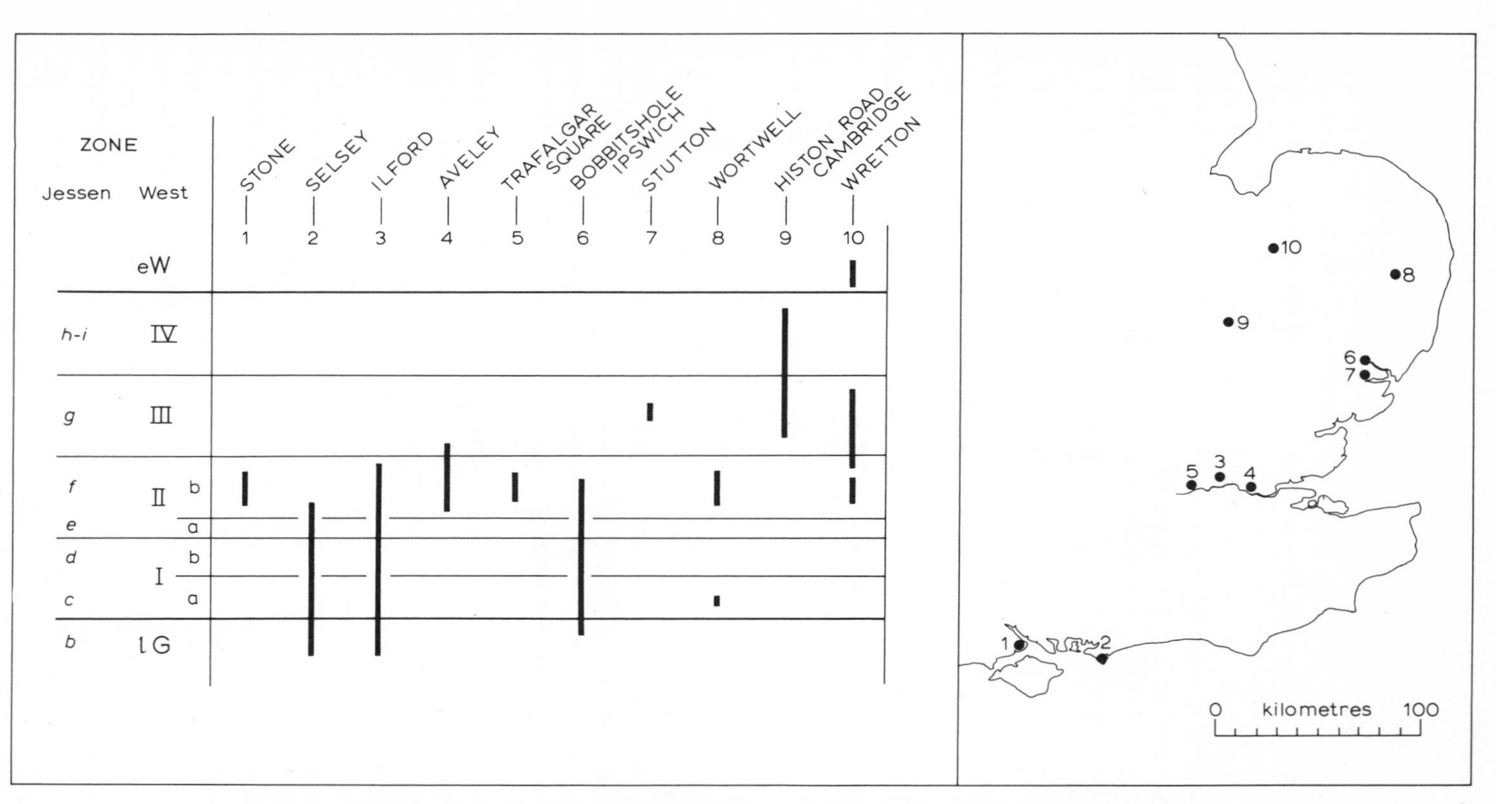

FIG. 10.1 Ipswichian interglacial sites in south-east England: location and time-span covered (after Sparks and West, 1970, fig. 1).

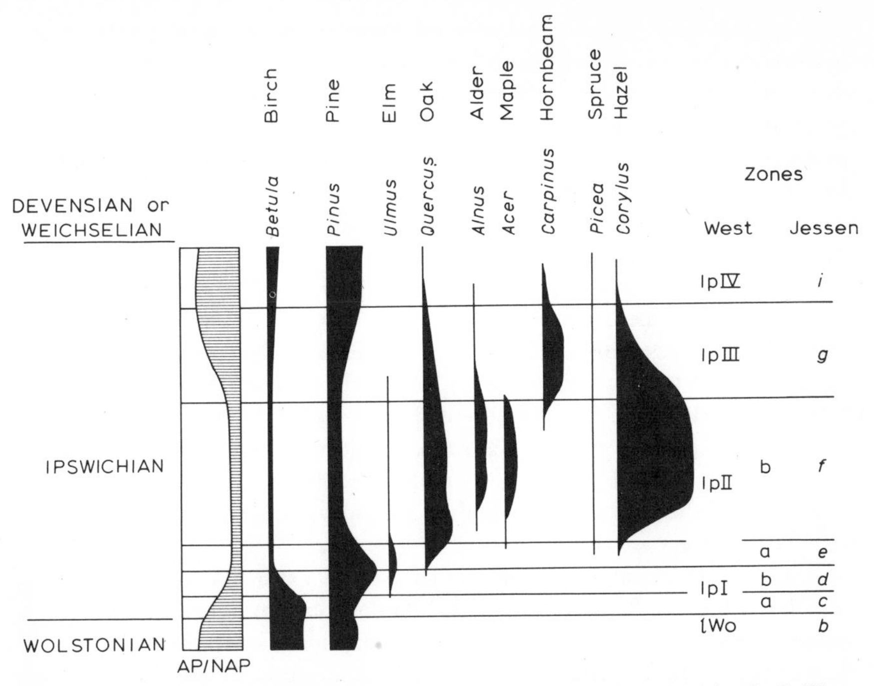

FIG. 10.2 Generalized pollen diagram through the Ipswichian interglacial (after R. G. West, *Pleistocene Geology and Biology*, London, Longmans, 1968, fig. 13.9).

A generalized pollen diagram through the interglacial, based upon these known British deposits, is shown in fig. 10.2. On this diagram are shown the modern general interglacial zones (see Chapter 6) and their subdivisions, as well as the older zones of Jessen and Milthers (1928). The vegetation succession of the interglacial may be interpreted from this and from the records of macroscopic plant remains.

The late-glacial phase of the Wolstonian, which appears at the base of the sections at Selsey, Bobbitshole and Ilford, is a period of predominantly open vegetation (80–90 per cent of the pollen total is non-tree pollen) with some *Betula* (birch), *Salix* (willow) and *Juniperus* (juniper).

These same three sections also show the best development of the pre-temperate zones, Ip Ia and Ip Ib. In zone Ip Ia there is a slow closure of the vegetation cover with a steady increase of the AP–NAP ratio, i.e. arboreal pollen–non-arboreal pollen ratio. The trees are dominated by *Betula* and *Pinus* (pine), but there are some regional differences. In the south, if we may judge from the Selsey pollen diagram, the two are almost equally represented, but farther north at Bobbitshole *Betula* is much more important than *Pinus*. In zone Ip Ib the forest cover is much more complete, as witnessed by the higher AP–NAP ratio, and its composition has changed.

Throughout Ip Ib *Betula* declines and *Pinus* increases until it reaches its maximum frequency at the boundary between Ip I and Ip II. In zone Ip Ib two important temperate forest trees appear, namely *Ulmus* (elm) and *Quercus* (oak), but they do not reach high pollen frequencies in this zone. Summing up, it can be said that this pre-temperate zone is dominated in turn by *Betula* and *Pinus*. Although designated pre-temperate, this must not be construed as implying a sub-Arctic climate, but as signifying a phase before the establishment of true temperate forest.

The early-temperate zone Ip IIa is essentially transitional. *Pinus* declines from its peak at the beginning of the zone; *Ulmus* increases to a maximum and then declines at Bobbitshole but not at Selsey; *Quercus* forms about 50 per cent of the tree pollen sum by the end of the zone; both *Acer* (maple) and *Corylus* (hazel) appear towards the end of the zone but not in great quantities. It is important to note that *Pinus* is still very important in the pollen total, though because of its high pollen production and because of the resistance of the pollen to weathering, this may exaggerate its real importance in the forest cover.

Zone Ip IIb is marked primarily by a large expansion of *Quercus* and *Corylus*, although *Pinus* remains important. *Acer* also reaches its peak frequency though it must have been somewhat local in its distribution: at Trafalgar Square its pollen forms about 30–40 per cent of the tree pollen; at Bobbitshole and Stone its maximum is about 10 per cent; at Ilford, Selsey and Wortwell there is only a trace. Before the Wretton deposit was discovered we had almost come to regard the Ipswichian as the alder-less interglacial, in great contrast to the Hoxnian and the Flandrian in both of which *Alnus* is abundant; and also in contrast to the Dutch Eemian (Ipswichian) deposits where *Alnus* pollen is more frequent. There is plenty of *Alnus* pollen at Wretton and also very frequent cones and fruits, so that it certainly occurs in the interglacial but seems to have been even more local than *Acer*.

The late-temperate zone Ip III is undivided in Britain. It is the most characteristic zone of the interglacial with the dominance of *Carpinus* (hornbeam) pollen, which forms as much as 60–70 per cent of the tree pollen at some sites, as its distinguishing feature. *Quercus* and *Corylus* both decrease through the zone and *Pinus* increases. It also seems that the forest cover opens out because the proportion of non-tree pollen rises. In the opinion of some palaeobotanists, this change may have been brought about not by climatic deterioration but by successful competition of *Carpinus* against the dominant trees of zone II and perhaps by changing soil conditions.

Finally in Britain, the vegetation cycle ends with Ip IV, the post-temperate zone, which is dominated by *Pinus* and *Betula* and has a high non-tree pollen ratio. Jessen and Milther's *Picea* (spruce) zone, which begins the correlative zone on the Continent, has not yet been found in Britain.

Thus, in general terms, though fortunately not in all details, the forest vegetation pattern is similar to that of other interglacials and the postglacial: a succession from birch and pine domination, with considerable open spaces, to a closed mixed oak forest and then back to conditions rather similar to those which started the interglacial. Even the local communities recognized from the assemblages of macroscopic plant remains seem very much like present-day communities.

Climate

It is difficult to translate the presence of plant remains into climatic terms, and just as difficult to interpret the animal evidence. It would probably be an accurate summing-up of the position to say that when one has a mass of biological evidence there is a basis for speculation, but not the material for an accurate deduction of all the features of the climate. In Britain the interglacials produced little lithological evidence for the interpretation of climate, unlike the glacials, so that all speculation is based on plants and animals.

One thing is certain: the British climate in the Ipswichian cannot have been greatly different from what it is now. That is shown very clearly by the general succession of vegetation zones. However, there were some detailed differences as can be detected from the presence of certain plants and animals at particular periods.

In the Wolstonian late-glacial (zone 1Wo.) are found remains of southern plants, e.g. ***Berula erecta***, ***Lycopus europaeus***, ***Potamogeton crispus***, ***Groenlandia densa***, ***Carex acutiformis*** and ***Typha latifolia***. The last is now confined south of the 14°C July isotherm in Scandinavia and is found below 1000 m in the Alps. The 14°C July isotherm does in fact go quite a long way north in Scandinavia. It passes through Trondheim on the middle Norwegian coast, but Archangel has almost 16° as its July temperature, and Haparanda, practically on the Arctic Circle, has 15° for July. Even the very coldest places in Siberia have mean July temperatures just above 14°. So the assumption of a continental climate for this phase would cover the temperature requirements and also square with certain other botanical evidence, namely the openness of the plant cover as shown by the high proportion of non-tree pollen. Yet open vegetation does not necessarily reflect severe climate: it might also be caused by a dry, windy climate. It may have been caused by waterlogging, a situation which could have arisen as the permafrost disappeared. It might have been due to the slow immigration rate of plants or to the slowness of soil development on raw environments failing to allow advanced plant communities to establish themselves. It could have been due to biotic factors: overgrazing by bison has been suggested for similar conditions in an intra-Devensian period at Upton Warren (Coope, Shotton and Strachan, 1961).

Assuming that the causes were climatic, do we have to infer a climate as continental as that of Archangel? The answer is probably not. The summer temperature may have been about 14° but the winter temperature is unlikely to have been as low as Archangel's −14° or Haparanda's −12°. Both of these places are in 65–66° latitude and the excess of winter radiation over insolation there can surely never have been equalled in 51–53° latitude. Yet, if we are to explain the treeless park–steppe climatically, the winter temperature must have been below 0°. Or is the answer to be found in assuming another feature of continental climate, namely a low total rainfall with a summer maximum, so that cold and drought acted to inhibit tree growth? Should the climatic analogy be with the steppes, but with a smaller temperature range? Of course, there is probably no precise analogy because of the different latitudes in question, a point that has been stressed before, and in fact few steppe plants are found in the deposits.

Turning now to the middle of the interglacial, we must try to assess the climate in zone IIb, which represents the climax of the temperate mixed oak forest. Both plants and animals rule out winters colder than we have now, but there is evidence for some small degree of extra summer warmth.

In the first place, although the *Acer* pollen which is abundant at some sites cannot be identified specifically, the winged fruits can and these are not those of the only British native species, *Acer campestre*, the field maple. It will be recalled that the sycamore, *Acer pseudo-platanus*, is not a native tree. The Ipswichian fruits turn out to be *Acer monspessulanum*, a species with a wide distribution in southern and central Europe. It will grow in Britain at the present time under protected conditions; for example there is a flourishing specimen in the Botanic Gardens at Cambridge though it does not set viable seed. Secondly, there are a few species which were present in zone Ip II which are not now native in Britain and have generally continental and southern distributions in Europe. Among them are *Najas minor* and *Salvinia natans*, both of which occur in zone Ip IIb, the former at Bobbitshole, Trafalgar Square, Selsey and Wortwell, the latter at Bobbitshole and Wortwell.

Other plants rarely fruit now and their fruits are much rarer in the postglacial than in the Ipswichian. Among these are *Hydrocharis morsus-ranae* (the frogbit) and *Lemna* cf. *minor* (duckweed). The latter now reproduces mainly vegetatively, but its seeds make up 3 per cent of the total of fruits and seeds at some levels in the Ilford deposit. The water soldier, *Stratiotes aloides*, has female plants only in the north of its range, both sexes in the middle, and only males in the south. At present Britain is in the north of the range with female plants reproducing vegetatively, but in the

Ipswichian both sexes must have been present to judge from the abundance of fruits in zone Ip IIb at Selsey.

Finally, *Trapa natans*, the water chestnut, a continental plant not now found in Britain, occurs at Trafalgar Square and Wortwell. It is also known from the earlier interglacial Cromer Forest Bed Series on the Norfolk coast.

These records, together with the zone Ip III dominance of *Carpinus*, a tree with a restricted distribution in south-east England and a wider distribution on the Continent, suggest an Ipswichian climatic optimum a little warmer and more continental than our present climate. It is difficult to say how much higher summer temperatures were – perhaps 2° or 3°C: this would agree very closely with Jessen and Milther's estimates of the increased warmth of the summers in the same period in Denmark. It would thus mean summer temperatures of about 19–20° in south-eastern England.

To a large extent the plant evidence is borne out by the animals. In this connection we must abandon any wild notions, engendered by the presence of such mammals as the hippopotamus, about tropical climates, and concentrate mostly on the evidence supplied by such lower orders as the beetles and the snails.

Beetles are known both from Bobbitshole and Trafalgar Square (Coope, 1965, 1967). At the former site the presence of the carabid ground beetle, *Oodes gracilis*, now extinct in Britain and typically central and southern European in its distribution, would seem to demand summers at least 1·5° warmer than those at present. At Trafalgar Square a large number of dung beetles, mainly scarab beetles, was found. At present the dung beetle fauna of Britain and northern Europe is dominated by the genus *Aphodius*, whereas in central and southern Europe the scarab beetles dominate the dung beetle fauna. Thus the beetle evidence points very clearly to the same sort of conclusion as the plants, and also agrees with what may be inferred from the presence of the European pond tortoise, *Emys orbicularis*, at Bobbitshole.

The snails and other non-marine molluscs reveal little about the early stages of the Ipswichian, because deposits of this age have so far yielded few faunas, the exception being at Wortwell where the species are all widely-distributed, tolerant European species.

In the middle of the interglacial, largely in zones IIb and III, is a number of characteristic species, now extinct or very much reduced in Britain (Sparks, 1964). Among these is a generally south-eastern and central European group which includes *Clausilia pumila*, *Helicella striata*, *Ena montana* and *Vallonia enniensis*: a southern group which includes *Belgrandia marginata*, *Helicodonta obvoluta* and *Corbicula fluminalis*: a continental species, *Fruticicola fruticum*, and two species which might be called south-western. One of these, *Potamida littoralis*, now lives in France and Spain, but it is known fossil from Germany and south-eastern Europe so that its present distribution may not indicate its potential range. The other is *Helicella crayfordensis*, a species now extinct, but placed on shell characteristics in the sub-genus *Candidula*, which is primarily south-western in Europe. But *Helicella* is notoriously difficult to diagnose on shell characteristics and the attribution of the species to *Candidula* could be in error and hence the distributional information drawn from it wrong. In any case, if one compares this primarily continental and southern batch of species with the characteristic species of the postglacial, the latter emerge as much more strongly oceanic, Mediterranean–oceanic, or central and south-western European. Man's clearance of the land in the postglacial should have increased the temperature range and the aridity of the climate near the ground, so that the generally more oceanic climate suggested by the snail distributions would probably under-represent the true climatic differences.

Although the representation of southern species declines towards the Devensian early-glacial phase (eD), they are nevertheless still found in small numbers. Some of the occurrences might be explained away as contamination from earlier deposits, but the feature is persistent and the species also occur in deposits which are full-glacial in type. *Belgrandia marginata*, for example, occurs in such deposits at Earith, Huntingdonshire; Wretton, Norfolk; and Sidgwick Avenue, Cambridge: *Segmentina nitida* is found at Wretton, *Truncatellina cylindrica* at Sidgwick Avenue, *Vertigo angustior* at both these sites, and *Corbicula fluminalis* (fresh-looking juveniles) at Earith. Thus, even at the end of the Ipswichian interglacial it would seem that there were either warm micro-climates or that there was still considerable summer warmth. Similar conclusions can be drawn from the survival of southern plants (Bell, 1969). All this is very difficult to square with the extreme periglacial climates suggested by some authorities to have preceded the onset of glacial periods.

It is unfortunate that the amount of evidence for the climate of interglacial periods is probably less than that for the Devensian (Weichselian) glaciation, where different lines of convergent evidence can be used (Shotton, 1962). The periglacial structures and patterns can be compared with the distribution of their present-day equivalents and the likely climate thus inferred; the snow line can be calculated from the position of corries and the amount of climatic change needed for its depression can be estimated; the beetle assemblages can be compared with the present beetle faunas of different Scandinavian vegetation zones (Coope 1965, 1967) to give a third means of estimating climate.

Meanwhile, all the evidence that we have of the Ipswichian climate seems to suggest a greater degree of continentality than we have now, especially early and late in the interglacial but still present in the middle. This is logical enough if we believe in anticyclonic conditions with easterly winds being more prevalent in cold periods and more normal westerly conditions being established in the middle parts of the interglacial. Yet the middle of the interglacial was apparently still a little more continental than either the Hoxnian or the postglacial.

Relief

It is always difficult to reconstruct that which has disappeared. The reconstruction of relief can only be made in very general terms. One may be able to arrive at the amount of erosion in a given period by comparing the landscape at the beginning of that period with the landscape at the end; unfortunately it is difficult to estimate the landscape at the beginning other than by adding the amount of erosion to the landscape at the end.

Presumably in the areas covered by the Devensian (Weichselian) ice sheet and at its margins the Ipswichian relief may have been very different from the present relief, but one is hampered, or perhaps aided, in one's estimates by the lack of evidence. There is little doubt that many of the sharper features of glacial erosion in the highlands of Britain were at least modified by the Devensian ice sheet. These probably include most of the glacial diversions of drainage in the highlands, whether caused by glacial breaching, overflow channels or sub-glacial channels. Still, we cannot really be sure of this, although it would have been very curious if the more extensive Anglian and Wolstonian ice sheets had failed to modify the drainage enormously. Therefore, though it might be expected that the Ipswichian landscape in the highlands was appreciably different from the present landscape, we have no reliable means of inference.

Outside the limits of the Devensian ice sheet we are in a better position to judge. All the Ipswichian deposits we have discussed lie in present valleys, usually as low terraces, or in present coastal locations, so that the main drainage patterns and valley depths seem to have been not very different from what they are now. The absence of Ipswichian interglacial deposits from the upper parts of valleys may indicate a scouring of those parts during the Devensian. In their generally valley situations the Ipswichian interglacial deposits contrast fairly strongly with the Hoxnian deposits. Outside the glacial limits the latter, e.g. the Swanscombe deposits on the Thames, are probably closely related to valleys, but within the area later covered by the Wolstonian ice sheet deposits at such places as Hoxne itself and Marks Tey near Colchester seem to be far less tied to the present relief, so that it may well be that considerable relief modification occurred between the Hoxnian and Ipswichian interglacials.

It is true that if one plots the position of Ipswichian terraces of rivers such as the Cam one may detect here and there lateral shifting of the streams of a kilometre or two in their wide valleys, but there have probably been few post-Ipswichian major relief changes.

This is also apparent from the general distribution of the boulder clay plateaux of the east Midlands and East Anglia, many of which are Wolstonian in age. The interfluvial areas, for example between the drainage to the Wash and the drainage to the east coast and the Thames, can have been little modified since the Ipswichian except perhaps at their edges. The latter assumption is necessary in the light of the amount of periglacial stripping of slopes suggested by Williams for southern England (see Chapter 4, p. 119). It seems likely that such erosion was confined to the edges of the plateaux and did not materially affect their central parts. It is also obvious that in these areas outside the Devensian ice sheet there can have been no further glacial modification of drainage patterns.

Finally, we must discuss the position of sea level in the interglacial as this must have been the basic control on erosion. In the early parts of the interglacial sea level was lower than now, but there is at present no known series of deposits which will allow an accurate determination of its height at a specific time. The first question to which some answer can be given is, at what stage did the interglacial sea level rise above the present level? The second is, what is the maximum height it reached?, and the third, when did it fall back below its present level?

The key area for approaching these questions is at present the English Channel. At Selsey in Sussex an enormous molluscan change to brackish water forms occurs at a level of −1·76 m O.D. (see p. 218) (fig. 7.5) in deposits of the early-temperate zone Ip IIb. The inference from this is that at this time the Ipswichian high-tide level rose above this height. If intertidal ranges have not changed much this probably represented a mean sea level of about −5 m O.D.

Immediately to the west of the interglacial deposit at Selsey marine gravels and brickearth in the low cliffs show the nature of the succeeding sedimentation. Low-level beach gravels are widely distributed over the Sussex coastal plain east of Selsey at varying elevations. They have been described as a '15-ft' (4·5 m beach) and seem to reach their highest point at Black Rock, Brighton, where the top of the beach shingle is known to reach almost 12 m, and even this is not quite at the inshore margin of the beach. If we assume that the inshore margin was at 12–13 m, that wave action extends about 1 m above high-tide level, and that the tidal range at Brighton was then about 7 m as it is now, then the highest mean sea level of the interglacial was probably about 8 m above present level. We are, of course, not sure whether the low raised beach deposits of the Sussex coast represent the transgression to or the regression from this level or both. Also, in Sussex there is at present no means of detecting the precise zoning of maximum sea level by pollen analysis.

At Stone in Hampshire confirmatory evidence of the Selsey section is provided by the occurrence of brackish Mollusca and littoral Foraminifera in a foreshore deposit, of zone Ip IIb age, which extends from −2·5 to −0·2 m O.D., and is backed by gravel cliffs which it has been suggested are formed of terrestrial gravels slightly re-worked by marine action.

Unfortunately, on the English side of the Channel there is no regression deposit which can be dated by pollen, but there is one such series of deposits on the French coast near Mulberry Harbour, the site of the 1944 Normandy landings. At Asnelles-belle-plage a change from marine to freshwater conditions occurs just below present mean sea level and deposits probably lying above this nearby at St Côme de Fresné have a pollen spectrum compatible with zone Ip IV. We cannot be absolutely sure of the date because similar spectra are known from Devensian interstadials.

All this evidence suggests a marine transgression in zone IIb, a maximum sea level of 8 m, and a regression in zone IV with the obvious implication that the highest Ipswichian sea level must have been in zone III. To regard zone III as the peak of the interglacial, as it were, would not conflict with the vegetation evidence, because it has already been suggested that zone III may represent not a climatic deterioration but a soil change. It also agrees reasonably well with von der Brelie's (1954) evidence from the Netherlands and north Germany of a transgression ranging in date between zones Ib and IIb and a regression in late zone III or early zone IV. The differences may be due to warping in the Netherlands but the general peak is similar in its placing in the interglacial.

By sheer luck the zoning of the peak interglacial sea level was obtained at Stutton in Suffolk. In the famous brickearth there is a minute number of the brackish snail species, *Pseudamnicola confusa*, at one level, while about 0·4 m below it the only pollen-bearing part discovered in the deposit gave a typical Ip III pollen spectrum, i.e. the *Carpinus* zone. Thus at Stutton the highest sea level cannot have been before zone Ip III and is unlikely to have been later. If we calculate the likely mean sea level at that time at Stutton it is about −1 m O.D. The difference of about 9 m between the levels at Brighton and Stutton represents the amount of downwarping of the East Anglian coast since zone Ip III, assuming the Sussex coast to have remained stable.

It is interesting to speculate where the hinge-line occurs. As stated above the inshore margin of marine action at Brighton is at 12–13 m O.D. In the central Fens the marine March gravels, which are generally thought to be Ipswichian though this has not been confirmed by pollen analysis, range up to about 12 m, while the Histon Road deposits at Cambridge, which are equivalent in age to Stutton, terminate at 12–13 m O.D. They are not marine but represent the high terrace of the Cam, surely at no great distance from the sea. Thus there seems to be no evidence for any appreciable warping between Sussex and the Cambridge region.

Again, the Ipswichian deposit at Ilford, Essex, is an aggradation reaching up to about 13 m O.D. Sea level cannot have been far away as there is a record of brackish snails from one of the older pits. At Aveley the aggradation goes up to 15 m and even a fraction higher at Crayford, though the latter has not been proved to be Ipswichian. Obviously there cannot have been any downwarping of this area compared with the Sussex coast. There may even have been some slight upwarping, as has been suggested from time to time. All these sites seem to lie west of the hinge-line which Wooldridge used to maintain extended roughly north–south through Braintree about 30 km east of the longitude of both Ilford and Cambridge.

In this connection the Ipswichian interglacial deposit at Wretton is important because it lies 25 km east of the Cambridge meridian and 5 km west of the Braintree meridian. Here brackish influences occur at intervals between 1·95 m below O.D. and 0·45 m above O.D. in zone Ip IIb deposits, i.e. at very much the same height as at Selsey and Stone and in the same zone, though we cannot be certain about relative position in the zone. But there is no trace of marine influence in zone Ip III deposits, which occur in an adjacent section at heights of −1·5 to about +1 m O.D. Unfortunately both zone IIb and zone III deposits are truncated a little above O.D. and succeeded by Devensian full-glacial deposits which extend up to about 4 m O.D. We do not know whether the interglacial deposits ever extended up to 12–13 m as they do at the other sites to the west and the south or not. What does seem reasonably certain is that the Devensian terrace here at 4 m is very similar in elevation to the Intermediate terrace of the Cam at Cambridge (Lambert, Pearson and Sparks, 1963) and the corresponding terrace of the Ouse at Earith (Bell, 1970). Not only are these features at the same elevation but they are also in the same geographical position where rivers debouch into the Fens, and they contain the same type of flora and fauna. Hence they are probably synchronous. Thus little downwarping has occurred at Wretton since the Devensian phase in question – there is a radiocarbon date of about 42,000 B.P. on the Earith deposit (Bell, 1969).

At Cambridge the zone II and III deposits lie in the terrace above the Devensian terrace, whereas at Wretton they are in the same terrace with the Devensian unconformably above them. At Wretton zone III is truncated at about 1 m O.D., while at Cambridge it is succeeded conformably at about 8 m by zone IV. This may indicate some downwarping at Wretton between Ip III and the Devensian, but we cannot be sure of this.

In thinking of this downwarping one must not exaggerate its magnitude. The differential movement between Cambridge and the East Anglian coast has averaged since the Ipswichian something like 8 mm per century.

There remains one final point about base-level changes in the Ipswichian: whether there was an oscillation of base level between zones IIb and III of which there is some evidence. Near Ipswich the two sites, Bobbitshole and Stutton, are only 8 km apart and roughly on the same meridian so that their involvement in the North Sea downwarping can be expected to be about equal. At Bobbitshole aggradation ends in the middle of zone IIb at about 0·4 m O.D. with no trace of marine influence (fragments of marine shells near the top were shown to be derived (Sparks and West, 1967) probably from the Red Crag). Yet at Stutton zone III aggradation – and it is well up in zone III – starts at about the same level. This strongly suggests a base-level oscillation in the middle of the interglacial. In fact, the way in which aggradation sequences of zone IIb deposits are almost everywhere truncated tends to confirm it. Similar evidence is found at Wretton where zone IIb aggradation ends in one section at about 0·8 m O.D. and is resumed, much later in the same zone, in another section at about −2 m O.D., and then continues up into zone III at about −1·4 m. We have considered the possibility of synchronous floodplain and channel deposits at different elevations, but rejected this as an explanation (Sparks and West, 1970). There seems to be genuine evidence of a mid-Ipswichian oscillation of base level, rather similar to van der Heide's (1957) findings in the Netherlands, but we would be the first to admit that some of the evidence is conflicting and our interpretation must be considered rather as a statement of the problem than as its solution.

These comments on the geography of the Ipswichian interglacial cannot be translated into map form. They are limited, tentative and hazy in their outlines. A map cannot express these attributes: it shows hard lines and is thus potentially one of the most misleading ways of describing reconstructions of the past.

References

Chapter 1

The following are mostly general texts which are applicable not only to this chapter but throughout the book.

BUTZER, K. W. (1965) *Environment and Archeology* (2nd ed. 1971). London: Methuen.

CHARLESWORTH, J. K. (1957) *The Quaternary Era*. London: Arnold.

DALY, R. A. (1934) *The Changing World in the Ice Age*. New Haven, Conn.: Yale University Press.

EMBLETON, C. and KING, C. A. M. (1968) *Glacial and Periglacial Geomorphology*. London: Arnold.

FLINT, R. F. (1971) *Glacial and Quaternary Geology*. New York: Wiley.

GROVE, A. T. and WARREN, A. (1968) Quaternary landforms and climate on the south side of the Sahara. *Geogr. J.*, **B4**, 194–208.

WEST, R. G. (1968) *Pleistocene Geology and Biology*. London: Longmans.

WOLDSTEDT, P. (1954) *Das Eiszeitalter*, Vol. I, 2nd ed. Stuttgart: Enke.

WRIGHT, W. B. (1937) *The Quaternary Ice Age*, 2nd ed. London: Macmillan.

ZEUNER, F. E. (1958) *Dating the Past*, 4th ed. London: Methuen.

ZEUNER, F. E. (1959) *The Pleistocene Period*, 2nd ed. London: Hutchinson.

Chapter 2

Most of the works listed at the end of Chapter 1 contain sections on the causes of the Ice Age and bibliographies of more detailed papers.

EWING, M. and DONN, W. L. (1958) A theory of ice ages. *Science*, **127**, 1159–62.

SIMPSON, G. C. (1934) World climate during the Quaternary period. *Q.J. Roy. Met. Soc., London*, **60**, 425–78.

Chapter 3

The literature on glacial landforms is enormous. Useful summaries are available, together with full bibliographies, in the following:

EMBLETON, C. and KING, C. A. M. (1968) *Glacial and Periglacial Geomorphology*. London: Arnold.

FLINT, R. F. (1957) *Glacial and Pleistocene Geology*. New York: Wiley.

SISSONS, J. B. (1967) *The Evolution of Scotland's Scenery*. Edinburgh: Oliver & Boyd.

SPARKS, B. W. (1960) *Geomorphology* (2nd ed. 1972). London: Longmans.

WEST, R. G. (1968) *Pleistocene Geology and Biology*. London: Longmans.

The following short list of papers and books covers the ones that have been mainly referred to in this chapter:

BULL, A. J. (1940) Cold conditions and landforms in the South Downs. *Proc. Geol. Assoc., London*, **51**, 63–71.

CAROL, H. (1947) Formation of roches moutonnées. *J. Glaciology*, I, 57–9.

CHARLESWORTH, J. K. (1953) *The Geology of Ireland*. Edinburgh: Oliver & Boyd.

CHORLEY, R. J. (1959) The shape of drumlins. *J. Glaciology*, **3**, 339–44.

DURY, G. H. (1953) A glacial breach in the North-Western Highlands. *Scot. Geogr. Mag.*, **69**, 106–17.

GREGORY, K. J. (1965) Proglacial lake Eskdale after sixty years. *Trans. Inst. Brit. Geogr.*, **36**, 149–62.

HAYNES, V. M. (1968) The influence of glacial erosion and rock structure on corries in Scotland. *Geografiska Annaler*, **50**, Ser. A, 221–34.

KENDALL, P. F. (1902) A system of glacier lakes in the Cleveland Hills. *Q.J. Geol. Soc., London*, **58**, 471–571.

LEWIS, W. V., ed. (1960) Investigations on Norwegian cirque glaciers. *Roy. Geogr. Soc. Res. Series*, No. 4 (among these papers we have referred especially to those by M. H. BATTEY, J. G. MCCALL and W. R. B. BATTLE).

LINTON, D. L. (1949 and 1951) Some Scottish river captures re-examined. *Scot. Geogr. Mag.*, **65**, 123–31, and **67**, 31–44.

NYE, J. F. (1952) The mechanics of glacier flow. *J. Glaciology*, **2**, 82–93.

PEEL, R. F. (1949) A study of two Northumbrian spillways. *Trans. Inst. Brit. Geogr.*, **15**, 75–89.

SPARKS, B. W. and WEST, R. G. (1964) The drift landforms around Holt, Norfolk. *Trans. Inst. Brit. Geogr.*, **35**, 27–35.

SUGDEN, D. E. (1968) The selectivity of glacial erosion in the Cairngorm Mountains, Scotland. *Trans. Inst. Brit. Geogr.*, **45**, 79–92.

Chapter 4

The following texts contain many references to further detailed work:

CAILLEUX, A. and TAYLOR, G. (1954) *Cryopédologie*. Paris: Hermann.

EMBLETON, C. and KING, C. A. M. (1968) *Glacial and Periglacial Geomorphology*. London: Arnold.

FLINT, R. F. (1971) *Glacial and Quaternary Geology*. New York: Wiley.

SPARKS, B. W. (1960) *Geomorphology* (2nd ed. 1972). London: Longmans.

TRICART, J. (1967) *Traité de géomorphologie:* Tome II, *Le Modèle des régions périglaciaires*. Paris: P.U.F.

WEST, R. G. (1968) *Pleistocene Geology and Biology*. London: Longmans.

The *Biuletyn Peryglacjalny*, published at Lódź in Poland, is devoted to periglacial studies, many of them in English.

Detailed papers specially referred to in this chapter are:

BULL, A. J. (1940) Cold conditions and landforms in the South Downs. *Proc. Geol. Assoc., London*, **51**, 63–71.

HOLLINGWORTH, S. E., TAYLOR, J. H. and KELLAWAY, G. A. (1944) Large-scale superficial structures in the Northampton ironstone field. *Q.J. Geol. Soc., London*, **100**, 1–44.

KENNEDY, B. A. (1969) *Studies of erosional valley-side asymmetry*. Unpublished Ph.D. thesis, University of Cambridge.

KERNEY, M. P., BROWN, E. H. and CHANDLER, T. J. (1964) The late-glacial and post-glacial history of the Chalk escarpment near Brook, Kent. *Phil. Trans. Roy. Soc., London*, **B**, **248**, 135–204.

MAARLEVELD, G. C. (1965) Frost mounds: a summary of the literature of the past decade. *Med. Geol. Stichting* (N.S.), **17**, 3–16.

MACKAY, J. R. (1963) *The Mackenzie Delta Area, N.W.T.*, Memoir No. **8**, Geographical Branch, Mines and Technical Surveys, Ottawa.

PISSART, A. (1963) Les traces de 'pingos' du Pays de Galles (Grande-Bretagne) et du Plateau des Hautes Fagnes (Belgique). *Zeits. Geomorph.* (N.S.), **7**, 147–65.

REID, C. (1887) On the origin of dry Chalk valleys and of coombe rock. *Q.J. Geol. Soc., London*, **43**, 364–73.

WATT, A. S., PERRIN, R. M. S. and WEST, R. G. (1966) Patterned ground in Breckland: structure and composition. *J. Ecol.*, **54**, 239–58.

WILLIAMS, R. B. G. (1964) Fossil patterned ground in eastern England. *Biul. Peryglacjalny*, **14**, 337–49.

WILLIAMS, R. B. G. (1968) Some estimates of periglacial erosion in southern and eastern England. *Biul. Peryglacjalny*, **17**, 311–35.

WOOLDRIDGE, S. W. (1950) Some features in the structure and scenery of the country around Fernhurst, Sussex. *Proc. Geol. Assoc., London*, **61**, 165–90.

Chapter 5

Over a hundred years of study of British Pleistocene stratigraphy has resulted in a very extensive literature on the subject. Many useful papers describing local Pleistocene sequences are to be found in the *Proceedings* of the Geologists' Association of London: the regional guides published by the same association, and the British Regional Geologies published by the Institute of Geological Sciences, are also very valuable. The following general texts give references to the regional detail of stratigraphy:

LEWIS, C. A. (ed.) (1970) *The Glaciations of Wales and Adjoining Regions*. London: Longmans.

PRAEGER, R. L. (1896) Report upon the raised beaches of the north-east of Ireland, with special reference to their fauna. *Proc. Roy. Irish Acad.*, **4** (3rd series), 30–54.

RAYNER, D. H. (1967) *The Stratigraphy of the British Isles*. Cambridge: Cambridge University Press.

SHOTTON, F. W. and WEST, R. G. (1969) Stratigraphical table of the British Quaternary. *Proc. Geol. Soc., London*, **1656**, 155–7.

SISSONS, J. B. (1967) *The Evolution of Scotland's Scenery*. Edinburgh: Oliver & Boyd.

WEST, R. G. (1967) The Quaternary of the British Isles. In *The Quaternary*, Vol. 2 (ed. K. RANKAMA). New York: Interscience Publishers.

WEST, R. G. (1968) *Pleistocene Geology and Biology*. London: Longmans.

WRIGHT, W. B. (1937) *The Quaternary Ice Age*, 2nd ed. London: Macmillan.

ZEUNER, F. E. (1959). *The Pleistocene Period*, 2nd ed. London: Hutchinson.

Chapter 6

The following general texts contain references to more detailed studies of the history of vegetation and of the British flora:

FAEGRI, K. and IVERSEN, J. (1964) *Textbook of Pollen Analysis*, 2nd ed. Oxford: Blackwell.

GODWIN, H. (1956) *History of the British Flora.* Cambridge: Cambridge University Press.

MATTHEWS, J. R. (1955) *Origin and Distribution of the British Flora.* London: Hutchinson's University Library.

PENNINGTON, W. A. (1969) *The History of British Vegetation.* London: English Universities Press.

WALKER, D. and WEST, R. G., eds. (1970). *Studies in the Vegetational History of the British Isles.* Cambridge: Cambridge University Press.

WEST, R. G. (1968) *Pleistocene Geology and Biology.* London: Longmans.

Chapter 7

These references are more detailed than those given in some of the earlier chapters, because much Ice Age zoology is contained in papers on individual deposits and in summary papers of the uses of certain types of animals.

BOYCOTT, A. E. (1934) Habitats of land Mollusca in Britain. *J. Ecol.*, **22**, 1–38.

BOYCOTT, A. E. (1936) Habitats of freshwater Mollusca in Britain. *J. Anim. Ecol.*, **5**, 116–86.

BROTHWELL, D. and HIGGS, E. (1969) *Science in Archaeology*, 2nd ed. London: Thames and Hudson. (This book contains a number of interesting essays on the use of animal remains in interpreting the past.)

COOPE, G. R. (1961) On the study of glacial and interglacial insect faunas. *Proc. Linn. Soc., London*, **172**, 62–5.

COOPE, G. R. (1962) A Pleistocene coleopterous fauna with Arctic affinities from Fladbury, Worcestershire. *Q.J. Geol. Soc.*, **118**, 103–23.

COOPE, G. R. (1965) Fossil insect faunas from late Quaternary deposits in Britain. *Adv. Sci.*, **21**, 564–75.

COOPE, G. R. (1967) The value of Quaternary insect faunas in the interpretation of ancient ecology and climate. *Proc. VII Congr. Internat. Assoc. Quat. Res.*, **7**: *Quaternary palaeoecology*, 359–80.

CORNWALL, I. W. (1956) *Bones for the Archaeologist.* London: Phoenix House.

ELLIS, A. E. (1951) Census of the distribution of British non-marine Mollusca. *J. Conch.*, **23**, 171–244.

ELLIS, A. E. (1962) British freshwater bivalve molluscs. *Synopses of the British Fauna*, **13**, Linn Soc.: London.

ELLIS, A. E. (1969) *British snails*, 2nd ed. London: Oxford University Press.

FISHER, M. J., FUNNELL, B. M. and WEST, R. G. (1969) Foraminifera and pollen from a marine interglacial deposit in the western North Sea. *Proc. Yorks. Geol. Soc.*, **37**, 311–20.

FREY, D. G. (1964) Remains of animals in Quaternary lake and bog sediments and their interpretation. *Arch. Hydrobiol. Beih.*, **2**, 1–114.

FUNNELL, B. M. (1961) The Palaeogene and early Pleistocene of Norfolk. *Trans. Norfolk Norwich Nat. Soc.*, **19**, 340–64.

FUNNELL, B. M. and WEST, R. G. (1962) The early Pleistocene of Easton Bavents, Suffolk. *Q.J. Geol. Soc., London*, **118**, 125–41.

GOULDEN, C. E. (1964) The history of the Cladoceran fauna of Esthwaite Water (England) and its limnological significance. *Arch. Hydrobiol.*, **60**, 1–52.

KERNEY, M. P. (1962) The distribution of *Abida secale* (Draparnaud) in Britain. *J. Conch.*, **25**, 123–6.

KERNEY, M. P. (1963) Late-glacial deposits on the Chalk of south-east England. *Phil. Trans. Roy. Soc., London*, **B**, **246**, 203–54.

KERNEY, M. P. (1968) Britain's fauna of land Mollusca and its relation to the post-glacial thermal optimum. *Symp. Zool. Soc., London*, **22**, 273–91.

MACFADYEN, W. A. (1932) Foraminifera from some late Pliocene and glacial deposits of East Anglia. *Geol. Mag.*, **69**, 481–97.

NORTON, P. E. P. (1967) Marine molluscan assemblages in the early Pleistocene of Sidestrand, Bramerton and the Royal Society borehole at Ludham, Norfolk. *Phil. Trans. Roy. Soc., London*, **B**, **253**, 161–200.

SAVAGE, R. J. G. (1966) Irish Pleistocene mammals. *Irish Nat. J.*, **15**, 117–30.

SHOTTON, F. W. (1965) Movements of insect populations in the British Pleistocene. *Geol. Soc. Amer., Special Paper*, **84**, 17–33.

SPARKS, B. W. (1961) The ecological interpretation of Quaternary non-marine Mollusca. *Proc. Linn. Soc., London*, **172**, 71–80.

SPARKS, B. W. (1964) Non-marine Mollusca and Quaternary ecology. *J. Ecol.*, **52** (Supplement), 87–98.

SPARKS, B. W. (1969) Non-marine Mollusca and archaeology. In BROTHWELL and HIGGS (see above).

SPARKS, B. W. and WEST, R. G. (1959) The palaeoecology of the interglacial deposits at Histon Road, Cambridge. *Eiszeitalter u. Gegenwart*, **10**, 123–43.

SPARKS, B. W. and WEST, R. G. (1970) Late Pleistocene deposits at Wretton, Norfolk. I, Ipswichian interglacial deposits. *Phil. Trans. Roy. Soc., London*, **B, 258**, 1–30.

SPARKS, B. W., WEST, R. G., WILLIAMS, R. B. G. and RANSOM, M. (1969) Hoxnian interglacial deposits near Hatfield, Herts. *Proc. Geol. Assoc., London*, **80**, 243–67.

SUTCLIFFE, A. J. (1964) The mammalian fauna. In OVEY, C. D., ed., The Swanscombe skull. *Roy. Anthrop. Inst., Occ. Paper* No. **20**, 85–112.

WEST, R. G. (1961) Vegetational history of the early Pleistocene of the Royal Society borehole at Ludham, Norfolk. *Proc. Roy. Soc., London*, **B, 155**, 437–53.

WEST, R. G. (1968) *Pleistocene Geology and Biology*. London: Longmans.

WEST, R. G. and SPARKS, B. W. (1960) Coastal interglacial deposits of the English Channel. *Phil. Trans. Roy. Soc., London*, **B, 243**, 95–133.

WRIGHT, W. B. (1937) *The Quaternary Ice Age*, 2nd ed. London: Macmillan.

ZEUNER, F. E. (1958) *Dating the Past*, 4th ed. London: Methuen.

ZEUNER, F. E. (1959) *The Pleistocene Period*, 2nd ed. London: Hutchinson.

Chapter 8

The following general texts contain many references to detailed work:

CLARK, GRAHAME (1967) *The Stone Age Hunters*. London: Thames & Hudson.

CLARK, GRAHAME (1969) *World Prehistory: A New Outline*. Cambridge: Cambridge University Press.

COLES, J. M. (1968) Ancient man in Europe. In *Studies in Ancient Europe*, ed. J. M. COLES and D. D. A. SIMPSON. Leicester: Leicester University Press.

COLES, J. M. and HIGGS, E. S. (1969) *The Archaeology of Early Man*. London: Faber & Faber.

OAKLEY, K. P. (1967) *Man, the Toolmaker*, 5th ed. London: British Museum (Natural History).

ROE, D. A. (1970) *Prehistory: An Introduction*. London: Macmillan.

Chapter 9

EPSTEIN, S., SHARP, R. P. and GOW, A. J. (1970) Antarctic ice sheet. Stable isotope analyses of Byrd Station cores and interhemispheric climatic implications. *Science*, **168**, 1570–2.

GODWIN, H. (1960) Radiocarbon dating and Quaternary history in Britain. *Proc. Roy. Soc., London*, **B**, **153**, 287–320.

GODWIN, H. (1970) The contribution of radiocarbon dating to archaeology in Britain. *Phil. Trans. Roy. Soc., London*, **B**, **269**, 57–75.

JELGERSMA, S. (1961) Holocene sea level changes in the Netherlands. *Med. Geol. Stichting*, Serie C–VI–No. **7**.

KIM, S. M., RUCH, R. R. and KEMPTON, J. P. (1969) Radiocarbon dating at the Illinois State Geological Survey. *Illinois State Geol. Survey, Environmental Geol. Notes*, No. **28**.

LIBBY, W. F. (1965) *Radiocarbon Dating*, 2nd ed. Chicago: Phoenix Science Series.

MONTFRANS, H. M. VAN (1971) *Palaeomagnetic Dating in the North Sea Basin.* Rotterdam: Princo N.V.

OAKLEY, K. P. (1969) *Frameworks for Dating Fossil Man*, 3rd ed. London: Weidenfeld & Nicolson.

OLSSON, I. U., ed. (1970) *Radiocarbon Variations and Absolute Chronology.* Stockholm: Almquist & Wiksell.

POLACH, H. A. and GOLSON, J. (1966) *Collection of Specimens for Radiocarbon Dating and Interpretation of Results.* Canberra City, Australian Institute of Aboriginal Studies, Manual No. 2.

STOKES, M. A. and SMILEY, T. L. (1968) *An Introduction to Tree Ring Dating.* Chicago, Ill.: University of Chicago Press.

WEST, R. G. (1968) *Pleistocene Geology and Biology.* London: Longmans.

ZEUNER, F. E. (1958) *Dating the Past*, 4th ed. London: Methuen.

Chapter 10

BELL, F. G. (1969) The occurrence of southern, steppe and halophyte elements in Weichselian (last-glacial) floras of southern Britain. *New Phytol.*, **68**, 913–22.

BELL, F. G. (1970) Late Pleistocene floras from Earith, Huntingdonshire. *Phil. Trans. Roy. Soc., London*, **B**, **258**, 347–78.

COOPE, G. R. (1965) Fossil insect faunas from late Quaternary deposits in Britain. *Adv. Sci.*, **21**, 564–75.

COOPE, G. R. (1967) The value of Quaternary insect faunas in the interpretation of ancient ecology and climate. *Proc. VII Congr. Internat. Assoc. Quat. Res.*, **7**: *Quaternary palaeoecology*, 359–80.

COOPE, G. R., SHOTTON, F. W. and STRACHAN, I. (1961) A late Pleistocene fauna and flora from Upton Warren, Worcestershire. *Phil. Trans. Roy. Soc., London*, **B**, **244**, 379–421.

DAVIS, A. G. (1955) The Mollusca of Bobbitshole interglacial beds, Ipswich. *Trans. Suffolk Nat. Soc.*, **9**, 110–14.

FRANKS, J. W. (1960) Interglacial deposits at Trafalgar Square, London. *New Phytol.*, **59**, 145–52.

HOLLINGWORTH, S. E., ALLISON, J. and GODWIN, H. (1950) Interglacial deposits from the Histon Road, Cambridge. *Q.J. Geol. Soc., London*, **105**, 495–510.

JESSEN, K. and MILTHERS, V. (1928) Stratigraphical and palaeontological studies of interglacial freshwater deposits in Jutland and north-west Germany. *Danm. Geol. Unders.*, **II** Raekke, Nr. **48**.

LAMBERT, C. A., PEARSON, R. G. and SPARKS, B. W. (1963) A flora and fauna from late Pleistocene deposits at Sidgwick Avenue, Cambridge. *Proc. Linn. Soc., London*, **174**, 13–30.

SHOTTON, F. W. (1962) The physical background of Britain in the Pleistocene. *Adv. Sci.*, **19**, 1–14.

SPARKS, B. W. (1957) The non-marine Mollusca of the interglacial deposit at Bobbitshole, Ipswich. *Phil. Trans. Roy. Soc., London*, **B**, **241**, 33–44.

SPARKS, B. W. (1964) The distribution of non-marine Mollusca in the last interglacial in south-east England. *Proc. Malac. Soc., London*, **36**, 7–25.

SPARKS, B. W. and WEST, R. G. (1959) The palaeoecology of the interglacial deposits at Histon Road, Cambridge. *Eiszeitalter u. Gegenwart*, **10**, 123–43.

SPARKS, B. W. and WEST, R. G. (1964) The interglacial deposits at Stutton, Suffolk. *Proc. Geol. Assoc., London*, **74**, 419–32.

SPARKS, B. W. and WEST, R. G. (1967) A note on the interglacial deposit at Bobbitshole, near Ipswich. *Trans. Suffolk Nat. Soc.*, **13**, 390–2.

SPARKS, B. W. and WEST, R. G. (1968) Interglacial deposits at Wortwell, Norfolk. *Geol. Mag.*, **105**, 471–81.

SPARKS, B. W. and WEST, R. G. (1970) Late Pleistocene deposits at Wretton, Norfolk. I, Ipswichian interglacial deposits. *Phil. Trans. Roy. Soc., London*, **B**, **258**, 1–30.

VAN DER HEIDE, S. (1957) Correlations of marine horizons in the Middle and Upper Pleistocene of the Netherlands. *Geol. en Mijnbouw.* (N.S.), **19**, 272–6.

VON DER BRELIE, G. (1954) Transgression und Moorbildung im letzten Interglazial. *Mitt. geol. (St) Inst. Hamburg*, **23**, 111–18.

WALKER, D. (1953) The interglacial deposits at Histon Road, Cambridge. *Q.J. Geol. Soc., London*, **108**, 273–82.

WEST, R. G. (1957) Interglacial deposits at Bobbitshole, Ipswich. *Phil. Trans. Roy. Soc., London*, **B, 241**, 1–31.

WEST, R. G. (1969) Pollen analyses from interglacial deposits at Aveley and Grays, Essex. *Proc. Geol. Assoc., London*, **80**, 271–82.

WEST, R. G., LAMBERT, C. A. and SPARKS, B. W. (1964) Interglacial deposits at Ilford, Essex. *Phil. Trans. Roy. Soc., London*, **B, 247**, 185–212.

WEST, R. G. and SPARKS, B. W. (1960) Coastal interglacial deposits of the English Channel. *Phil. Trans. Roy. Soc., London*, **B, 243**, 95–133.

Index

1 Scree development, Suilven, Sutherland. Scree at the foot of the free face developed in nearly horizontal Pre-Cambrian Torridonian Sandstone: hummocky glaciated Lewisian Gneiss to the right.

2 Corries near Col de l'Iseran, French Alps. Two corries separated by jagged arêtes hanging above a deep glacial trench at the head of the Val d'Isère.

3 Llyn Cau, Cader Idris, North Wales. View of the corrie lake at about 450 m from a cleft in the back wall at about 750 m. The corrie is excavated in mudstones, while the surrounding walls are mainly composed of igneous rocks.

4 Arêtes and peaks, Northern Arran. Caisteal Abhail in right foreground, Cir Mhòr (pointed summit) in right middle distance and Goat Fell to the left.

5 Glacially smoothed rock, Foel Penolau, near Trawsfynydd, North Wales. The up-glacier side of a roche moutonnée developed in Cambrian Barmouth Grit.

6 Glacially plucked outcrop, Roman Steps, near Harlech, North Wales. The down-glacier side of an outcrop of jointed Cambrian Rhinog Grit (compare the angular blocks and opened joints with the closed joints and smoothed rock of pl. 5).

7 Glaciated Lewisian Gneiss, Wester Ross. Knock and lochan landscape seen from the jointed Pre-Cambrian Torridonian Sandstone of Stac Polly.

8 Glacial trough, Glen Dee, Aberdeenshire. A typical Scottish glaciated valley with a corrie (left middle distance) and a breached watershed (centre middle distance).

9 Hanging valley, Stelvio Pass, Italian Alps. A high-level glacial trough hanging above the cloud-filled valley below.

10 Loch Nevis, near Mallaig, Inverness-shire. A Scottish sea loch with glaciated highland at the back.

11 Drowned drumlins, Clew Bay, Co. Mayo, Eire. Note the considerable variation in shape, though some modification by marine processes is apparent in the foreground.

12 Kettle moraine, Ehenside, Cumberland. Note the various stages of plant colonization of the kettle hole lakes, the two in the foreground still possessing open water.

13 Moraine complex, Strathardle, Perthshire. Confused morainic country of Devensian (last glaciation) age.

14 An esker in Finland.

15 Complex esker, Muir of the Clans, Inverness–Nairn border. Note the differences in distinctness of form between the cultivated and uncultivated areas.

16 Interbedded till and outwash, Upper Sundon, near Luton. Weathered till in outwash gravels which include a seam of fine sand above.

17 Disturbed very chalky till, Upper Sundon, near Luton. Individual chalk pebbles shattered by frost.

18 Outwash gravels, Upper Sundon, near Luton. Well-sorted, steeply-dipping gravels with seams of sand.

19 A glacial meltwater channel, Glen of the Downs, Wicklow. Note the typical straight side slopes of this channel which cuts across the relief trend.

20 A stone stream, Dovedale, Derbyshire. Probably formed by the selective removal of finer material from a flow which spread out as a small fan at the foot of the slope.

21 The upper periglacial complex at Hoxne, Suffolk. A layer of coarse solifluxion gravel intercalated in finer sediment probably deposited in a lake. A small ice-wedge cast occurs to the right of the top step. The meticulous nature of proper archaeological excavation (by Mr J. J. Wymer) is apparent.

22 Solifluxion gravel, Hoxne, Suffolk. Detail of pl. 21 to show the unworn, unbedded and unsorted nature of a layer of solifluxion gravel.

23 Brickearth overlying chalk solifluxion, Angmering-on-Sea, Sussex. Note the rough vertical fissility of this loess deposit.

24 Ice mound features, Walton Common, Norfolk. Ramparts of cryoturbated chalk marking the sites of former ice mounds. The irregular pattern suggests either irregularly-shaped segregations of ice or the repeated development and thawing of segregations in the same area.

25 Involutions in dark, organic layer, Wretton, Norfolk. An involuted organic bed, just below the level of the figure, in Devensian (last glaciation) sands and gravels.

26 Small ice wedges, Wretton, Norfolk. Larger wedge to the left and two smaller wedge casts near the shovel to the right. Both the terrace deposits and the wedges are Devensian (last glaciation) in age.

27 A problem, Upper Sundon, near Luton. Are these ice-wedge sites filled with surface decalcified and oxidized gravels or are they giant forms of the surface solution effects shown in pl. 28? The bed of till just below the dark upper gravel seems to turn down into the forms and so roughly does the bedding of the oxidized gravel, features which one would expect of filled wedge casts. Problematic cases of this type constantly occur in the field.

28 Weathered surface layer in gravel, Upper Sundon, near Luton. Dark surface layer of decalcified, oxidized gravel, probably formed by last interglacial weathering. The irregular depth of weathering is caused by permeability differences and can produce a superficial resemblance to poor ice-wedge casts.

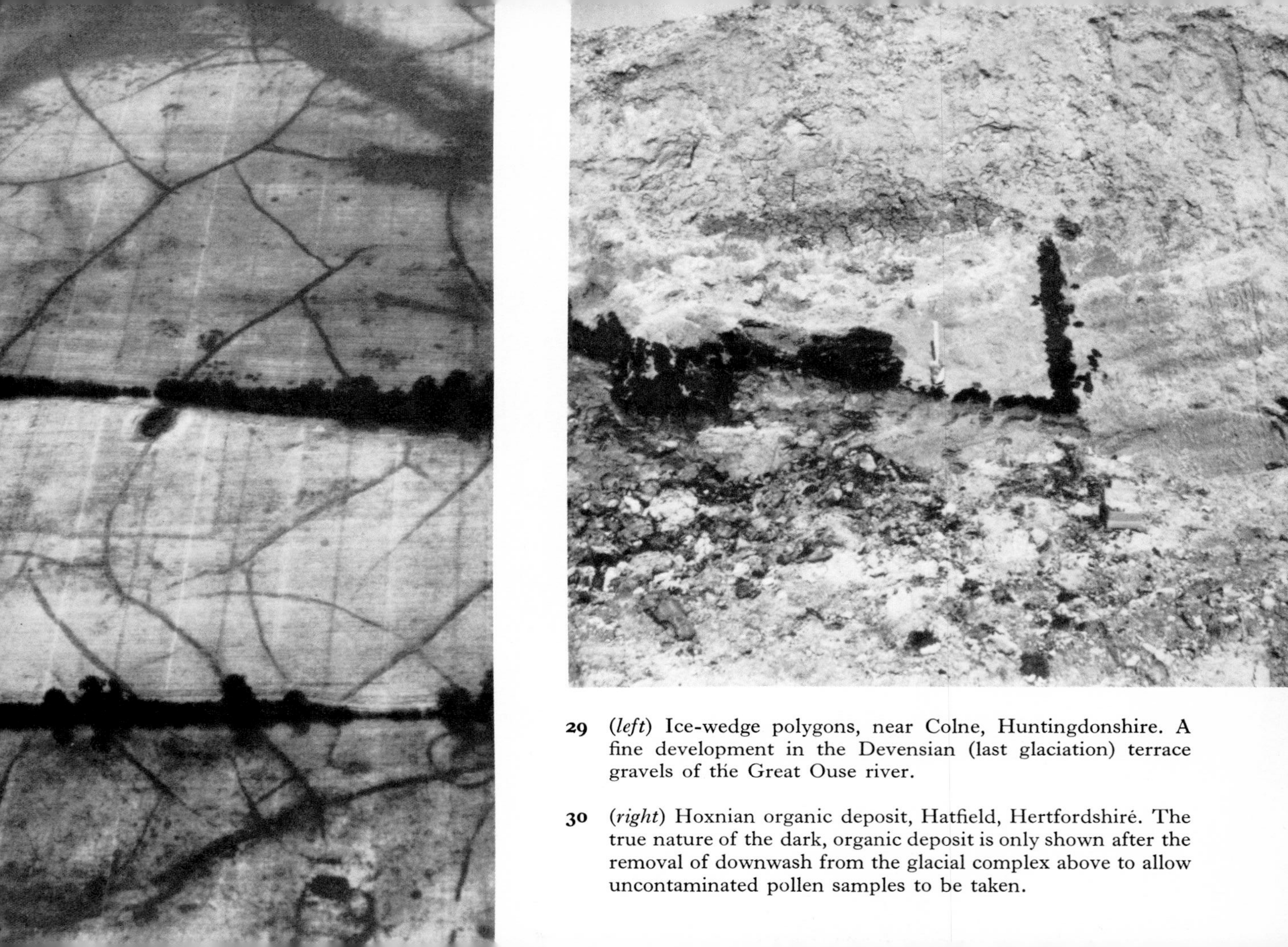

29 (*left*) Ice-wedge polygons, near Colne, Huntingdonshire. A fine development in the Devensian (last glaciation) terrace gravels of the Great Ouse river.

30 (*right*) Hoxnian organic deposit, Hatfield, Hertfordshiré. The true nature of the dark, organic deposit is only shown after the removal of downwash from the glacial complex above to allow uncontaminated pollen samples to be taken.

31 Breckland stripes, Duke's Drive, near Elveden, Norfolk. Alternating stripes of sandy and chalky material on a gentle slope with elongated polygons near row of pines in the background.

32 An organic interglacial deposit, Hoxne, Suffolk. The interglacial deposit is here covered by only a thin layer of made ground, the junction showing clearly on the far face of the excavation. The excavation was made under Mr J. J. Wymer to recover artefacts and vertebrate remains: the site is being surveyed and is pegged out for a metre grid to locate finds precisely.

33 Raised bog, Wedholme Flow, Cumberland. This dome-shaped bog has been extensively cut and drained for peat extraction, but its original shape is still clear from the contrast between the cultivated fields and the bog vegetation of *Sphagnum*, *Calluna* and *Eriophorum*.

34 Eroded blanket bog, near Plynlimmon, Cardiganshire.

35 (*left*) Varved sediments, Finland. Each pair of layers represents the deposition of one year, the coarser being deposited in summer and the finer in winter.

36 (*right*) Coring the Histon Road interglacial at Cambridge. Taking a 4-inch core by hand in 1956: the procedure is now mechanized.

37 A peat section at Godwin's Piece, Westhay, Somerset Levels in 1948. The sequence of peat types reveals the history of peat-forming plant communities at the site: f. humified forest floor of recent vegetation; e. fresh *Scheuchzeria* peat with aquatic *Sphagna* formed between the hummocks and eventually growing over them as water level rose; d. humified *Eriophorum-Calluna* peat, resulting from the growth of hummocks on the bog surface; c. fresh *Sphagnum* peat with aquatic *Sphagna* and *Scheuchzeria*: a flooding horizon; b. humified *Sphagnum–Calluna–Eriophorum* peat, resulting from slower peat growth under drier conditions; a. fresh *Sphagnum* peat with aquatic *Sphagna* and *Andromeda*: a flooding horizon.

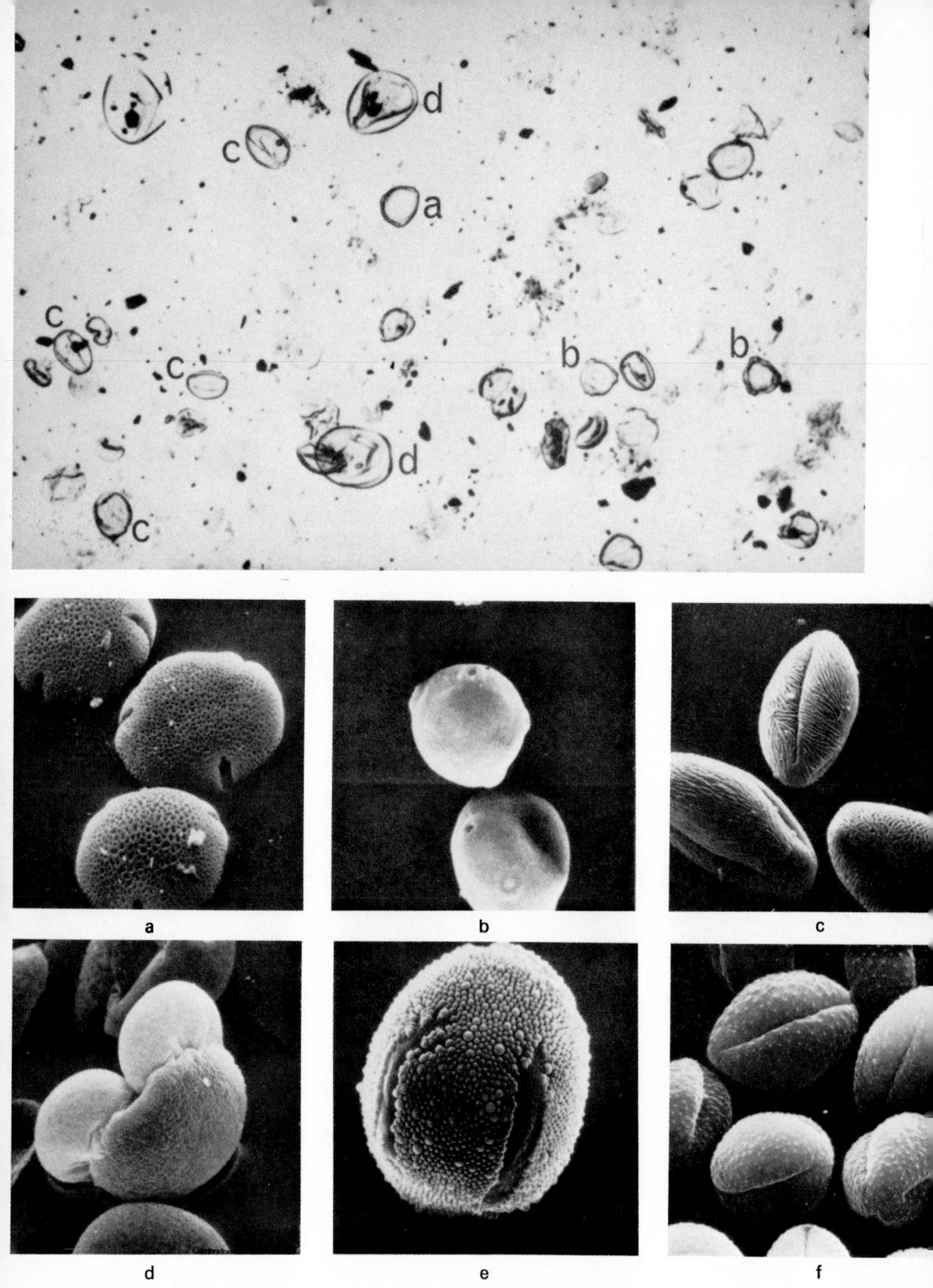

38 Pollen grains. *above,* A preparation of fossil pollen grains on a slide. The field of view shows several types of pollen grains and spores, as well as much unidentifiable debris. a, *Corylus*; b, *Betula*; c, *Quercus*; d, fern spore (x 200).
below, Pollen grain photographs taken with a 'Stereoscan' microscope (*by courtesy of Cambridge Scientific Instruments Ltd*), a, *Tilia* (x 900); b, *Carpinus betulus* (x 600); c, *Acer* (x 1000); d, *Pinus sylvestris* (x 500); e, *Linum usitatissimum* (x 1000); f, *Artemisia* (x 1000).

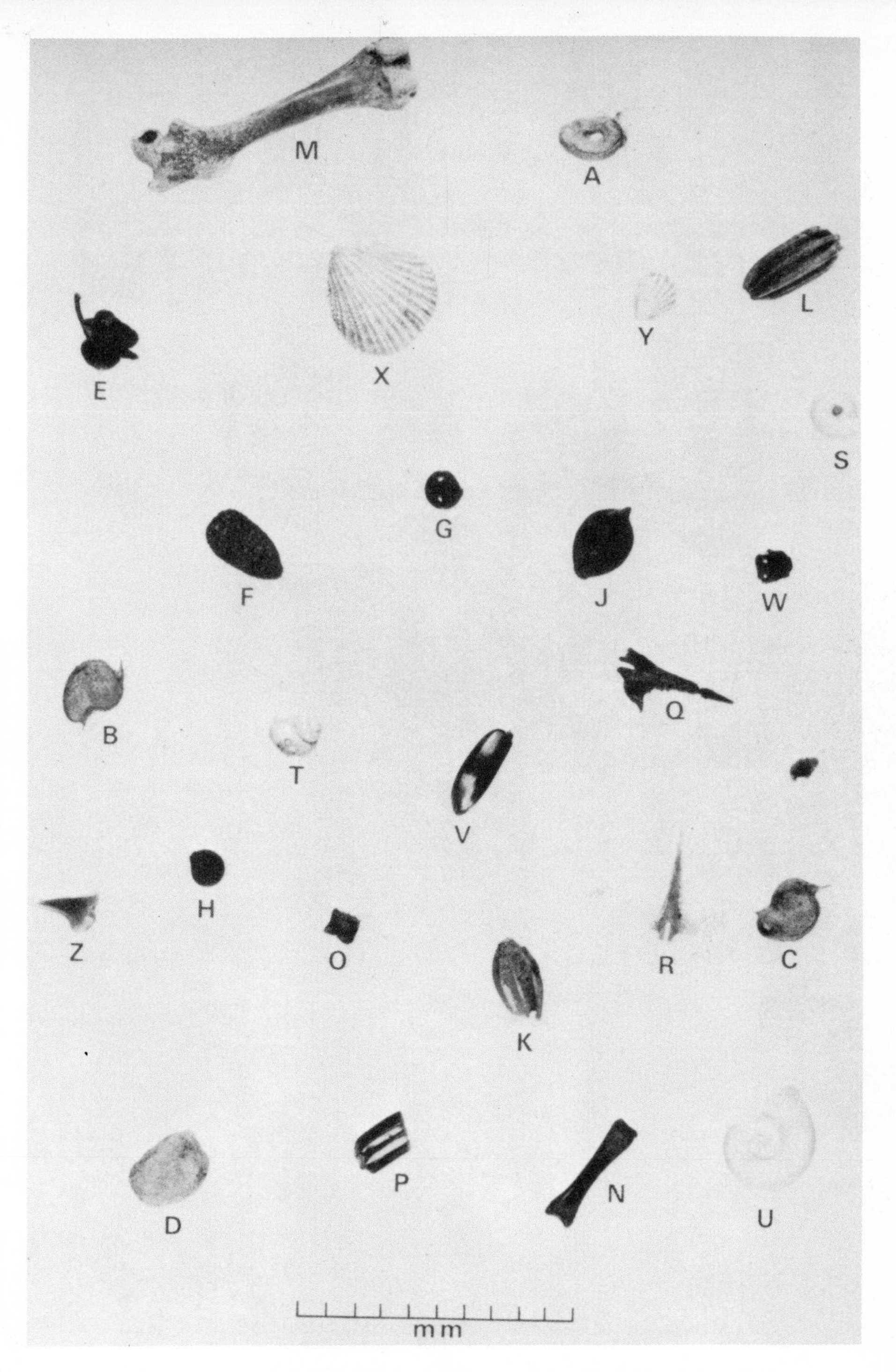

39 Macroscopic plant and animal remains. The plate shows the range of organic remains often found in deposits and includes:

Fruits and seeds, including *Potamogeton* (A, B, C, D), *Rumex* (E), *Sambucus* (F), *Chenopodium* (G, H), *Scirpus* (J), *Sparganium* (K), an umbellifer (L).

Vertebrates, including bones (M, N), vertebra (O), rodent tooth (P), stickleback spines (Q, R). Snails (S, T, U). Beetle wing case (V), and weevil head (W). Derived Cretaceous fossils including brachiopods (X, Y) and a fishtooth (Z).

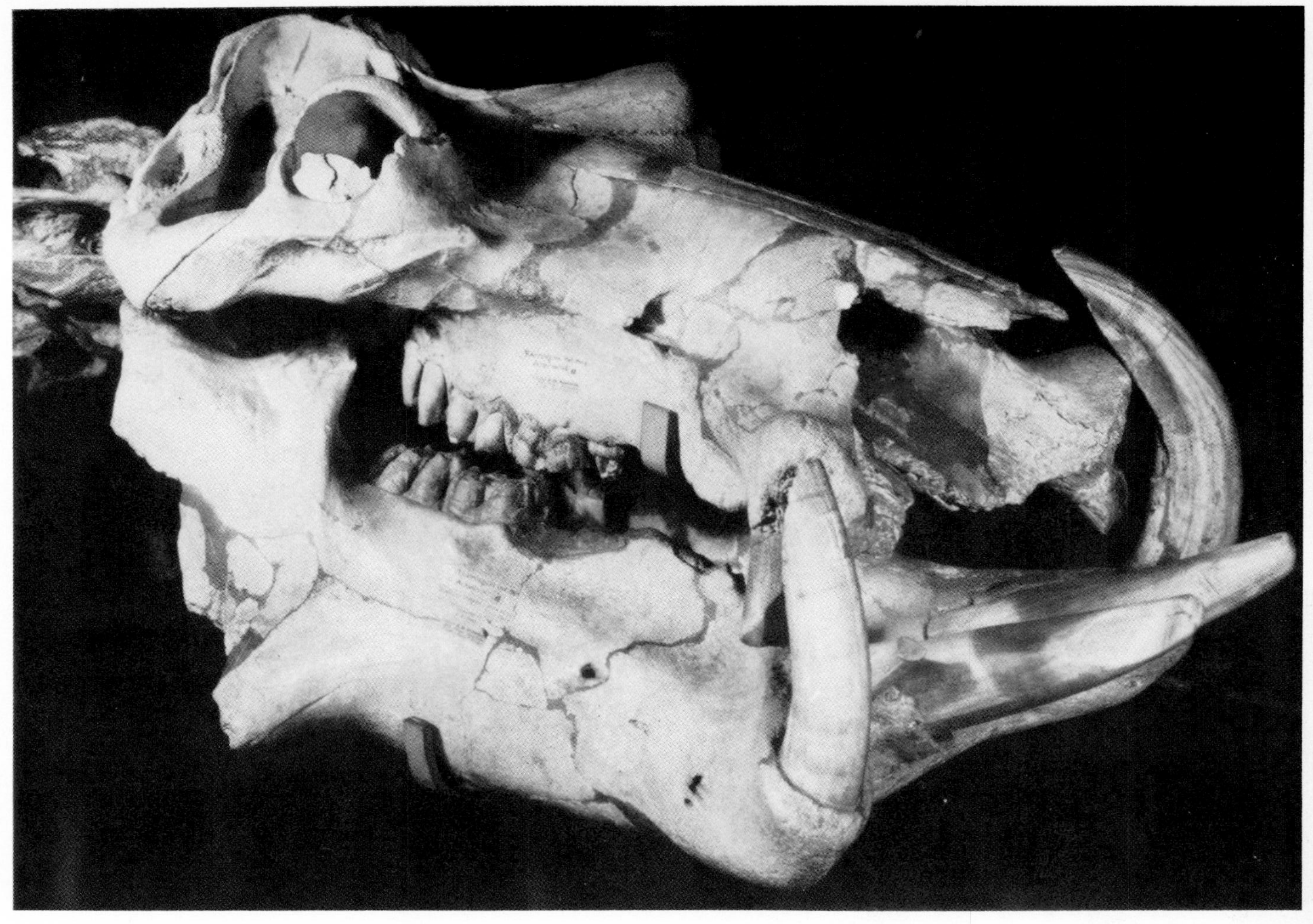

40 Hippopotamus. The skull of a complete skeleton, preserved in the Sedgwick Museum, from the interglacial deposit at Barrington, Cambridgeshire, which has never been properly dated.

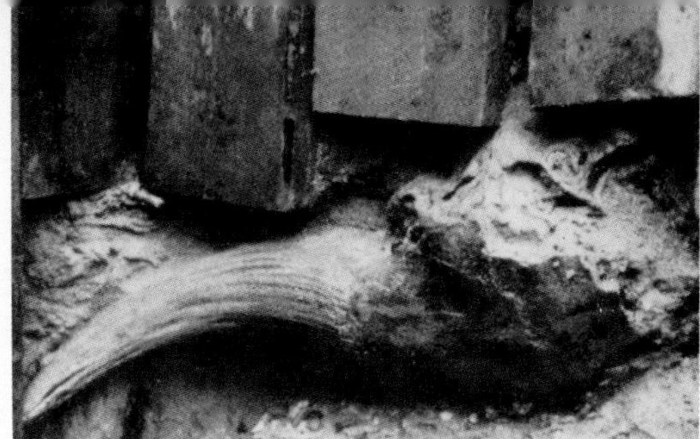

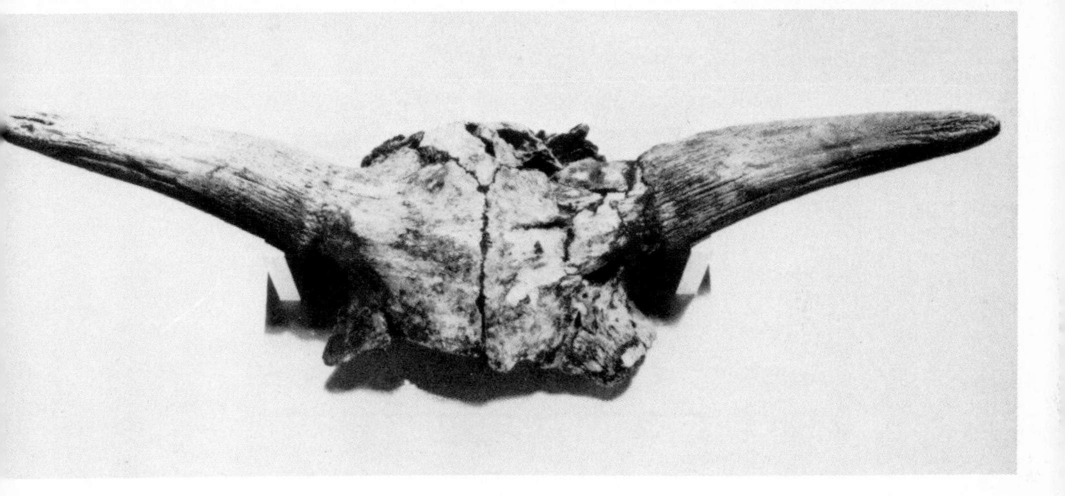

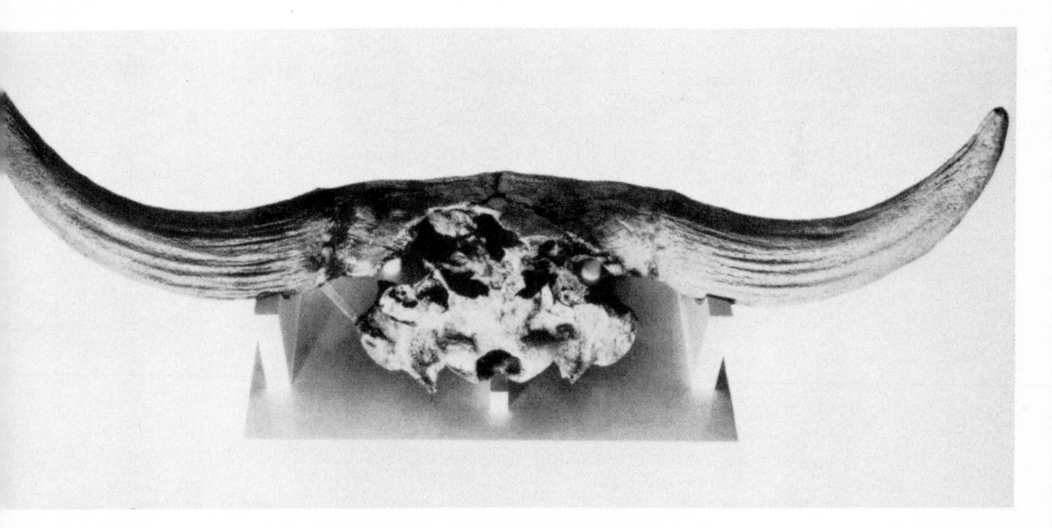

41 A Bison skull of Devensian age.

Top left: as first seen in excavations for a Lecture Block, Sidgwick Avenue, Cambridge, September 1957; Top right: identity revealed as material is removed; Middle and bottom: front and rear views of the skull pieced together and mounted in the Sedgwick Museum.

42 Non-marine Mollusca from the British Pleistocene.

A. *Hydrobia ventrosa*. Interglacial. A brackish water snail; B. *Agriolimax* sp. Cold and interglacial deposits. A slug; C. *'Succinea' arenaria*. Usually cold deposits. A marsh snail; D. *Potamida littoralis*. Interglacial. A freshwater mussel now extinct in Britain; E. *Vallonia enniensis*. Interglacial. A land snail, found in marshy places, now extinct in Britain; F. *Belgrandia marginata*. Interglacial. A freshwater snail, now extinct in Britain; G. *Columella columella*. Usually cold deposits. A dry-land snail, now extinct in Britain; H. *Planorbis vorticulus*. An interglacial freshwater snail, now with a restricted distribution in Britain.

The scales are of 1 mm except for *Potamida*, which is 1 cm.

43 The Neolithic Honeygore track, Somerset Levels. This track, excavated in 1948, has longitudinal timber of birch, with an alder stump (left base) growing *in situ* overturned and incorporated in the trackway. The track was probably built as a response to flooding of the fen, as shown by its position above fen peat but below raised bog peat. The radiocarbon date for the wood is 4750 B.P.

44 Radiocarbon Dating Laboratory, University of Cambridge. On the left is the vacuum line, in which carbon dioxide is brought to a high state of purification before it is led into the proportional gas counters. The counters, and their surrounding anticoincidence shield, are placed within the massive lead castle in order to shield them from unwanted soft gamma radiation.